On Demand Compu

Technologies and Strategi

IBM Press Series—Information Management

On Demand Computing Books

On Demand Computing: Technology and Strategy Perspectives
Fellenstein

Grid Computing
Joseph and Fellenstein

Autonomic Computing
Murch

Business Intelligence for the Enterprise
Biere

DB2 Books

DB2 Universal Database v8.1 Certification Exams 701 and 706 Study Guide
Sanders

DB2 for Solaris: The Official Guide
Bauch and Wilding

DB2 Universal Database v8.1 Certification Exam 700 Study Guide
Sanders

DB2 Universal Database v8 for Linux, UNIX, and Windows Database Administration Certification Guide, Fifth Edition
Baklarz and Wong

Advanced DBA Certification Guide and Reference for DB2 Universal Database v8 for Linux, UNIX, and Windows
Snow and Phan

DB2 Universal Database v8 Application Development Certification Guide, Second Edition
Martineau, Sanyal, Gashyna, and Kyprianou

DB2 Version 8: The Official Guide
Zikopoulos, Baklarz, deRoos, and Melnyk

Teach Yourself DB2 Universal Database in 21 Days
Visser and Wong

DB2 UDB for OS/390 v7.1 Application Certification Guide
Lawson

DB2 SQL Procedural Language for Linux, UNIX, and Windows
Yip, Bradstock, Curtis, Gao, Janmohamed, Liu, and McArthur

DB2 Universal Database v8 Handbook for Windows, UNIX, and Linux
Gunning

Integrated Solutions with DB2
Cutlip and Medicke

DB2 Universal Database for OS/390 Version 7.1 Certification Guide
Lawson and Yevich

DB2 Universal Database v7.1 for UNIX, Linux, Windows and OS/2—Database Administration Certification Guide, Fourth Edition
Baklarz and Wong

DB2 Universal Database v7.1 Application Development Certification Guide
Sanyal, Martineau, Gashyna, and Kyprianou

DB2 UDB for OS/390: An Introduction to DB2 OS/390
Sloan and Hernandez

On Demand Computing
Technologies and Strategies

CRAIG FELLENSTEIN

IBM

PRENTICE HALL PTR

PRENTICE HALL PROFESSIONAL TECHNICAL REFERENCE
UPPER SADDLE RIVER, NEW JERSEY 07458
www.phptr.com

Editorial/production supervision: *Kathleen M. Caren*
Cover design director: *Jerry Votta*
Cover design: *Nina Scuderi*
Manufacturing buyer: *Alexis Heydt-Long*
Acquisitions editor: *Jeffrey Pepper*
Editorial assistant: *Linda Ramagnano*
Marketing manager: *Robin O'Brien*
IBM Consulting editor: *Susan Visser*

PRENTICE HALL PTR

Published by Pearson Education, Inc.
Publishing as Prentice Hall Professional Technical Reference
Upper Saddle River, NJ 07458

Prentice Hall books are widely used by corporations and government agencies for training, marketing, and resale.

For information regarding corporate and government bulk discounts please contact:

Corporate and Government Sales (800) 382-3419 or corpsales@pearsontechgroup.com

Printed in the United States of America

10 9 8 7 6 5 4 3 2 1

ISBN 0-13144024-1

Pearson Education LTD.
Pearson Education Australia PTY, Limited
Pearson Education Singapore, Pte. Ltd.
Pearson Education North Asia Ltd.
Pearson Education Canada, Ltd.
Pearson Educación de Mexico, S.A. de C.V.
Pearson Education — Japan
Pearson Education Malaysia, Pte. Ltd.

Contents

Preface

A lot of very sophisticated technology is involved in the journey to becoming an on demand business. This book de-mystifies these complexities, places boundaries in areas that have none, and brings clarity to areas that may have (until now) seemed vague to many people. This book describes on demand business strategies and technologies in terms that are concise, hard-hitting, and to the point.

Becoming an on demand business operation is absolutely a business conversation that many companies today should be having. Being an on demand business means not operating in the same old mode.

This book introduces a way to understand the real potential of the networked world, that is, where business strategy and processes are inextricably linked with technology strategy and deployment. Where thinking and lines of business intersect in a surprisingly seamless manner. As your company has started to navigate this intersection of thinking, business, and technology, the IBM Corporation has focused on and invested in becoming the strongest partner to support you in this On Demand Journey.

Can You See It?

It is important to know that on demand business is an evolution, not a revolution, and *yes,* it is a fundamental shift in the way the best companies operate, anywhere in the world. An on demand business is constantly thinking about:

- Allocating precious resources
- Structuring new and more efficient processes
- Competing and leading from the front
- Interacting with partners, employees, and customers with improved efficiencies
- Realizing lower operational costs and capital expenditures

Becoming an on demand business is a response to a world that has become more volatile than ever before. This is not just about finding new ways to manage one's business through a weak economy. An on demand business mode of operation applies whether the economy is strong or weak. In fact, growth can drive as many challenges as a downturn to a supply chain, to customer relationships, or to the resilience of your underlying infrastructure.

An on demand business is not necessarily easy to see at first glance. As with any evolution, it is time itself that is the determining factor.

What is driving this worldwide operational shift? The consistent theme is that notable forces are converging on organizations, today, where both business forces and technical possibilities can easily intersect. These forces are driving different choices about business designs and underlying computing infrastructures. Of course, these forces are not new, but in an advanced networking services world, you feel these pressures more acutely and in real time.

Because of the global marketplace and the Internet, every institution has far greater contact with the world: access to more markets and information, exposure to more threats, and a rapid-fire, competitive environment. Those companies that lead their industries are the ones best able to adapt and build the right partnerships at this intersection of thinking, business, and technology.

At the highest level of this evolution, there is a relationship between business and IT. First, technology was applied as an integrated solution to a business problem. In this case, the problem was the inefficiency of manual back-office functions such as accounting and payroll. This inefficiency solution space yielded several benefits, but these benefits of yesterday remain today somewhat inflexible.

Next, there was the issue of exploding technology into every corner of the business, onto desktops, networks integrated into homes, even into our pockets and the backpacks of our children. There was a lot of value in empowering these networks into departments within businesses and also to private individuals, but proliferation spawned complexity and islands of autonomic functions.

Now we are well along in extending applications and business processes to the "Net." However, an on demand business is not just connected; it is completely integrated, end-to-end, to drive organizational productivity and cost efficiencies. Our approach in the IBM Corporation not only includes a view showing that integration is an incremental process, but also one that delivers fundamental leaps in efficiency and responsiveness, management and control, as well as the empowerment of virtually all lines of business. This is a robust approach that brings a new agenda for a new global economy. This book presents this "new agenda" in a concise, hard-hitting, and to-the-point fashion.

In this environment, what does any business need from the IT industry? We think it is a partner that can help leverage competitive advantages, optimize existing IT infrastructures, and improve the variability and management of cost structures. This partner must be able to do this all within the context of what matters in a particular industry. This is, in part, what drove IBM's acquisition of PriceWaterhouseCoopers Consulting, and it is why our client teams are more specialized than ever in the competitive pressures and dynamics across more than 17 industries. This is so that we at IBM can clearly communicate with our customers, having equal insights about the business and technical implications of (for example):

- A very large telecommunications company planning for a next-generation network. The new networking services and infrastructures have to give the company the ability to experiment with new, and most oftentimes very innovative, services, causing it to scale up (or down) and do so without heavy up-front capital investments.
- A very large retail outlet chain planning to use technologies that will let it "*see*" how much product is in transit, or on store shelves. All this will allow the chain to become more informed about buying patterns and respond to unforeseen events like inventory discrepancies, theft, security breaches, delays, and unpredicted sales spikes.
- A very large home and commercial appliances company trying to automate the management of a computing environment that includes 17,000 clients and 700 distributed servers.

- A very large moving and storage company launching a new division to focus on inventory management and home delivery of high-value consumer goods.
- A large food and beverage retail company needing an IT environment that will integrate worldwide operations while creating more flexibility in IT costs.

Our work at IBM with thousands of companies around the world tells us that leaders in all industries think about three things and the relationships among them. These three items of thought are:

1. The design of business models and processes
2. The implications for the supporting technology environments
3. The most efficient and effective ways to acquire and manage processes and technologies

One point to keep in mind as you read this book is that businesses will start this "On Demand" journey from any one of multiple entry points, and from anywhere within a business model, as seen in Figure P.1.

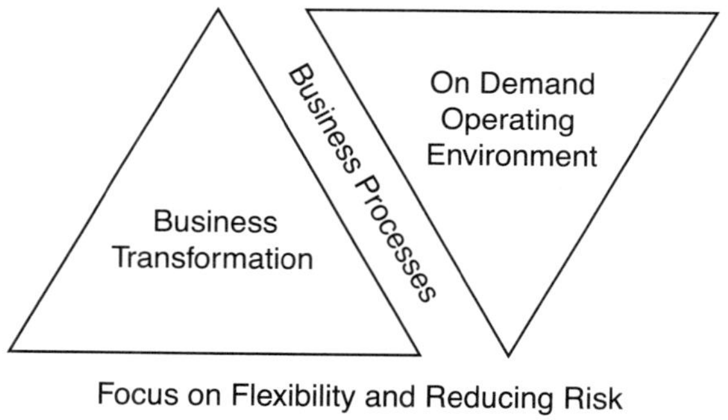

FIGURE P.1 Increasing flexibility and reducing risk is the key—business models, processes, infrastructure, plus fiancing and delivery.

As illustrated in Figure P.1, entry points can be obtained from many positions. There is no set way to engage in this transformation process other than to engage. IBM has indeed proven that on demand business is a profitable and world-class means in which to operate a business, and we have helped many of our largest customers to achieve this state of operation in many of the areas shown in the figure.

As the author of this book, it is my hope that you find reading this book as interesting and challenging as I did writing it. I spent countless hours (and air miles) just trying to summarize and legitimize in my mind exactly what best describes the technologies and strategies of on demand business. This effort, however, is not the reflection of any single mind. The technologies and strategies described in the book are composed and represented by a vast number of individuals, many of whom are deeply involved in worldwide IBM on demand business implementations.

The IBM Corporation has absolutely demonstrated that it has both the capabilities and the resources to be a role model for on demand business efficiencies. We at IBM invite any interested businesses with a desire to achieve these same efficiencies realized by an on demand business enterprise to reach out and contact us. We are anxious to share with you our insights and achievements so that you too can realize the tremendous benefits afforded by on demand business operations. On demand business is an achievement that does not necessarily benefit the IBM Corporation; rather, it benefits worldwide business enterprise operations, which will ultimately impact our cultural economies in many ways yet to be realized.

Finally, one overriding conclusion of this book is that on demand business (regardless of a company's mission) is an evolution of many bright minds, a refined language with a defined set of autonomic business practices, and a tremendous critical skills teaming environment, all which yield the result of redesigning business operations. This is a realization that many worldwide corporations have already experienced, are currently challenged with, or are about to engage in for the first time. *On demand business is the new agenda for our new world industrial economy.*

An Open Invitation for On Demand Business *Computing* Author Dialogue ...

As the author of this book and a practitioner of many strategic implementations in on demand business global operations (and other distinguished engineering topics), I welcome your thoughts, proposals, challenges, and comments related to topics in this book. I can be contacted through the Prentice Hall PTR publications team at:

Prentice Hall PTR
c/o Craig Fellenstein (Author: *On Demand Computing*)
One Lake Street
Upper Saddle River, NJ USA 07458

Part 1
On Demand Business

The on demand business era is here. This era introduces a current place in time where the stakes are higher than ever, the perils are significant, and the business opportunities are enormous. It is a time when everyone who makes your business run—customers, suppliers, partners, and shareholders—will expect everything just as you do ... to be delivered better, faster, and at far less cost.

ON DEMAND BUSINESS ...
As will be discussed throughout this book, an on demand business is an enterprise whose business processes—integrated end-to-end across the company and with key partners, suppliers, and customers—can respond with agility and speed to any customer demand, market opportunity, or external threat.
Achieving the state of on demand business is a journey, an evolutionary transformation process consisting of several levels of business transformation. We will describe these levels in later sections.
This book explores this fascinating journey and delivers concise messages regarding critical technologies and strategies to consider along the way.

So how do we get to this state of on demand business? The answers are straightforward and unveiled throughout this entire book.

This part of the book discusses aspects of the on demand business challenge and introduces the on demand Operating Environment. We will explore interesting intersections of business and thinking, and reflect on how the world has awakened to the values that this new operational state of on demand business delivers.

1 Introduction to IBM On Demand Business

The end of the 20th Century set a precedent for a new era, where the structure of global business began to fully demonstrate the changes brought about by the Internet. Toward the end of the 1990s, we notably could see a shift in both computing practices and the social and business implications of this phenomenon.

Although global businesses have successfully adopted many tools of the information age while at the same time enhancing their employees' efficiencies, many businesses have not yet had to fundamentally change their competitive and operational strategies. With economic and business pressures now more intense than ever before, emerging technology enablers are finally able to allow substantive changes more practically. Global businesses have reached a transformational stage where the need to execute has run directly into the ability to deliver, dislodging a complex, steady state of business operations to one of *transformation*.

E-BUSINESS

IBM defines e-business as "an organization that connects its core business systems to key constituencies using Intranets, extranet and the Web; the process of building and enhancing business relationships through the thoughtful use of network-based technologies; and the leveraging of Internet technologies to transact and interact with customers, suppliers, and partners and employees in order to achieve and sustain competitive advantage."

As a result, global business leadership teams are engaging in new forms of business operations called *"On Demand Business,"* where their companies are moving beyond simply integrating their various business processes. On demand is where portions of Web applications might begin to fail and are immediately self-provisioned to other machinery for continued operations. On demand business is where networks are self-provisioning and self-healing—they are self-managed. On demand business is a business strength, which allows a business to respond to any market threat..

Today, we operate in a world where business needs to be able to sense and respond to fluctuating market conditions. This requires more than simple process integration; in almost every case, it requires process transformation across all the core competencies.

Target, Equifax, John Hancock, Nordea, and ING are just a few examples of on demand business initiatives IBM launched in 2003. Later in the book we will explore many more examples of on demand business transformations. In later chapters of this book we will also explore the on demand Operating Environment in great detail. It is important to note that the differences between "On Demand" and an "e-business on demand" are subtle. An on demand Operating Environment maintains a multitude of tools and utilities for providing on demand business computing disciplines, tools, and services.

ON DEMAND BUSINESS

As we will see in the following chapters of this book, IBM has proven that roadmaps for the journey to on demand business can be created in flexible, bite-sized chunks. And because of this, we are able to help our clients become an on demand business.

At the IBM Corporation, we recognize that achieving on demand operations is a customer state and not any particular company's solution. We at IBM work to help enable and accelerate our clients through the stages of on demand business adoption. An on demand business:

- Is *responsive*—It responds almost intuitively to dynamic, unpredictable changes in demand, supply, pricing, labor, competitors' moves, capital markets, and the needs of all its constituencies—customers, partners, suppliers, and employees.

- Uses *variable* cost structures and adapts processes in a flexible manner—This flexibility enables an on demand business to reduce risk, and to accomplish business at high levels of productivity, cost control, capital efficiency, and financial predictability.
- Is *focused* on its core competencies and differentiating tasks and assets—This implies that tightly integrated strategic partners manage selected tasks, everything from manufacturing, logistics, networking services, fulfillment, human resources, and financial operations.
- Is *resilient* enough to manage changes and threats—From computer viruses and intrusions, to earthquakes, to power outages, to spikes in usage, every on demand business must promise consistent availability and security.

An on demand business, while efficiently providing critical on demand business services, affords the opportunity for its business leaders to see and manage the company as an integrated whole, even if that requires that other, outside companies handle critical aspects of the business. Instituting an on demand business is not an overnight transformation; it requires several levels of transformation that must occur over time. Elements of Autonomic Computing, which we will explore in detail in Chapter 3, can be key to an on demand business.

Autonomic Computing can be a key part of any on demand business. Autonomic Computing is an approach to self-managed computing systems that operate with a minimum of human interference. The term derives from the body's autonomic nervous system, which controls key functions without conscious awareness or involvement.

The following levels of business transformation are critical for an enterprise to factor into its strategy. These particular levels of business transformation deal with varying states of Autonomic Computing, which are:

- *Level 1:* Basic—The starting point where most systems are today, this level represents manual computing in which all system elements are managed independently by an extensive, highly skilled information technology (IT) staff. The staff sets up, monitors, and eventually replaces system elements.
- *Level 2:* Managed—Systems management technologies can be used to collect and consolidate information from disparate systems onto fewer consoles, reducing administrative time. There is greater system awareness and improved productivity.

- *Level 3:* Predictive—The system monitors and correlates data to recognize patterns and recommends actions that are approved and initiated by the IT staff. This reduces the dependency on deep skills and enables faster and better decision-making.
- *Level 4:* Adaptive—In addition to monitoring and correlating data, the system takes actions based on the information. Service level agreements (SLAs) enhance IT agility and resiliency with minimal human interaction.
- *Level 5:* Autonomic—Fully integrated systems and components are dynamically managed by business rules and policies, enabling IT staff to focus on meeting business needs with true business agility and resiliency.

Autonomic Computing sometimes helps to enable on demand business. Autonomic Computing, as stated by Ric Telford, Director of Architecture and Technology for Autonomic Computing at IBM, "*helps make an on demand Operating Environment work, but it alone does not define on demand business.*" Grid Computing is another environment that oftentimes can contribute to very powerful and cost-effective on demand business initiatives, and can work in conjunction with Autonomic Computing. In simple terms, Grid Computing is analogous to an electric "utility" power network where power generators are distributed in a grid-like fashion and the users are able to access the electric power without having to know anything about the source of the power and its locations. We will explore, in later sections of this book, specific details of both Grid Computing and Autonomic Computing.

While the definition of Autonomic Computing will likely transform as contributing technologies mature, the following list describes eight defining characteristics of an autonomic system:

1. An autonomic computing system needs to "know itself" - its components must also possess a system identity. Since a "system" can exist at many levels, an autonomic system will need detailed knowledge of its components, current status, ultimate capacity, and all connections to other systems to govern it. The system will need to know the extent of its "owned" resources, those it can borrow or lend, and those that can be shared or should be isolated.
2. An autonomic computing system must configure and reconfigure itself under varying (and even unpredictable) conditions. System configuration or "setup" must occur automatically, as well as dynamic adjustments to that configuration to best handle changing environments.
3. An autonomic computing system never settles for the status quo—it always looks for ways to optimize its workings. It will monitor its con-

stituent parts and fine-tune workflow to achieve predetermined system goals.

4. An autonomic computing system must perform something akin to healing—it must be able to recover from routine and extraordinary events that might cause some of its parts to malfunction. It must be able to discover problems or potential problems, and then find an alternate way of using resources or reconfiguring the system to keep functioning smoothly.
5. A virtual world is no less dangerous than the physical one, so an autonomic computing system must be an expert in self-protection. It must detect, identify, and protect itself against various types of attacks to maintain overall system security and integrity.
6. An autonomic computing system must know its environment and the context surrounding its activity, and act accordingly. It will find and generate rules for how best to interact with neighboring systems. It will tap available resources, even negotiate the use by other systems of its underutilized elements, changing both itself and its environment in the process—in a word, adapting.
7. An autonomic computing system cannot exist in a hermetic environment. While independent in its ability to manage itself, it must function in a heterogeneous world and implement open standards—in other words, an autonomic computing system cannot, by definition, be a proprietary solution.
8. An autonomic computing system will anticipate the optimized resources needed while keeping its complexity hidden. It must marshal IT resources to shrink the gap between the business or personal goals of the user, and the IT implementation necessary to achieve those goals—without involving the user in that implementation.

In this chapter, we will introduce on demand business. We will also take a look back at a vision of today from the not-too-distant past.

The successful execution of creating an on demand business involves a series of activities that an enterprise can plan and execute over some period of time. This happens by executing tactical activities as a surgically planned strategic approach and, in turn, yields the characteristics described in the previously stated Levels 1–5. You may now be asking: What is required to become an on demand business that demonstrates an on demand responsiveness to a variety of situations, leveraging variable cost structures that are tightly focused across all core competencies while demonstrating extreme resilience? Simply stated, a widespread business transformation must occur to become an on demand business. This transformation can actually be progressively executed across five different levels of business transformation, which we will fully treat later in this book. To effectively become an on demand business involves three fundamental initiatives that will formulate the on demand business apex; these are:

1. *Transforming your business processes* to amplify the core competencies of the business. This is critical. Seek to efficiently streamline labor-intensive processes, and then wherever possible, explore Autonomic Computing products and disciplines targeting the enrichment of core internal business operations and capabilities.
2. *Embracing new mechanisms for delivery* of services and pricing is paramount. Throughout the industries of the world, this is referred to as "*Utility Computing*." Seek to develop (or purchase) an internal systems management infrastructure that leverages many types of systematic *utilities*, with common goals of notably improving efficiencies and reducing costs. Establish a tactical plan for maximum reuse of this utility fabric.
3. *Sustaining a flexible operating environment* is mandatory. A rigid and cost-prohibitive, overly complex infrastructure has today been surpassed with the introduction of open standards. Seek to embellish these new open standards approaches. Then, utilizing these newly integrated open standards (discussed later in this book), look outside the organization for best-of-breed solutions, resisting the urge to always try to develop them with internal resources. This, as discovered by many service providers, will enable an enterprise to deliver "next-generation services."

Initially, there was a general misconception in the marketplace that "on demand" meant simply utility computing. We at IBM believe that on demand business has a much deeper meaning. For example, Flexible Financing and Delivery Options helps you to determine exactly how you can acquire the capabilities you need to deliver, and realize the results you expect.

At IBM, we know your current operating environment may be composed of pieces and parts of solutions from multiple vendors, so we take a holistic approach to enabling your long-term on demand Operating Environment. Or, you may not have yet built your infrastructure, and you are wondering how you might be able to do this and become an on demand business as a result.

We are not exempt from this situation, nor are our customers. In order to assist, we have innovative financial plans allowing one to finance non-IBM products, including vendor (ISV) software. We can also manage your existing assets, no matter how complex, complicated, or customized these assets may be. We can do all of this, while at the same time, deploying new on demand capabilities for you—for example, you may want to start with something as straightforward as server consolidation. We will always look for ways to help you free-up cash, so this additional cash can then be leveraged to fund your on demand initiatives. Other areas to start may be providing billing for on demand business capabilities, application provisioning, self-healing, and self-managed network capabilities. Becoming an on demand business will not happen overnight, but it will happen.

The goal is to help you realize measurable benefits as quickly as possible and progress. In this process, IBM can help you to mitigate risks by:

- Shifting to IBM responsibility for services such as guaranteed service level management, event correlation, problem management, self-managed networks, and more
- Enabling variable cost structures

This lets you pay for only what you use—as you use it. This also allows your customers to do the same. And, yes, this is exactly what an on demand business should experience, and be able to provide.

The IBM Corporation has fully realized what it means to be a world-class leader, establishing the global on demand business ecosystem. That is, we now expect (and deliver) more information simply because it is there. We are resilient. We expect near-complete automation simply because it is possible. We have invented and provided new mechanisms for the delivery of information. And, as we will soon illustrate, IBM has mastered a powerful combination of two innovative computing disciplines: Autonomic and Grid Computing. We have, indeed, established a flexible on demand Operating Environment. On demand Operating Environments are capable of operating and sustaining both Grid Computing environments and Autonomic Computing environments.

Grid Computing is a form of networking where—unlike conventional networks that focus on communication among devices—grid computing harnesses unused processing cycles of literally thousands of computers. The intention in creating this grid of many endpoint computer systems for solving complex problems is that a Grid Computing network is established for solving problems too intensive for any stand-alone machine.

A well-known Grid Computing project is the SETI (Search for Extraterrestrial Intelligence) initiative. This is commonly referred to as the SETI@Home project, in which computer users worldwide donate unused processor cycles to help search for signs of extraterrestrial life. This is accomplished in the grid by analyzing signals coming from outer space. This grid project relies on individual computer users to volunteer to allow the project to harness the unused processing power of the user's computer. This method saves the project both money and resources.

Grid computing does require special software that is unique to the computing project for which the grid is being used. We will explore this software toolkit (Globus), and many key development considerations as it applies to creating an on demand Operating Environment. The Autonomic Computing discipline combined with Grid Computing capabilities will absolutely deliver rich on demand Operating Environments. In fact, we have proven that together, these two disciplines have helped deliver some of the most responsive and resilient on demand Operating Environments in the world. That is not to say, however, that one must always implement both disciplines to transform into an on demand business. These operating environments can also be implemented independently for a variety of other types of implementations. We will further discuss throughout this book these on demand business perspectives surrounding services, technologies, and strategy perspectives across several different levels.

Now that we have introduced a few perspectives of what on demand business will deliver, let us take a look back to 1999. In this next discussion, as stated in a past *IBM Systems Journal* article, "*Turning points in information technology,*" published by Dr. Irving Wladawsky-Berger of the IBM Corporation, one will note a credible turning point in global business practices.

This next discussion appropriately states that "Business has never lived within the 'brick and mortar' confines of the enterprise, but in the whole web of relationships among employees, customers, business partners, suppliers, shareholders—all those on whom the enterprise relies for success. Yet in the heterogeneous environment that evolved through the years from competing architectures and operating systems, these relationships were unreachable by IT approaches in any integrated sense, dooming business to fragmented, discontinuous implementations that resulted in isolated 'islands of automation'."

Not surprisingly, a vision was forming of a larger role for technology and business. As industrial developments furthered the progress of IT and its use in business, various individuals and groups contributed new ideas and expertise. To their credit and our benefit, these men and women were trying to define a role beyond individual products—to provide an architectural context for the systematic integration of IT into business and a conceptual framework for thinking about the business itself. In reaching the current state of IT, some of these developments influenced past progress to such a degree that they are considered to be turning points.

The *IBM Systems Journal* article states from historical and visionary perspectives that "In fact, business technologies and techniques will become so thoroughly integrated into our lives that more than business will be transformed. Society in general and our day-to-day lives will be automated to a degree undreamed of in the past. Still, we would all do well to bear in mind one critical principle: In a business, few are enthralled by the prospect of being surrounded by technology; what does thrill us is the potential for transformation." This notion was published in 1999 and it certainly sets some form of precedence for the on demand business transformations our customers are experiencing today.

THE NEXT SECTION IS A REPRINT OF [WLADAWSKY-BERGER].

This *IBM Systems Journal* article, "*Turning points in information technology,*" exactly as published by Dr. Wladawsky-Berger in 1999, sets a precedent for several of the topics that will be discussed throughout this book.

The following is an excerpt from the *IBM Systems Journal*, Vol. 38, Nos. 2 and 3.

Turning points in information technology

Historically, the driving force behind the adoption of any technology has been its potential for transforming the way we live and work. Nowhere has that been more evident than in the progress of computer technology, and particularly in IBM's contributions through the years. A selection of papers representing those contributions is presented in this retrospective issue {*IBM Systems Journal, Vol. 38, Nos. 2 and 3*}, providing examples of turning points documented by the *IBM Systems Journal* since 1962.

In the early years, OS/360 gave customers of IBM the ability to write applications and run them on a consistent operating system, making genuine business computing available for the first time. IBM made history by automating the back offices of *Fortune* 500 companies. All those humdrum basic business applications like payroll and BICARSA—billing inventory control, accounts receivable, and sales analysis— could now be carried on more efficiently, accurately, and quickly than in the old manual mode.

In the 1970s, IBM married data to networks, uniting IT with communications, and created major products such as CICS* (Customer Information Control System), IMS* (Information Management System), and SNA (Systems Network Architecture) that brought on the age of real-time transaction processing. This further transformed business by basing relationships with customers on "live" information. An airline could make, change, and confirm bookings at a moment's notice. A bank teller could instantly inform a customer about an account balance and, eventually, with the development of the automatic teller machine, the relationship between bank and customer would become automated and immediate.

Business has never lived within the "brick and mortar" confines of the enterprise, but in the whole web of relationships among employees, customers, business partners, suppliers, shareholders—all those on whom the enterprise relies for success. Yet in the heterogeneous environment that evolved through the years from competing architectures and operating systems, these relationships were unreachable by IT in any integrated sense, dooming business to fragmented, discontinuous implementations that resulted in isolated "islands of automation."

Not surprisingly, a vision was forming of IT's larger role in business. As developments furthered the progress of IT and its use in business, various individuals and groups contributed ideas and expertise. To their credit and our benefit, these men and women were trying to define a role beyond individual products—to provide an architectural context for the systematic integration of IT into business and a conceptual framework for thinking about the business itself. In reaching the current state of IT, some of the developments influenced past progress to such a degree that they are considered to be turning points.

In this section of this issue, several reprinted papers describing some of these turning points are included as representative selections of IT topics that were published in the *IBM Systems Journal*. As the author of the first of these papers, which is entitled "A Framework for Information Systems Architecture," J. A. Zachman is well-known for the architectural framework that helps to place business elements in a structure that organizes information to better utilize IT. In the second paper, "Strategic Alignment: Leveraging Information Technology for Transforming Organizations," J. C. Henderson and N. Venkatraman explain how the goals of a business can be aligned with IT so that the operation of the business is more productive. W. H. Davidson carries this idea further, showing how IT can be used to completely transform a business, in "Beyond Reengineering: The Three Phases of Business Transformation," also reprinted here. In addition to the papers appearing here, a selection of related papers is presented in the bibliography.

The formation of the vision of IT's larger role, though compelling, would remain Utopian in the absence of standards-based technologies. Meanwhile in the university and research community, the Internet had been germinating out of public sight since 1969. Not long after development of the World Wide Web in 1991 and later Mosaic, the first graphical user interface for the Web, business began to see the Internet's technologies and standards as a way to connect all the information systems in which it had invested (by some estimates more than $1 trillion) over the years.

Businesses began integrating Web standards and technologies into their existing information technology. In the process, a new, extended IT infrastructure emerged along with a new on demand business model based on the ability to reach anyone inside or outside a company any time of the day or night. The transformation activity introducing on demand business for a large number of companies has been dramatic and continuing. An example of this historical progression is noted in the published paper by Abad Peiro et al[(1)] that appeared in Volume 37, Number 1, 1998, of the *IBM Systems Journal*. That entire issue as well as Volume 37, Number 3, 1998, focuses on computing on the Internet and on related technologies.

As the use of IT becomes even more important to business, the nature of work and commerce is changing. These changes are now described, and some of the directions in which developments in computing are leading business are presented.

Changing the Nature of Work

The innovative use of Web standards and technologies enables thousands of businesses to establish intranets linking geographically dispersed workers and information within an enterprise. Lockheed Martin Corporation and The Boeing Company, collaborating over an intranet, developed the Darkstar aircraft in 11 months with 50 people, a process usually requiring hundreds of designers and years of work.

In turn, intranets are being linked to extranets connecting a firm to all its trading partners. These internal and external links are delivering what was thought impossible a short time ago: efficient, timely collaboration within the enterprise and between firms separated from each other by thousands of miles and many time zones. For the first time, a worldwide enterprise, and its trading partners, can act as a unified, global team, because workers can leverage a shared base of knowledge delivered from anywhere.

A combination of intranets and extranets enables ABB, the Swiss transnational, to integrate over 60,000 users in a worldwide corporate network spanning more than 80 countries and to connect over 100 external companies—both customers and business partners. IBM increasingly relates to the world through its external Web site. Meanwhile, its internal site is a source for all of its employees on everything from strategy to the Tax Deferred Savings Plan (401K plan).

Changing the Nature of Commerce

The global reach of heretofore-local companies may startle some firms that thought they were established in their markets. Likewise, Web-based networks are permitting new competitors with very clearly focused core competencies to enter established industries and then develop on-line partnerships to complement their operations. Moreover, with the use of the Web, power is migrating to the buyer who, armed with all sorts of information on price, quality, and availability, can now easily compare a global universe of suppliers.

Among all the Web-generated changes in commerce, however, the most exciting is the growing degree of personalization facilitated by this new medium. In the past, business focused on the product, designing it to appeal to a mass market. Now with the colossal amounts of information and processing power available to an on demand business, suppliers can gain a far more comprehensive knowledge of their customers and focus on them as individuals. Instead of ending after the transaction, the relationship with the customer can span an entire lifetime because on demand business technologies can build a corporate memory of each person's needs and wants. With growing personalization, suppliers are beginning to differentiate their brands as much on knowledge of the customer as on price and quality.

The Next-Generation On Demand Business

No end to the on demand business transformation appears to be in sight, as many powerful forces conspire to drive it forward.

- Driven by the imperatives of electronic commerce, companies are automating and integrating all their business processes from the customer at one end to all their suppliers at the other. In this e-commerce environment, one transaction (one click) by a customer will trigger multiple transactions throughout the enterprise. As a consequence, the next generation of transaction processing built on incredibly scalable, reliable, manageable, and secure hardware and software servers are emerging before our eyes.

- *Knowledge management* technologies are beginning to bring employees together across time and space, making them more effective, responsive, and innovative [(2)]. Knowledge management is already making distributed learning a reality, and soon synchronous communications will make real-time video collaboration available for education, training, and general on demand business communication.
- *Deep computing* directly descended from Deep Blue*, an IBM supercomputer noted for its chess-playing ability, is already being used to transform information into useful, valuable knowledge. It is being employed by customers as diverse as the San Diego Supercomputing Laboratory and *The New York Times*, in the first case to support researchers from all over the world and in the latter to target advertising to precisely the right people.
- *Pervasive computing* is in the process of "Web-enabling" everybody and everything that can benefit from information technology. Everything from our cars to our washing machines may be connected to the Web and automated to one degree or another.
- *Digital media* will enrich the Web experience with video, sound, graphics, animation, and all sorts of effects. As a result, the Web will begin to reach more and more people because it will approach us on our own terms, that is to say, through our eyes and ears.
- Undergirding all this will be a renewed Internet infrastructure, phenomenally reliable and secure, a hundred to a thousand times faster than the Internet today, and providing the most sophisticated applications imaginable.

In fact, on demand business technologies and techniques will become so thoroughly integrated into our lives that more than business will be transformed. Society in general and our day-to-day lives will be automated to a degree undreamed of in the past. Still, we would all do well to bear in mind one critical principle: In a business (or in life for that matter), few are enthralled by the prospect of being surrounded by technology; what does thrill us is the potential for transformation.

*Trademark or registered trademark of International Business Machines Corporation.

Cited References in this reprint:

1. J. L. Abad Peiro, N. Asokan, M. Steiner, and M. Waidner, "Designing a Generic Payment Service," *IBM Systems Journal* 37, No. 1, 72–88 (1998).
2. K.-T. Huang, "Capitalizing on Intellectual Assets," *IBM Systems Journal* 37, No. 4, 570–583 (1998).

Bibliography

1. G. Gordon, "A General Purpose Systems Simulator," *IBM Systems Journal* 1, 18–32 (September 1962).
2. P. L. Kingston, "Concepts of Financial Models," *IBM Systems Journal* 12, No. 2, 113–125 (1973).
3. L. Bronner, "Overview of the Capacity Planning Process," *IBM Systems Journal* 19, No. 1, 4–27 (1980).
4. M. M. Parker, "Enterprise Information Analysis: Cost-Benefit Analysis and the Data-Managed System," *IBM Systems Journal* 21, No. 1, 108–123 (1982).
5. R. L. Katz, "Business/Enterprise Modeling," *IBM Systems Journal* 29, No. 4, 509–525 (1990).

Beginning the *On Demand Business* Journey

As we begin to unfold this book, it is key to understand that on demand business is not a revolution, but an evolution. As mentioned previously, on demand business (and IT transformation) has today manifested in an intersection of critical thinking and business operations.

This is a transformation of IT across industry, which has been brought about by virtue of advanced networking services combined with unprecedented computing power. This has opened doors to a vast number of new business opportunity possibilities.

The strategies presented in this book are compelling, and are designed to present discussions toward advanced levels of thinking, to consider aspects of business operations perhaps not yet considered by business leaders and senior technologists.

Subsequent chapters in this book will explore the various technology and strategy underpinnings of an on demand business, which are basically decomposed into four areas:

1. The on demand Operating Environment

2. Autonomic Computing
3. Grid Computing
4. Flexible hosting approaches to delivering service provider environments

In the next chapters, we will provide full treatment of the five levels of Autonomic Computing, which can be critical in any transformation process. An on demand business transformation will not occur overnight, but it can occur in a reasonable period of time. We will also more fully explore the *Autonomic Computing Blueprint* and the future of *Grid Computing*. These focus areas, along with the surrounding details of each, will provide a telling story of how any business interested in achieving an on demand business state of operations can more directly address this transformation.

Finally, the book will close by discussing critical industry issues driving the transformation activities of many worldwide companies. We will profile customer stories that describe the results of various companies' transformation efforts such that one will understand the driving industry and economic issues stimulating these transformations, the specifics of what was addressed in each transformation, specific elements in the solutions, and exactly how each of the customers started its own transformation.

The On Demand Operating Environment

2

The IBM Corporation is the worldwide leader in Autonomic Computing disciplines, products, and services. Autonomic Computing can be an important component of an on demand Operating Environment (odOE)—it is an enabler for on demand business. This chapter focuses on world-class examples of odOEs, and many strategic perspectives of IBM business leaders of on demand business.

An odOE defines a set of integration and infrastructure management capabilities that customers can utilize in a modular and incremental fashion to become an on demand business.

An on demand Operating Environment

An odOE exhibits four characteristics:

- Integrated
- Open
- Virtualized
- Autonomic

These characteristics play a key role throughout any on demand business "transformation," which will be further explored in this (and other chapters) of this book.

An odOE, as described by Bart Jacob [Jacob02] of the IBM Corporation's International Technical Support Organization, is "an operating environment that unlocks the value within the IT infrastructure." It allows for the deployment of powerful on demand values to be applied across the enterprise for the purpose of solving complex business problems in very efficient and cost-effective manners.

There are several well-known examples of odOEs today, already in action. Perhaps one of the most well known examples is the popular online auction site, eBay. The eBay worldwide online auction site is an international example of an odOE. The eBay site delivers a variety of on demand computing business services, all related to its domain-specific online business environment, to millions of online end-users, 24 hours a day, 365 days a year. There are subscription services, billing services, graphic management services, buyer services, seller services, and consumer protection services, just to name a few. A "buyer" or "seller" on eBay is able to conduct a variety of business transactions, down to the last second, and do so in an end-to-end, service-oriented manner. Virtually every service that anyone might require to transact goods, products, or services on eBay, at any particular moment in time, are available.

All these respective services required to accomplish on demand business are located on (or from) the eBay site itself. This odOE is a world-class example, renowned for delivering outstanding services. For instance, any end-user can very quickly establish his or her own online business using eBay on demand business services. This is accomplished by utilizing eBay's advanced capabilities and services, to establish a point of commerce across the world.

To become an on demand business, you need an odOE. Let's begin by exploring the odOE in this chapter, and then follow up with other on demand business topics in the next few chapters.

AN odOE IS UNIQUE AND POWERFUL

An odOE defines a set of integration and infrastructure management *capabilities* that enterprises can utilize, in a modular and incremental fashion, to become an on demand business. These are each unique services that work together to perform a variety of on demand business functions. The odOE is an example of a Services-Oriented Architecture (SOA), which is addressed in specific areas of this book. It is important to understand that it is an odOE that allows for an on demand business to operate.

Kim Kemble, Program Manager for the odOE in IBM Marketing, contributes to many of the discussions in this chapter related to the odOE.

In October 2002, IBM announced its vision of the next major phase of on demand business adoption in the company. In IBM, this concept of on demand business is not simply a vision; it is a statement of IBM's belief of how businesses will absolutely need to transform themselves to be successful in today's highly networked global environment. Global and local businesses will have to adapt and transform to cope with the ever-increasing pressures from competition and other factors associated with the global economy. This implies a transformation to a fully integrated business across people, processes, and information, including suppliers, distributors, customers, and employees.

As we will continue to point out in this book, IBM defines an on demand business as an enterprise whose business processes—integrated end-to-end across the company and with key partners, suppliers, and customers—can respond with agility and speed to any customer demand, market opportunity, or external threat. As we will also continue to reinforce in this book, there are four key attributes of an odOE:

- *Responsive*—Able to sense and respond to dynamic, unpredictable changes in demand, supply, pricing, labor, competition, capital markets, and the needs of customers, partners, suppliers, and employees.
- *Variable*—Able to adapt processes and cost structures to reduce risk while maintaining high productivity and financial predictability.
- *Focused*—Able to concentrate on core competencies and differentiating capabilities.
- *Resilient*—Able to manage changes and external threats while consistently meeting the needs of all constituents.

These attributes define the business itself. For a business to successfully attain and maintain these attributes, it must build (or enhance) an IT infrastructure that is designed to specifically support the business' goals. The next section describes the characteristics of an odOE and how such an environment can be delivered.

The on demand Operating Environment (odOE)

An odOE defines a set of integration and infrastructure management *capabilities* that customers can utilize, in a modular and incremental fashion, to become an on demand business. These *capabilities* guide what is needed for customers to increase business flexibility and simplify their IT infrastructure in a way that aligns with their business objectives.

An odOE supports the needs of a business, allowing the business to become and remain responsive, able to accommodate many variables. It makes the business tightly focused, responsive to change, and highly resilient. An odOE unlocks the capabilities within the IT infrastructure to allow these resources to be applied to solving on demand business problems. It can be viewed as an integrated environment, based on open standards, that enables rapid deployment and integration of on demand business applications and automated processes. An odOE must be:

- Flexible
- Self-managing
- Scalable
- Economical
- Resilient
- Based on open standards

To assist in this global challenge, IBM delivers numerous odOE capabilities to customers, which can be categorized into two primary areas:

- *Integration*—Provides the facilities to gain a unified view of processes, people, information, and systems—in a simple and cost-effective manner.
- *Infrastructure management*—Provides the facilities to reduce the complexity and simplify the management of an IT infrastructure, while at the same time, aligning it to business goals.

The value of an odOE is its ability to dynamically link business processes and policies with the allocation of IT resources. In an odOE, resources are allocated and managed without intervention, enabling them to be used efficiently based on business requirements. Having flexible, dynamic business processes increases the ability to grow and manage change within the business. Figure 2.1 provides an overview of the key components of an odOE:

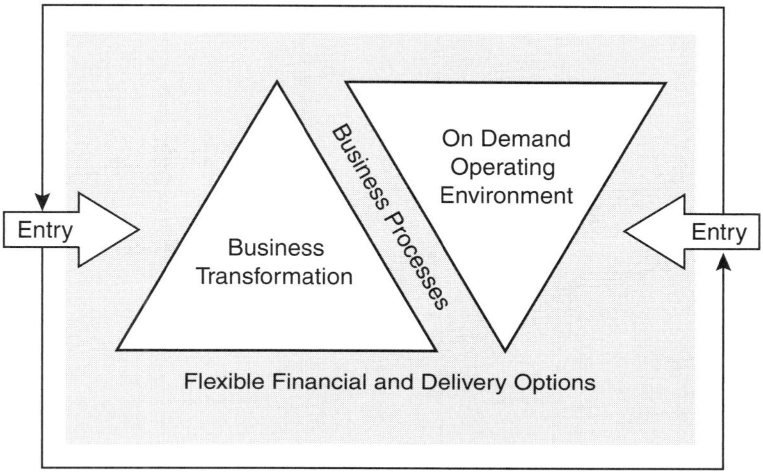

FIGURE 2.1 Where you start depends on *your* organization's priorities. Increasing flexibility and reducing risk is the key—business models, processes, infrastructure, plus financing and delivery.

Integration

Integration is the efficient and flexible combination of resources to optimize operations across and beyond the enterprise. Integration involves the resources of people, processes, and information, which are detailed in the next subsections.

People

The IT environment must enable the business to interact with employees, customers, business partners, distributors, and suppliers by supporting integrated business processes, collaboration, and data sharing.

No matter the role of the user, the operating environment must:

- *Simplify the end-user experience*—An odOE provides the tools and facilities to make employees more productive, and to make it easier for customers, suppliers, and partners to do business with the company.
- *Provide secure, role-based interactions*—To enable people with what they need, when they need it, requires a solution that includes integrated, role-based enforcement of policies. Such facilities help ensure privacy and the protection of data and resources while meeting the dynamic demands of the users.
- *Standardize access to applications*—This is required as the business changes to meet market pressures. New applications must be made available on demand, and users must be able to become productive without extensive training or skills. By providing standardized inter-

faces to applications, and using facilities such as portals to gain access to applications, not only can users quickly adapt to new applications, but resources within the company can quickly be redeployed as necessary to meet the changing demands of the business.

- *Allow users access to processes and data*—User communities require access to process enforcement and data anytime and anyplace, whether using a desktop system, laptop, or some other device such as an automated teller machine (ATM), personal digital assistant (PDA), or cell phone. It is now a fact that vast numbers of users worldwide demand access to applications and information from anywhere and at any time, day or night.

Processes In an odOE, an organization can no longer afford to develop and maintain isolated, vertical business processes. An odOE must manage and coordinate the entire enterprise horizontally, and the IT infrastructure needs to be an enabler, not a barrier. Business processes, IT infrastructure, and critical thinking regarding business operations must all intersect. To sustain this intersection, an odOE will have to support several very valuable attributes, including:

- *Consistent modeling of business processes*—This is accomplished by providing a consistent model of optimized business processes, where one can more easily adapt applications as business needs change. It is worthy to note that consistent modeling is independent of underlying product implementations; therefore, the model can persist even when more cost-effective products might be chosen in which to build the solution.
- *Integration of applications*—It is no longer necessary, practical, or cost-effective to build vertical, one-off solutions (in a vacuum) that are independent of, and without consideration for, other applications in other parts of the business. Just as a business must integrate all its business processes to run more efficiently and be responsive to changing demands, an odOE must provide the infrastructure needed to allow applications to be integrated by using common standards and open technologies. As new, previously unforeseen opportunities and requirements surface, applications that previously might have seemed unrelated will need to be quickly and efficiently integrated.
- *External connectivity*—As previously mentioned, by conforming to open standards, applications and processes can be connected to other external applications and processes in simpler, more cost-effective ways. As the business changes, new partners arise or mergers occur; then, the cost and time associated with rewriting applications or rede-

signing processes to be compatible with existing tools is unacceptable in an odOE. By planning for and enabling the integration and interconnectivity of applications, the business enables dynamic integration capabilities. This way, as a business continues being transformed (or is already operational as an on demand business), it becomes more responsive to changes and market dynamics as they occur.

In addition, by providing a consistent and integrated environment, a more economical approach can be engaged for information attainment and delivery. We will explore this critical aspect in greater detail in the next discussion.

Information Information integration enables real-time access to diverse and distributed information in transactions across and beyond the on demand enterprise. Information can reside in multiple source systems (e.g., IBM DB/2, Oracle databases, spreadsheets, flat files, etc.) and can be distributed across a variety of operating platforms (e.g., Linux, AIX, UNIX, z/OS, Microsoft Windows NT, etc.).

One strategic interest area of many industries, especially the telecommunications industry (also referred to as "service providers"), is reducing costs and increasing revenues. An indirect enabler of this notion is to form a tighter integration of information systems, and to do so in a timely, more cost-effective way. As an example, why should it cost (sometimes) millions of dollars to integrate a (let's say) billing system into the odOE infrastructure when a computer end-user can plug a credit card-size networking card into a slot in his or her computer, and as a result, be on the network in less than 60 seconds?

This tighter integration of information, in turn, better serves customers and suppliers, assisting them in making better decisions and allowing them to respond more quickly to new opportunities or competitive threats.

An odOE provides a multitude of information strategy perspectives, including:

- *A greater range of information access strategies to best match business values*—An odOE must allow companies to either consolidate information onto fewer platforms or immediately access data in a variety of ways, depending on business needs.
- *Access diversity*—Key to an odOE is the ability to access data required by diverse and distributed entities as though it was being obtained from a single source. Needless to say, an odOE must be capable of quickly locating data and content to be accessed independently of its location or platform. This is referred to as *federated data*, utilized by an on demand business. This federated approach allows for access to real-

time information, which most often resides outside the enterprise. An odOE requires the ability to access data from third parties without impact to the existing IT infrastructure. This enables corporations to gain greater return on existing information assets.

- *Data consolidation*—This is where data needs to be consolidated and/or moved for reasons of greater performance and availability. An odOE provides data placement and self-management techniques (discussed in later sections of this book), and for these reasons, having open standards related to data acquisition is mandatory. The self-management capability alone allows information and data to be acquired, consolidated, moved, or cached when and where operational business requirements demand it.

The following illustration that shows the integration layers across people, processes, and information is not a new concept. However, the values of investing in concepts of on demand business approaches to develop these types of critical operational intersections are new. The ability to apply the technologies and strategies described in this book truly enables such integration. This, for many, has never been more attainable than it is today.

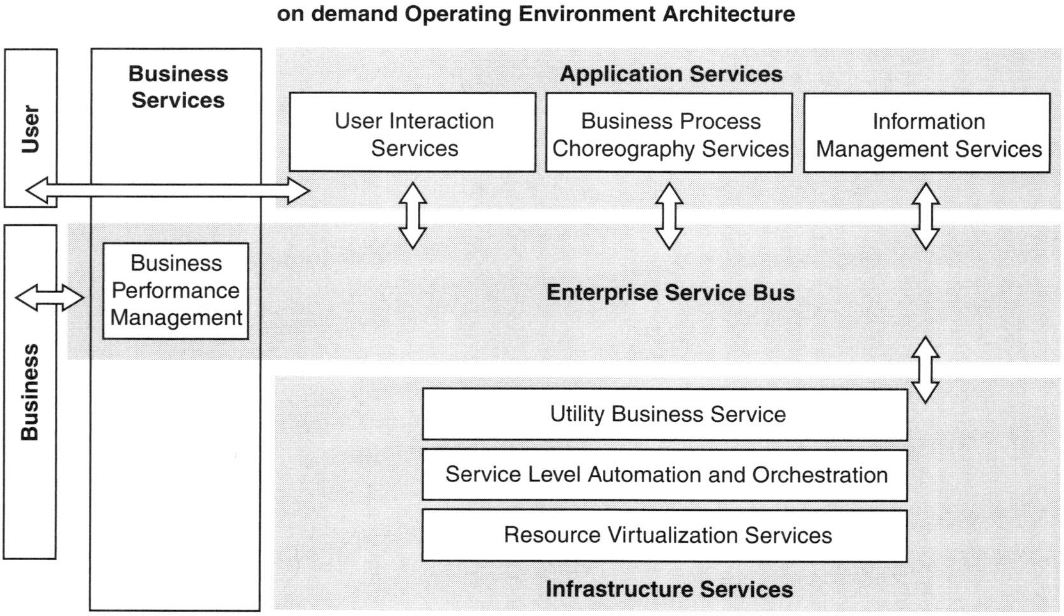

FIGURE 2.2 The on demand Operating Environment is based upon the concept of a Service Oriented Architecture (SOA). Each element of the architecture is a "service" that together implements the Operating Environment.

Autonomic Computing and Automation There has been a significant amount of discussion within the industry about Autonomic Computing. It is a new computing discipline openly embraced by many individuals. This next discussion introduces automation as one of the key goals of an e-business odOE. It is natural to ask the following question: What is the difference between Autonomic Computing and automation? The term "autonomic" deals with the types of components in the infrastructure; "automation" is the delivered feature and end-result.

AUTONOMIC vs. AUTOMATION ...

Autonomic technologies represent the components "inside"; the actual automation is the end-result that is realized "outside" to the customer or on demand business enterprise.
The result also complements, and is enhanced by, the Grid Computing side of the odOE.

IBM Autonomic Computing technologies and services address the need for autonomic features that can manage and improve an infrastructure's own operations with minimal human intervention. As IT systems become more complex, more difficult to maintain, and more expensive to manage, IBM seeks to resolve these problems by virtue of the Autonomic Computing disciplines. IBM has made significant progress in this area, as represented by a wide range of products and services specifically targeting this subject.

Indeed, automation is one fundamental goal of Autonomic Computing, but at the same time, it is the end-result and prime benefit that the enterprise will achieve. So, one could say that Autonomic Computing is both the goal and the benefit. It delivers many value-added benefits. Autonomic Computing is a discipline with capabilities that can deliver significant enabling technologies that establish the on demand business in an odOE. Autonomic Computing is oftentimes the driver for many on demand business offerings and transformation endeavors.

Autonomic Computing utilizes several technologies to enable information systems to become self-managing. These self-managing characteristics combine themselves to deliver the critical business automation required in an odOE. Autonomic technologies are the components "inside"; the actual automation is the end-result that is realized "outside" to the customer or on demand business enterprise.

Infrastructure Orchestration *Infrastructure orchestration* allows an IT infrastructure to *sense, trigger, and respond according to business goals. Orchestration* provides an end-to-end IT service that is dynamically linked to key business policies, thus allowing the IT infrastructure to adapt to changing business conditions as they occur. This is a critical element of an odOE.

Focusing on the fact that this concept is directed at achieving the state where each individual element of an IT system will have to respond to change is definitely a fundamental step toward policy-based orchestration. Realizing this goal, however, is to truly become an on demand enterprise, which requires orchestration of the automation of multiple elements so that the entire IT infrastructure can respond appropriately to changes in business policies or conditions.

For example, if a customer's order entry application suddenly finds itself among exponential and unexpected networking service traffic loads, simply allocating more machine capabilities may not be enough; the networks may also require additional (immediate) storage capabilities, more network bandwidth capacity, perhaps additional servers, new circuits to route traffic, or additional support personnel to handle the sudden increased demands. All these changes must be immediately orchestrated so that the dynamic allocation of these multiple resource elements occurs seamlessly and according to the autonomic business policies linked to the desired quality of service (QoS) levels.

Infrastructure orchestration is a fundamental success factor in an on demand business.

Provisioning *Provisioning* makes the right resources available to the right processes and people at the right time. It is the end-to-end capability to automatically deploy and dynamically optimize resources, products, and/or services. The act of autonomic provisioning is performed in response to complex business objectives and policies in heterogeneous, autonomic environments, which are intended for, or being directed by, on demand operational activities.

Provisioning enables an on demand business to respond to changing business conditions by dynamically allocating entities to the processes that most require them at a particular moment in time. This is being driven by business policies and the orchestration of seamless technical transactions in an automated fashion. Provisioning individual elements involves activities such as: service subscription identities, storage resources, servers, applications, operating systems, network circuits, workflow capabilities, third-party connections, and various types of software middleware. "Provisioning" is more of a generic term, yet it applies across several complex dimensions of technology and services.

Provisioning is a multi-faceted and extremely critical step in being able to orchestrate autonomic resources across the odOE while responding to on demand business transformation and operational efforts.

Availability *Availability* helps ensure the health and appropriate functioning of IT environments. Systems must be available 24 hours a day, 7 days a week, 365 days a year (in some customer-defined form) to meet the requirements of an on demand business. To meet these availability requirements without employing huge amounts of human capital, autonomic products and services can assist in many ways. For instance, autonomic products from IBM can help by monitoring enterprise systems and automatically taking actions to maintain high availability without human intervention before issues become problems. This can be performed according to automated policies.

To provide the kind of infrastructure that supports an on demand business requires a complex set of underlying technologies, both hardware and software. However, to support a flexible and responsive business, components must be able to be atomically and automatically reconfigured, self-managed, and applied to the appropriate business objectives according to policies. This task is immensely complex, and absolutely cannot be accomplished in a timely, cost-effective manner (in today's complicated world of technologies) without automation.

Optimization *Optimization* helps customers make the best use of their existing resources while helping to ensure all resources are running at peak performance and efficiencies. It is critical to make the best use of the resources the enterprise already has, to focus on the heterogeneous mix of equipment and cross-platform support that integrate with what is already installed. Constant pressures to diminish IT budgets, combined with the need to respond quickly to meet customers' demands and needs, cause customers to insist that their current investments be optimized. This is not a one-time action, but rather a consistently evergreen and ongoing need for continuous improvement.

As with integration and virtualization (described next), optimization is not a new concept in the IT industry. An odOE requires many new and innovative levels of automation. These levels of automation must consistently provide the flexibility and responsiveness to support the odOE. Automation is achieved through a plurality of autonomic solutions being deployed into the existing infrastructure.

Security

Security helps to ensure that information assets remain confidential and that data integrity is protected. This keeps on demand business systems protected from threats, provides functions for a greater user experience in accessing applications and data, and keeps out unwelcome users.

Business Service Management

Business service management helps to visualize an IT environment in business terms and manage service levels to business objectives. Business-driven service management capabilities provide the necessary tools to manage service levels, meter system usage, and bill customers for that usage, as well as model, integrate, connect, monitor, and manage business processes comprehensively for thorough linkage of IT and business processes.

Resource Virtualization

Resource virtualization provides a single, consolidated, logical view of and easy access to all available resources in a network (including servers, storage, and distributed systems). It is the process of presenting computing resources in ways that users and applications can easily ascertain values from many types of integrated implementations rather than presenting them in a way dictated by a specific implementation, geographic location, or their physical packaging. A meta-level of implementation views can in some sense describe the overall concept of virtualization. Simplifying the management of infrastructures and utilizing virtual meta-level data across multiple heterogeneous machines and applications (over the Internet) is one significant challenge in implementing virtualization.

In an odOE, where resources are used efficiently based on business requirements, virtualization can improve working capital and asset utilization. Virtualization enables the sharing of resources in several areas, which are discussed in the following subsections.

Data

Data virtualization includes the capability to view data from many different global sources, without knowing where the data actually resides. This implies that data virtualization involves software- and hardware-independent implementations, leveraging open standards of some sort (e.g., the need for consolidating storage from multi-vendor backup systems).

Associated with data are database management systems. Database virtualization allows database applications to access database systems without being aware of the machines where the databases are geographically located. This allows any application to view data elements from disparate database management systems. The database servers can also be of different technology origins—for example, IBM DB2, Oracle, Informix, Ingres, Microsoft SQL Server, etc.—as long as the database systems adhere to open technology standards such as Open Database Connectivity (ODBC), Java Database Connectivity (JDBC), or Object Linking and Embedding Database (OLE DB).

For this reason, it is an absolute requirement that database virtualization must utilize open database standards, which simplifies the management of databases shared across global, heterogeneous systems.

Storage Storage virtualization involves providing access to data, regardless of its physical or geographic locations or file structures, and provides economies of scale through cost-effective storage media and more efficient data-sharing concepts.

The Storage Networking Industry Association's European division (SNIA-E) defines storage virtualization as "pooling data from disparate devices, so it can be managed from a single, centrally managed console." This pooling (federation) of data can be accomplished regardless of the approach used to store the data: Storage Area Network (SAN[1]), Network-Attached Storage (NAS[2]), Direct Access Storage (DAS[3]), or some hybrid approach.

Distributed Systems Taking advantage of advances in distributed systems such as Grid Computing allows resources across the enterprise (i.e., individual systems, servers, clusters, and storage devices) to be shared and dynamically allocated to meet business needs. This dynamic allocation allows for a higher utilization of resources, resulting in cost savings as well as increased capability to meet unforeseen processing requirements by allocating available resources on demand.

Networking As the world of business operations has become pervasive and networked together through the Internet, it is now critical to be able to manage and control portions of the network. Networking services are often

1. SAN is a network of storage devices using a high-speed *fiber channel* to switch between host and storage.
2. NAS is a storage array device that is connected to a storage input/output (I/O) that interconnects to the host.
3. DAS is a storage device, either a redundant array of independent disks (RAID) or disk type, attached directly to the application server that uses data on the storage unit (most often a SCSI [Small Computer Systems Interface] connection).

shared among many different enterprises as individual or virtual networks. This includes networking technologies such as virtual private networks (VPNs), virtual local area networks (VLANs), wide area networks (WANs), Internet Protocol (IP) virtualization, wireless, satellite, and many other approaches.

In an odOE, virtualization of networking services and computing resources within the IT infrastructure provides many critical benefits. Where the underlying hardware and system software are hidden from the users and applications, an open standards-based infrastructure can and should be deployed to simplify:

- *System administration*—Rather than having individual servers dedicated to specific applications, and therefore specialized support staffs, virtualization allows for a common environment that can simplify overall administration requirements while allowing resources to be allocated as needed by the business. This is accomplished according to autonomic policies and controlled by the same.
- *Asset portfolio management*—An odOE provides a common ecosystem for running applications while maintaining independence from underlying hardware and systems software. In this operational environment, assets are tracked, managed, changed, and reallocated without requiring changes to the applications. Likewise, applications can be modified and enhanced to meet business needs without necessarily requiring changes to the underlying systems. If additional computing power or storage capabilities are required, these can be provided through available resources by enacting autonomic policies.
- *Cost structure*—By utilizing virtualization to dynamically allocate systems and storage resources, the overall cost of hardware and software can be normalized, and thus better controlled.

Overall, a virtualized environment provides simplified forms of access to data and IT resources in an odOE. As an example of one of the benefits, idle capacity can be used to meet unforeseen end-user traffic demands, and to reduce the need to purchase additional hardware and software. Staying with this example, savings can then be realized from reduced capital and operational expenditures; these savings can then be reinvested in other functional disciplines to help transform other areas of the business.

Virtualization in different forms has been present for at least some 40 years, dating back to the earlier mainframe computing days of the IBM System 360 (i.e., the early "virtual machines"). However, with the maturing of Grid Computing technologies, Autonomic Computing disciplines, and SANs, we are now transitioning from the days of virtual storage, or even virtual machines, to entire virtual computing environments, allowing businesses to gain even larger financial and business advantages.

On Demand Technologies

In previous sections, we discussed many strategy perspectives, which will also be explored in much greater detail throughout the remainder of this book. We presented some of the characteristics of an odOE and briefly discussed the business benefits that could be derived by implementing these types of environments.

In this section, we will introduce some of the technologies intersecting within the on demand ecosystem, and later in Chapter 7, we will discuss these technologies within the context of "service provider" environments.

There are many technologies, both new and evolving, that create an on demand ecosystem. It is important to understand that any odOE is always based on the implementation of a broad set of open standards, working together to provide a consistent and comprehensive set of fully integrated facilities.

In the next subsections, we will highlight these technologies as we describe how they apply to on demand computing and Web services.

Web Services

Web services provide the online facilities needed to access corporate data, legacy transaction systems, and new business applications. Web services are always accessed through a multitude of open standards Web browsers.

Based on open standards such as Java II Enterprise Edition (J2EE), extensible markup language (XML), and Simple Object Access Protocol (SOAP), advanced Web services are becoming the standard way to develop and deploy new applications in a timely and cost-effective manner. An interesting observation here is that Grid Computing standards and Web services standards seem to merging together.

Because of the standard Web viewing interfaces (i.e., Web browsers), many forms of advanced Web services can be combined into a single (virtualized) view. This is accomplished to generate new integrated applications quickly, while leveraging existing traditional systems with on demand business components.

The discussions surrounding advanced Web services are highly complex across several dimensions. This topic will be further explored in Chapter 7, as it is related to several different types of "service provider" environments.

Grid Computing As we will discuss in great detail in several chapters of this book, Grid Computing enables the virtualization of widely distributed Web computing resources, such as machine processing capabilities, advanced storage and capacity management capabilities, data storage and disposition capabilities, network provisioning, and bandwidth management capabilities. This is to present the appearance of a single computing environment. These integrated grid capabilities, combined in unison as a holistic on demand business solution, are utilized to create a single virtual system while allowing the capability of granting users and applications seamless access to vast IT resources across a widely dispersed geographical environment.

In simple terms, Grid Computing enables customers to get more business value out of what they own, by leveraging other resources outside their own odOE in concert with their own assets.

Through open technologies such as the Globus Toolkit for Grid Computing and emerging standards such as the Open Grid Services Architecture (OGSA) and Open Grid Services Infrastructure (OGSI), Grid Computing is no longer applicable to only scientific and research projects. Many businesses worldwide seriously look at Grid Computing solutions to realize higher infrastructure utilization capabilities while sustaining their own computing and storage resources, thus providing efficient and cost-effective management of their operational environments. For example, Grid Computing enables the dynamic reallocation of machine and networking resources to immediately manage end-user traffic volume spikes inline with unpredictable demands placed on the networks, or dynamically changing business requirements. This is a simple example, but it also presents a simple-to-understand scenario.

Autonomic Capabilities

As you might imagine, Autonomic Computing capabilities are being developed and integrated with hardware and software products across several important IBM product lines. Autonomic capabilities are considered to be fundamental to the odOE, and are utilized to automate many types of critical operational activities for very specific, desired business results.

Autonomic capabilities are (generally speaking) categorized into four very specific areas that involve high amounts of automation. These autonomic areas are discussed in the following subsections.

Self-Configuring

Self-configuring is the capability to dynamically configure a machine anywhere and at any moment in time. This also includes the capability for the machine to initialize itself in the context of the larger system and/or odOE. It also includes the ability to influence relevant environmental changes in other products in the overall ecosystem, (again) anywhere and at any moment in time. This is accomplished without human intervention.

Self-Healing

Self-healing is the capability to recover from a failing component (or portion of a larger system, or portions of [or entire] Web services) by first detecting improper operations. This is accomplished either proactively through predictions, or otherwise inferred through "intelligent" methods. This process initiates corrective actions without disruption (or minimal causal impacts) to the resources/applications in the ecosystem. This is accomplished without human intervention.

Self-Optimizing

Self-optimizing is the capability of systems or components in an odOE to efficiently maximize resource allocation and utilization to meet end-user demands. This is accomplished without human intervention.

Self-Protecting

Self-protecting is the capability of a component in an on demand ecosystem to detect hostile or intrusive systematic behaviors, often introduced by a security breach or an organized attack. This act of cyberdefense detection is accomplished as it occurs and takes immediate autonomous actions to protect the odOE. This is accomplished by executing multiple systematic, autonomic defensive actions, which in turn make the component safe and/or less vulnerable, minimizing risks to the overall odOE. This is accomplished without human intervention, yet it will almost always notify responsible personnel when activated in security situations.

Information Integration

Information integration assumes many forms in an on demand ecosystem, from federated databases, to virtualization, to content management, to all the information required to perform Autonomic and Grid Computing functionalities.

Federated databases allow data stored in various databases and file systems to be accessed through a common interface, such as the structured query language (SQL), hiding the complexity of the environment and actual location of the data. Using such technologies addresses the integration of applications and services while providing location transparency.

Content management systems enable users and applications to easily find and manage the data they need to perform their tasks.

The entire on demand ecosystem depends on information, as well as the engineering and processing of this information, to make it available when it is needed and in the form in which it is required.

Blade Computers

Blade computers allow large numbers of server-class computing resources to be tightly packed into very small and compact machine footprints. This technology provides more efficient and cost-effective utilization of space, making it easier to physically manage a large number of very powerful servers. Using blade computers is one form of server consolidation, which in the overall strategic and tactical objectives simplifies aspects of the physical management of hardware and software. By utilizing various virtualization and autonomic technologies, blade computers can be efficiently reconfigured and reallocated to meet changing demands quickly and simply.

Summary

In this chapter, we highlighted key technologies that are utilized to provide a flexible and powerful odOE. It is important to understand, however, that an odOE is not simply about individual technologies. It is about applying integration and infrastructure management capabilities toward the overall strategic nature of the business: This is first and foremost. These advanced capabilities can be tightly integrated so that each technology supports another in such a way that the final result is greater than the sum of the individual parts.

In this chapter, we explored odOE concepts that help to enable on demand business. We also discussed some of the fundamental technology underpinnings that deliver Autonomic Computing disciplines, which can contribute to odOEs in very significant ways.

Part 2
Autonomic and Grid Computing

This part of the book discusses several aspects of Autonomic Computing and Grid Computing. We will explore many fascinating aspects of autonomic capabilities, including a "Blueprint" for Autonomic Computing. We will explore Grid Computing, the future for Grid Computing, and the powerful strategies surrounding the Grid Computing discipline.

This part of the book will provide full treatment of Autonomic Computing, which is actually what makes Business On Demand work. We will explore how this fascinating computing strategy does this, and how it even mirrors the human body and many of its own complex functions, such as "self-healing." We will contrast Autonomic Computing to the human neural system.

This part of the book will also provide full treatment of Grid Computing, which is a worldwide initiative involving many companies and academic institutions, all of which have achieved some outstanding results and application deliveries. The IBM Corporation is a world leader in the Grid Computing discipline. Grid Computing deliverables are capable of providing some very powerful solutions in any Business On Demand strategy. This part of the book will explore these Grid Computing topics in great detail by describing both architectural and service infrastructure accomplishments from around the world.

In this part of the book, we will treat many application strategies in depth, and explore incredible aspects of the strategic thinking surrounding the Autonomic Computing and Grid Computing concepts. We will, at the same time, illustrate the power of becoming an on demand business.

Autonomic Computing Strategy Perspectives

3

This chapter will address one of the fundamental on demand business strategy perspectives, *Autonomic Computing*. This chapter's focus on Autonomic Computing will emphasize both strategic and technological perspectives. The overall on demand business strategy focuses on two computing areas of interest: *Autonomic Computing* and *Grid Computing*. It is the Autonomic Computing disciplines that provide the necessary efficiencies required to conduct on demand business.

Previously, we discussed the need for critical intersections to occur among IT, business operations, and autonomic transformations. We introduced the three high-level aspects of on demand business and the concepts driving these types of environments. In this chapter, we will provide further treatment of the strategic perspectives defining the on demand business world of operations, including the IBM Autonomic Computing strategy.

Here is another example of Autonomic Computing: A computer intermittently freezes up. No customer transactions can be processed for several seconds, costing thousands of dollars in business and customer loyalty. Today, the IT support staff might not even find out about the problem for more than a day; and when they do, it may take a couple days to figure out what the problem is and which of the multiple scenarios matches the problem.

With an autonomic system in place—with real-time event monitoring and auto-tuning analysis—the freeze-up is detected the very first time it happens and matched against historical problem data. The settings are reset automatically, averting any real loss of revenue and customer loyalty. A report then shows the actions that were taken to resolve the issues.

And yet another example of Autonomic Computing: If an airline starts a fare sale and is hit by a high volume of customer inquiries, it would take less then a minute for the autonomic software to determine that more power is required and add another computer. The system can also turn off computers as they are no longer needed.

Autonomic Computing introduces *autonomic* efficiencies into the overall scheme; however, Autonomic Computing alone does not entirely constitute an on demand business. As we discussed in Chapter 1 of the book, on demand business involves three fundamental activities to transform to an efficient on demand Operating Environment.

ON DEMAND BUSINESS...

The IBM Corporation has defined *on demand business* as an enterprise whose business processes—integrated end-to-end across the company and with key partners, suppliers, and customers—can respond with agility and speed to any customer demand, market opportunity, or external threat.

Respective to the Autonomic Computing vision, are capabilities like maintaining security against undesirable intrusions, enabling fast automatic recovery following system crashes, and developing "standards" to ensure interoperability among myriad systems and devices. Systems should also be able to fix server failures, system freezes, and operating system crashes when they occur; or better yet, prevent them from developing in the first place.

An on demand business is a way of doing business (a strategy), and the on demand Operating Environment consists of the systems and procedures that are used to do business (executing the strategy). As a review of our previous discussion, and to set the stage for this discussion on Autonomic Computing, let us again consider these three key activities. Each activity is paramount to the successful transformation of any corporation, enterprise, or organization with a desire to be an on demand business. These three activities are:

1. Engaging in several "levels" of streamlining and transforming one's overall business enterprise processes.

2. Embracing new mechanisms for the delivery of services and pricing, which is also referred to in the industry as "*Utility Computing*."
3. Ensuring the enterprise is delivering and sustaining a flexible operating environment.

As described throughout much of this chapter by David Bartlett (IBM's Director of Autonomic Computing), Autonomic Computing plays an important role in several on demand business activities. Several fundamental questions arise as one begins to consider what is involved in the on demand transformation, as illustrated Figure 3.1:

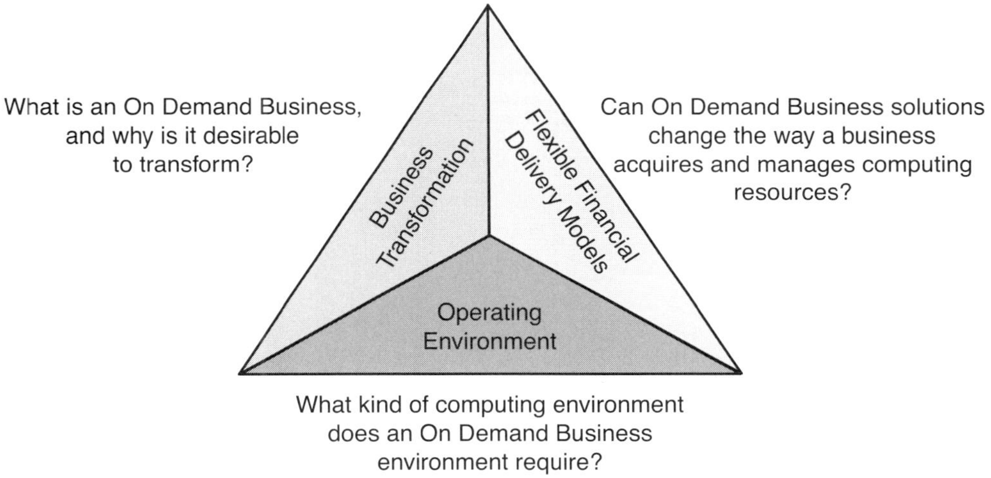

FIGURE 3.1 The key focus areas in Autonomic Computing, which must be addressed while creating an on demand business environment. The questions unveil critical thinking related to a business' transformation.

Exploring these three key questions begins to unveil what is necessary to build an on demand Operating Environment. Autonomic Computing is all about business transformation, which requires engaging in several "levels" of streamlining and transforming one's overall business enterprise processes. This will involve the shift of IT infrastructure from a reactive, human-intensive management paradigm to an adaptive, technology-balanced approach, supported by Autonomic Computing technologies. *Utility Computing* is also involved; this embraces new mechanisms for the delivery of services and pricing methods. The Utility Computing approach also involves the integration of attractive, new, flexible financial delivery models.

Let us explore a bit more the sections illustrated in the pyramid of Figure 3.1:

Business transformation involves transforming an organization's strategy, processes, technology, and culture by applying deep business process insight with some combination of advanced technologies to increase business productivity and enable flexible growth.

The *operating environment* is typically an approachable, adaptive, integrated, and reliable infrastructure for delivering on demand services to an on demand business operation.

Flexible financial and delivery offerings embrace the delivery of business processes, applications, and/or infrastructure On Demand, with usage-based charges around IT and/or business metrics. This is a new approach for most IT departments and service providers, either working independently or together. It is paramount to more effectively align IT assets with business priorities, provide reliability through more granular SLAs (Service Level Agreements), and achieve cost savings through increased utilization and proactive management.

Autonomic Computing, with its focus on reducing management costs, leverages the Utility Computing approach by enabling infrastructure and on demand business process services to subscribing clients at the lowest possible cost. Autonomic Computing, quite obviously then, has the most obvious and direct connection to ensuring that one is delivering and sustaining a very flexible operating environment.

Consider the following question, which is being asked in Figure 3.1: What kind of computing environment does on demand require, and how do I build one? The answer will ultimately include functions and features such as the following: Monitoring, workload management, provisioning, dependency management, and policy-based computing. All these features are at the core of both the on demand Operating Environment and the Autonomic Computing framework.

Establishing the fundamental on demand business intersections for any company involves the implementation of standardized and automated processes, applications, and infrastructures over networks and advanced services with business and IT functionality. As published in the *Silicon Valley Business Ink* in April 2003 (by Alan Ganek [Ganek01], Vice President of Autonomic Computing for IBM Corporation's Software Group), the following article describes some innovative thinking and strategy perspectives regarding Autonomic Computing.

In this reprinted article, Alan Ganek states:

> Technology is like a race car going 200 miles per hour on the fastest track on Earth. Systems crash. People make mistakes. Computers need a lot of maintenance to keep them up and running. This is life. At the same time, consumers have become increasingly intolerant of computer failures, while placing ever-greater demands on the technology they utilize.

The high-tech industry has spent decades creating systems of marvelous and ever-increasing complexity, but like Charlie Chaplin falling into the machine in "Modern Times," complexity itself is the problem.

Consider this fact: One-third to one-half of a company's total IT budget is spent preventing, or recovering from system crashes, according to a recent study by the University of California: And this is no wonder. A system failure at a financial firm or online retail store can cost millions of dollars in lost business.

Approximately 40 percent of computer outages are caused by operator errors. That is not because operators are not well-trained. It is because the technology is difficult to figure out, and IT managers are under pressure to make decisions in seconds.

So we are headed for a wall.

Businesses cannot roll-in processors and storage fast enough to avoid meltdowns when usage spikes, fend off viruses and hacker attacks, or manage the different operating systems that access information. People are good, but they are not that good.

Since no one is close to writing defect-free software or hardware, we need computers that are capable of running themselves, with far greater levels of intelligence built into the technology.

We are not talking about computers that can write the next "Ninth Symphony." We are talking about the same kind of intelligence we take for granted in our own bodies. We walk up three flights of stairs and our heart rate increases. It is hot, so we perspire. It is cold, so we shiver. We do not tell ourselves to do these things, they just happen.

If systems and networks adopt these attributes, managers could set business goals and computers would automatically set the IT actions needed to deliver them. For example, in a financial-trading environment, a manager might decide that trades have to be completed in less than a second to realize service and profitability goals. It would be up to software tools to configure the computer systems to meet those metrics. The implications for this "autonomic" business approach are immediately evident: A networking service of organized, "smart" computing components that give us what we need, when we need it, without a conscious mental or even physical effort.

The goal is to increase the amount of automation that businesses need to leverage, extend, and sustain. Because the more that you can get human error out of the loop, the more efficient your business will become—whether you are a financial institution, a shipping company, an automotive manufacturer, an airline, or an online retailer. The beauty of it is that all of these complexities are hidden from the user.

The logic is compelling: Relief from the headaches of technology ownership and maintenance; an improved balance sheet; and, a much greater flexibility in meeting the demands of running a business.

However, in the end, perhaps the greatest benefit would be the freedom it would unlock. Sure, it will create enormous efficiencies. But the game-changing impact will be freeing up all companies—whether just starting out or well-established. The following figure shows the high-level problems being addressed and solved by Autonomic Computing.

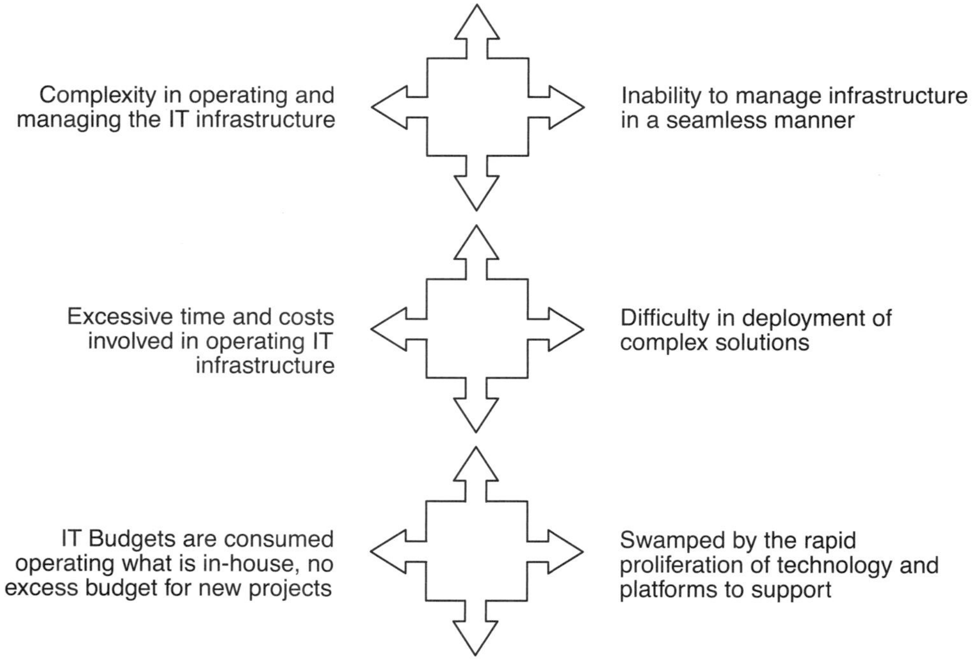

FIGURE 3.2 Where you start depends on the on demand business priorities. Increasing flexibility and reducing risk is the key—business models, processes, infrastructure, plus financing and delivery.

These are all outstanding insights that provide some of the strategic perspectives toward removing the mystery of Autonomic Computing.

The Autonomic Goal Is Simplicity

"The goal is to increase the amount of automation that businesses need to sustain. Because the more that you can get human error out of the loop, the more efficient your business will become—whether you are a financial institution, a shipping company, or an online retailer. The beauty of it is that all of these complexities are hidden from the user."

—Alan Ganek, Vice President of Autonomic Computing for IBM Corporation's Software Group.

Perhaps you are now thinking about what would be involved for your company or organization to become an on demand business operation, with a multitude of Autonomic Computing transformation efforts successfully completed. Or, perhaps you are wondering what this would mean for the many employees in your company, should you execute such a transformation. If not, then you might be wondering how effectively you have been able to accomplish what the contents of this book address—establishing on demand business operations, including critical Autonomic Computing and maybe some key Grid Computing success stories.

Alan Ganek, in one of his public speaking engagements related to Autonomic Computing, mentions: "The goal of our Autonomic Computing initiative at IBM is to help you build more automated IT infrastructures to reduce costs, improve up-time, and make the most efficient use of increasingly scarce support skills to get new projects on line more rapidly." This strategic perspective summarizes the fundamental values of Autonomic Computing.

To transform any organization or company to an on demand business operation, developers and IT professionals will be expected to help transform key areas of the technical infrastructure to support the integration of on demand processing. Business leaders will need to foster a way of thinking across the workforce to focus skills and critical thinking toward a common transformation goal. Managing innovation will become a staying thread for the marketing and sales teams, technology practitioners, and leading strategic thinkers. This is not an impossible task, nor is it a revolution, simply a transformation toward a more effective means of conducting business.

In the on demand business *flexible hosting* model, customers leverage on demand services only as they are required. The major resulting benefit from this flexible hosting-based approach is that the customer pays only for the "flexible hosting service" by the transaction, according to usage amounts, and is insulated from the core infrastructure used to deliver these types of flexible hosting services. These flexible hosting approaches allows customers the ability to quickly and easily leverage many on demand business products and services to conduct their on demand business activities in a seamless manner.

Customer benefits realized in the on demand business flexible hosting model strategies include the ability to "pay as you go" for services. This improves the ability to better forecast technology needs, to be able to activate a service on demand at the point of need, and to access additional capacity and resources for only short periods of time. For example, a company could request autonomic capacity provisioning or server utilities if it suspects having to sustain increased levels of Web traffic (at any unknown time); later, this autonomic flexible hosting service capability could be turned off, as the increased Web traffic is realized and begins to reduce in intensity.

Although this flexible hosting approach to computing is very important and noteworthy to understand in the context of on demand business, this chapter will focus on the major fundamental concepts in the IBM on demand business strategy, Autonomic Computing, and less on IBM's flexible hosting computing concepts.

Consider the fact that a realistic vision is an achievable mission. And, not surprisingly, many of today's existing IT infrastructures are not staged for the kind of dynamic, responsive, integrated operating environment required for being a true on demand business enterprise. The IBM Corporation is, however, operating in an on demand Operating Environment today, and has carefully positioned the entire company to help others interested in achieving this same type of on demand Operating Environment. IBM continues to help many global customers develop a realistic vision, to create and execute on their own achievable on demand business mission.

At IBM, we recognize four essential transformation characteristics for the on demand Operating Environment. These characteristics are key for any company to consider, as it enters the on demand strategic transformation process. These on demand business transformation characteristics are defined as:

- *Integrated*—This is the key to the castle. Data must maintain its integrity and will require transaction processing of the highest order across custom applications throughout the enterprise. Instead of integrating "vertically" (within the operating system of the computer), applications will integrate "horizontally," freeing them from the restrictive underpinnings of their underlying infrastructure.

- *Virtualized*—Companies will be able to access and pay for computing the same way they get electricity, telecommunications, or water—by the flip of a switch, the push of a button, the turn of a knob, or the click of a mouse. When traffic volumes or transactions unexpectedly spike, capacity can be added automatically over the 'Net. When things are quiet, your company pays less, and capital is then freed up to invest back into your business.
- *Open*—With most companies already having made huge investments in technology, the ability to "rip and replace" an entire system is just not an option. Open technical interfaces and agreed-upon standards are the only realistic way that many business processes, applications, and devices will be able to connect.
- *Autonomic*—The term "Autonomic Computing" is from the human anatomy's autonomic nervous system. The same way we take for granted the human body's management of breathing, digestion, and fending off viruses, companies will one day take for granted the network's ability to manage, repair, and protect itself. Figure 3.3 shows control measures with resources.

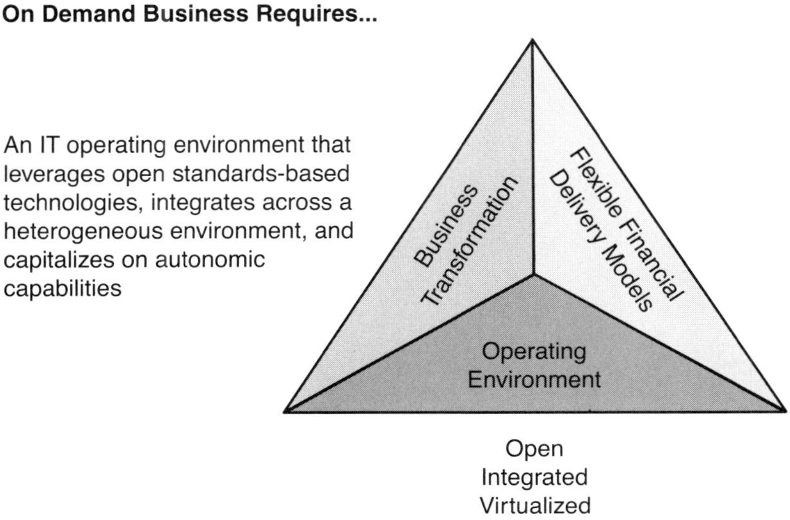

FIGURE 3.3 This illustration shows automonic control measures as applied to resources (year 2000).

IBM has realized that *networking services* are the key enablers of all on demand business ecosystems, and these complex, autonomic networking services grids must be carefully considered in the evolutionary on demand business transformation. A key consideration here is the fact that networks and the complex services they provide are critical to the success of any company in the on demand transformation. In on demand Operating Environments, especially concerning elements of Autonomic and Grid Computing, networking services quickly become a major dependency in the success of the overall mission, and are no longer a hidden assumption in the overall IT equation.

THE COSTS OF DOING BUSINESS

"A decade ago, hardware soaked up 80 percent of the cost of a data center," said Alan Ganek, Vice President, Autonomic Computing, IBM Software Group. *"Now, half the money spent goes for the people to run the center."*

More intuitive computers will provide a buffer for more complex IT systems, Ganek said.

The IBM Corporation's vision of Autonomic Computing embraces the development of intelligent, open systems and networks capable of running themselves, adapting to varying circumstances in accordance with business policies and objectives, and preparing resources to most efficiently handle the workloads placed on them. These Autonomic Computing systems manage complexity, "know" themselves, continuously tune themselves, and adapt to unpredictable conditions, all while seeking to prevent and recover from failures.

Just as the human body grows throughout life, the movement to a fully self-managing Autonomic Computing environment can only be realized on a gradual deployment model. IBM has been working for several years to make this long-range vision a reality with the introduction of advanced technologies in many current IBM products and services. These innovative developments in Autonomic Computing strategies and technologies deliver computing systems that offer both IBM and our global customers improved resiliency, accelerated implementation of new capabilities, and increased ROI in IT, while providing safe and secure operational environments.

ON DEMAND BUSINESS

To transform into an on demand business operation exploiting Autonomic Computing, one must manage transformation and effectively foster a new way of thinking. Senior technologists and key business leaders become the core of this evolutionary thinking process: exploration by many strategic thinkers of how Grid Computing and Autonomic Computing become fundamental in many aspects of the operational model. This leads to a fundamental operational graph, which everyone must be able to envision within his or her own domain(s).

A new "success" agenda for the entire workforce must be cascaded across the organization. "Lead from the front" is the single most important message. This helps to build and substantiate the *on demand business roadmap* for success in your organization.

The *on demand business* evolution, incorporating Autonomic Computing, is not an overnight transformation in which system-wide, self-managing environments suddenly appear inside the infrastructure. Autonomic Computing must be a gradual transformation to deliver new technologies that are adopted and implemented at various stages and levels. These transformation levels are born at the most basic levels of the infrastructure and business, and transition across defined levels to fully autonomic operations.

Autonomic Computing is focused on the most pressing problem facing the IT industry today: the increased complexity of an IT infrastructure that typically accompanies the increased business values delivered through computing advancements. This problem is contributing to an increasing inability of businesses to absorb new technology and solutions.

The scope of Autonomic Computing in any operating environment must address the complete domain of IT management complexities—marginal improvements will be afforded by individual technologies—or single operating environments will not be sufficient enough to sustain the computing advances that the IT user community requires and expects.

All elements of the IT system must be included. The deployment of business solutions invariably involves the coordination of all IT resources, including the following: Servers, storage, clients, middleware, applications, and network services. Furthermore, Autonomic Computing initiatives must insure growth in the levels of autonomic compliance in each of these areas. They must also be able to support all the platforms and/or vendors that supplied the IT infrastructure elements.

This multi-level autonomic journey will be an evolutionary process, while allowing each business to adopt Autonomic Computing capabilities in a non-obtrusive manner and at its own speed. Autonomic Computing delivers quantifiable savings and other qualitative values to customers as functions/features in the new version of products (both computing elements and management offerings). Multiplicative values will be realized when the Autonomic Computing capabilities of each product are integrated to deliver a fully autonomic system. Existing IT implementations will require a prescribed migration path/plan through which a fully Autonomic Computing environment and its associated benefits will be realized. The realization of an on demand Operating Environment must be driven from industry collaboration. The industry-wide advancement of Autonomic Computing technologies is based on open and de facto industry standards, which is the most feasible approach to transformation.

Paul Horn issued a challenge to the IT industry in 2001 to work together to solve the growing problem of complexity faced by IT customers. IBM has stepped up to that challenge and taken on the industry-wide leadership. This industry-wide leadership is what differentiates IBM from other suppliers' individual autonomic, adaptive, or organic computing initiatives. IBM's leadership makes sense, given the breadth and depth of IBM's experience across the most extensive product and services portfolio(s) in the industry, and IBM's open standards-based approach and commitment to work with its products and services across global heterogeneous environments.

IBM's mission, in addition to leading the industry in on demand business, is to establish IBM products as the best possible examples for Autonomic Computing that deliver distinct values and IBM services; and leads the facilitation for the adoption of Autonomic Computing technologies. This will be accomplished by dedicated focus and execution at an unprecedented level of integration across IBM, in response to Sam Palmisano's (IBM's Chairman and Chief Executive Officer) *"On Demand"* call to action in 2002.

The Autonomic Computing business strategy for supporting this transformation includes several very important Autonomic Computing initiatives. These perspectives are not only focused on IBM transformation accomplishments, but also are extensible to the entire IT industry, worldwide. These transformation initiatives are as follows:

1. An overarching architectural framework, or blueprint, and corresponding open standards that underpin the industry's development of autonomic technologies
2. An IT deployment model that defines each progressive level of autonomic maturity
3. New, integrating core Autonomic Computing technologies that when combined with existing products, serve to fulfill the architecture and deliver on the vision of the on demand Operating Environment
4. Autonomic Computing technological enhancements to existing product lines that conform to the architecture and standards that enable level progression
5. Autonomic Computing service offerings that can define the roadmap of required on demand business initiatives for any given business and deliver the technology and services as prescribed by those initiatives
6. Customer/partner programs to co-create standards and technologies and validate the openness of our initiative to get the most relevant and complete self-managing systems to market sooner

The following points explore each of these Autonomic Computing perspectives, which are extensible across all global industries. Consider deliverables such as the following in your organization:

1. *An overarching architectural framework, or blueprint, and corresponding open standards that underpin the industry's development of autonomic technologies*

   Architecture, technology, and standards deliverables drive the industry toward a common technical vision of Autonomic Computing. This *Autonomic Computing Blueprint* will be an overarching view of the technology underpinning Autonomic Computing. These technological underpinnings depict how the various key technologies, interfaces, services, functions, and capabilities all fit together to form an end-to-end technical framework. The purpose of the framework is to set the context of what is meant, technically, by the terms of Autonomic Computing, and to show relationships. This framework is an active enabler in the quest to realize the on demand business vision. This is discussed in more detail in later sections of this chapter.

2. *An IT deployment model that defines each progressive level of autonomic maturity*

   As we introduce the need for business transformation in these early chapters, let us again review this evolution. The following list and Figure 3.4 prescribe these five levels of transformation toward achieving a refined state of Autonomic Computing, which is required in every on demand Operating Environment. These levels are as follows:

   - *Level 1*: Basic—The starting point where most systems are today, this level represents manual computing, in which all system elements are managed independently by an extensive, highly skilled IT staff. The staff sets up, monitors, and eventually replaces system elements.
   - *Level 2*: Managed—Systems management technologies can be used to collect and consolidate information from disparate systems onto fewer consoles, reducing administrative time. There is greater system awareness and improved productivity.
   - *Level 3*: Predictive—The system monitors and correlates data to recognize patterns and recommends actions that are approved and initiated by the IT staff. This reduces the dependency on deep skills and enables faster and better decision-making.
   - *Level 4*: Adaptive—In addition to monitoring and correlating data, the system takes action based on the information. This can be mapped to SLAs, thereby enhancing IT agility and resiliency with minimal human interaction while insuring that the SLAs are met.

- *Level 5*: Autonomic—Fully integrated systems and components are dynamically managed by business rules and policies, enabling IT staff to focus on meeting business needs with true business agility and resiliency.

	Basic Level 1	Managed Level 2	Predictive Level 3	Adaptive Level 4	Autonomic Level 5
Characteristics	Multiple sources of system generated data	Consolidation of data and actions through management tools	Systems monitors, correlates and recommends actions	System monitors, correlates and takes action	Integrated components dynamically managed by business rules/policies
Skills	Requires extensive, highly skilled IT staff	IT staff analyzes and takes actions	IT staff approves and initiates actions	IT staff manages performance against SLAs	IT staff focuses on enabling business needs
Benefits	Basic requirements Met	Greater system awareness Improved productivity	Reduced dependency on deep skills Faster/better decision making	Balanced human/system interaction IT agility and resiliency	Business policy drives IT management Business agility and resiliency
	Manual				Autonomic

FIGURE 3.4 The various transformation levels of Autonomic Computing.

Although we will go into more detail on these autonomic levels of transformation in the forthcoming section titled "An Architecture Blueprint for Autonomic Computing," let us begin to explore these levels now to better understand operational positioning. Consider the following as a map, or route, for the future of driving efficiencies into business enterprise operations. To assist one in this assessment, IBM is also publishing guidelines and "adoption models." In summary, these five levels show this evolution (not revolution) based on one's need for investment protection.

Delivering autonomic IT infrastructures is, indeed, an evolutionary process that is enabled by technology; however, this transformation is ultimately implemented by each enterprise through the adoption of these technologies, along with supporting business processes and the proper critical skills.

IBM views these five levels as a map of the future state of business for any enterprise engaged in the on demand business journey. The representation of these levels starts at the basic level and progresses through the managed, predictive, and adaptive levels, and finally to the fully autonomic level.

The first level, the *basic level*, represents the starting point where many IT infrastructures are today. At this level, one will note that IT professionals who set it up, monitor it, and eventually replace it will typically manage each element in the infrastructure independently.

At the second level, the *managed level*, systems management technologies can be utilized to collect information from disparate systems onto fewer consoles, thus reducing the time it takes for the administrator to collect and synthesize information as the systems become more complex to operate.

At the third level, *predictive*, as new technologies are introduced that provide correlation among several elements, the infrastructure itself can begin to recognize patterns, predict the optimal configuration, and provide advice on what course of action the administrator should execute.

As these technologies improve and as people become more comfortable with the advice and predictive power of this infrastructure, we can progress to the fourth level, the *adaptive level*. At this level, the elements themselves, via *closed loop automation*, can automatically take the right course of action based on the information available to them, and the knowledge and state of what is actually occurring in the infrastructure.

Finally, in this five-level transformation process, to achieve the fully *autonomic level*, the fifth level, the IT infrastructure must be governed by business policies and objectives.

Many industry analysts deem this multi-level "evolution" as the right approach to Autonomic Computing, ultimately delivering the arrival of companies to fully enabled on demand Operating Environments.

3. *New, integrating core Autonomic Computing technologies that when combined with existing products, serve to fulfill the architecture and deliver on the vision of the on demand Operating Environment*

   Integrating core Autonomic Computing technologies will serve as the fundamental common "building blocks," ensuring consistent implementations and behaviors of autonomic systems. This will also facilitate the integration and interoperability of many heterogeneous components. After surveying what autonomic functionalities exist in the marketplace, a pattern to meet one's business requirements will become obvious across most technologies. This pattern will be related to individual products, services, or operating environments.

To realize the autonomic vision in the majority of customer IT implementations that are heterogeneous in nature (i.e., multiple products across multiple operating systems), the strategy must include architectures and standards for the creation of the integrating core Autonomic Computing technologies; for example, technologies allowing end-to-end self-management capabilities across the heterogeneous environment. Furthermore, the very business processes that necessitated the need for the heterogeneous mix of technology to begin with must now ultimately drive these management actions. The evolution of technology also requires the evolution of processes, skills, and their respective linkages to achieve the on demand Operating Environment. Figure 3.5 shows a management interface.

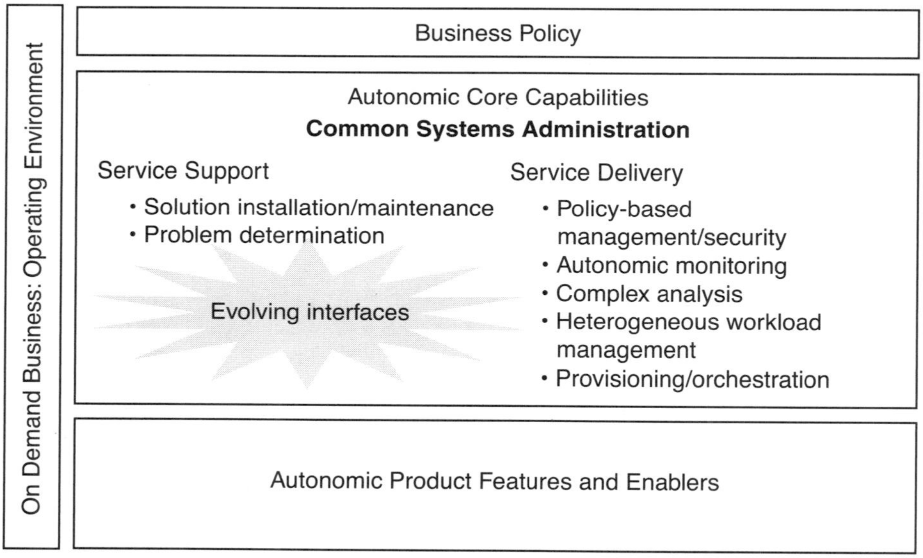

FIGURE 3.5 Core capabilities for enabling Autonomic Computing.

The following examples of Autonomic Computing core technologies illustrate why they are so critical as key foundational elements of the overall strategic approach.

The first example is a set of integrating core technologies required in the area of problem determination. The industry today is largely represented by a vast number of individual, product-unique approaches to problem identification and resolution. Even within individual products, problem determination is often inconsistent and incomplete in its approach and implementation. In most cases, a product's problem determination constructs have the appearance of an afterthought, as opposed to the product of a well-thought–out, robust, and extensible design.

This "afterthought" problem is a systemic and pervasive concern across many industrial solutions. So often, the focus is on advancing, new, and leading-edge technology or functionality. And then, in retrospect, the field support of that technology remains as an afterthought, perhaps based on the assumption that the technology will be developed without fault.

Consider for a minute some comments made by a college student at one of the top engineering universities in the U.S. This discussion concerns the student's participation in a robotics project as part of an international competition. The robot was able to perform a number of very impressive maneuvers, employing some truly innovative ideas that were implemented by the development team. As we discussed the project, there was a lot of excitement and passion for the newly created abilities of the robot, but far less enthusiasm when the conversation turned toward robot maintenance.

The need, therefore, is to define, standardize, and integrate an industry-wide approach to problem determination to achieve *self-healing* in multi-component environments. This will always consist of a standards-based approach to data capture, analysis, and remediation to realize the self-healing aspects, and would practically be achieved by a phased approach, over time, represented by incremental levels of increasing autonomic maturity.

The first step is to get the right data from the system, in a consistent, standards-based format. The next step consists of putting a set of symptoms and corresponding actionable causes in a consistent format, and building tools that can correlate the data to match against a cross-product, standards-based problem/symptom database: in other words, autonomic event correlation.

We ultimately want to automate fixing defects by being able to automate the provisioning of an application with standardized fixes (or temporary workarounds) based on the business policies that govern each application.

While we evolve IT infrastructures toward self-healing, there are many benefits that can be realized and are already having an impact. Here are two examples: First, the common format for log entries, submitted as a standard, is dramatically reducing training time for administrators and providing a consistent format to evaluate multiple logs together. Second, automated correlation engines are reducing manual analysis by providing a programmatic method to correlate the logs that are adapted to the common format. Figure 3.6 shows autonomic nervous system compared to the autonomic model.

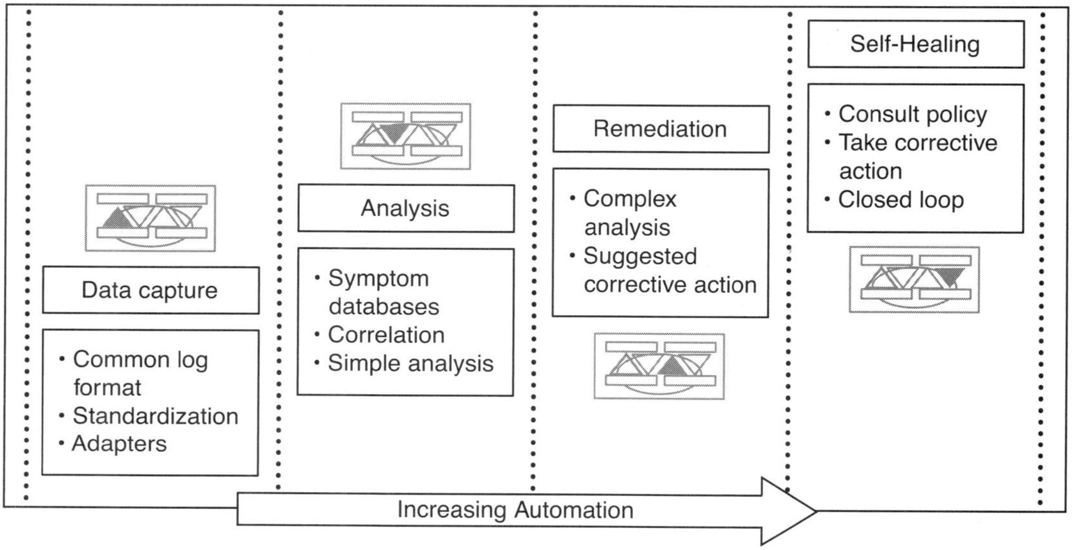

FIGURE 3.6 The various levels involved in creating self-healing systems.

Other examples, which are discussed in more detail later in this chapter, include:

- An integrated solutions console for common system administration that addresses the complexity of operating multiple heterogeneous products, each with its own end-user interface. The idea here is to provide one consistent administration user interface for use across an IT infrastructure product portfolio using WebSphere Portal Server as a basis. This would include the provision of a common runtime infrastructure and development tools based on industry standards.
- Consistent software installation technology across all products that provides solution packaging standards for defining complete installation end-to-end solutions and consistent and up-to-date configuration and dependency data, which are key to building self-configuring autonomic systems.
- Policy tools for policy-based management that provide uniform cross-product policy definition and management infrastructure needed for delivering system-wide self-management capabilities that map automated actions to the needs of the business. This would include the

development of the ability to describe business needs, rules, and policies in system-readable form.

- Workload balancing across heterogeneous systems requires infrastructure enablement, allowing business transactions to operate in a resource-balanced way across server pools (e.g., in heterogeneous environments).
- Policy-based, intelligent orchestration and provisioning that drive infrastructure provisioning, capacity management, and service-level delivery across the automation environment and enable the ability to rapidly deploy a complete application environment. This allows one to re-purpose computing resources for uses such as application staging, load testing, and support of anticipated peaks in application demand.

Integrating such core technologies will have a profound effect on the reduction of complexities (and costs) for the customer, and also for the supplier, which can take advantage of component reuse as well as having a dramatic impact on service costs. These cost savings can then be applied to continue to fund the advancement of autonomic technologies and their respective implementations.

The successful adoption of these technologies will also require corresponding adjustments to the IT processes and skills that are impacted by these core technologies. This must be a conscious effort that begins at the early design stage of both the development of the technologies as well as the customer adoption of these technologies. The creation of flow sequence diagrams is required to document the interaction patterns between humans and products in the accomplishment of service and system management. These sequence diagrams will be used to identify requirements for IBM and industry products to improve the value of Autonomic Computing core technologies and Autonomic Computing functions and features.

4. *Autonomic Computing technological enhancements to existing product lines that conform to the architecture and standards that enable level progression*

The work must take place across IBM and the industry to deliver industry-leading Autonomic Computing functions and features while simultaneously insuring existing product offerings can be easily integrated within an autonomic system environment. This includes driving continued implementation of specific self-management product capabilities into product offerings along four dimensions, as depicted in Figure 3.7:

- *Self-configuration* features to provide greater ease of use and increased availability

- *Self-healing* functions to prevent customer downtime
- *Self-optimizing* functions to provide the highest utilization of customer resources
- *Self-protecting* functions to safeguard customer access to system resources and protect customer data

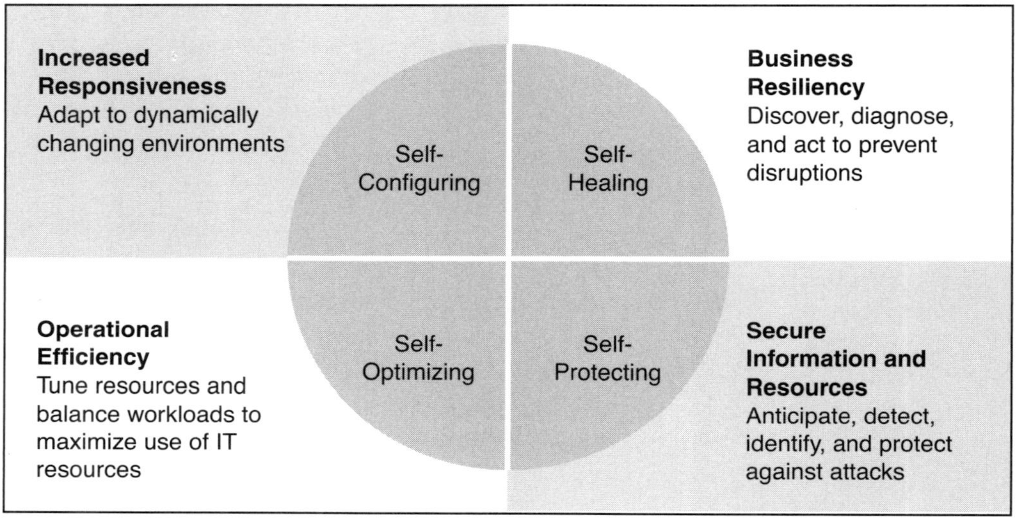

FIGURE 3.7 The various Autonomic Computing self-managing capabilities.

Additionally, it is important to enable the integration of product elements within an autonomic system environment by ensuring adoption of common standards. This is accomplished by applying autonomic core technologies and compliance with strategic Autonomic Computing architecture guidelines for evolving autonomic system behaviors.

Here are some examples of how Autonomic Computing technologies enable IBM brand offerings:

Servers—Servers that respond to unexpected capacity demands and system errors, without human intervention, yielding dramatic improvements in the server's reliability, availability, and serviceability while simultaneously significantly reducing both downtime and cost of ownership.

Storage—Storage systems that utilize hot spot detection and online data migration techniques to guide object placement (e.g., data and applications) in a shared storage pool. The details of low-level object-to-device assignment, disk space provisioning, automated failure detection and recovery, and storage network traffic management are handled invisibly by the system while providing an accurate depiction of storage capacity and bandwidth trends.

Software—Self-diagnosing, self-repairing, and self-managing behaviors in software that will yield more cost-effective customer service and prevent unethical hacking attacks, as well as IT systems that operate together more efficiently in the accomplishment of business objectives. For example:

- *Integration middleware software*—The system will allow users to express policy and what they believe should happen, and then the software will drive the execution. Pre-emptive diagnostics will automatically recognize and solve problems such as configuration settings, software updates, provisioning, and load balancing across application server clusters, and intercept/block unauthorized system calls.
- *Data management software*—Sophisticated database management tools such as performance and health monitors, recovery experts, and configuration advisors are complemented by capabilities that result in operating parameters changing in real time to better address changes in the environment or business priorities.
- *Systems management software*—Automatically deploy, update, track, repair, and manage equipment, software, and configuration changes to provide better alignment between business priorities and IT.

Systems Management	• Access/Identify/Risk Managers • Storage Resource Manager • Service Level Advisor
Client	• ImageUltra • Rapid Restore PC • Embedded Security Subsystem
Application	• Prioritization of User Transactions • EdgeServer Performance Optimization • Problem Analysis and Recovery
Database & Collaboration	• DB2 Query Patroller • DB2 Configuration Advisor • Tivoli Analyzer for Domino
Servers	• Dynamic Partioning • IBM Director • BladeCenter • Electronic Service Agent
Storage	• Intelligent Cache Configuration • Predictable Failure Analysis • Dynamic Volume Expansion

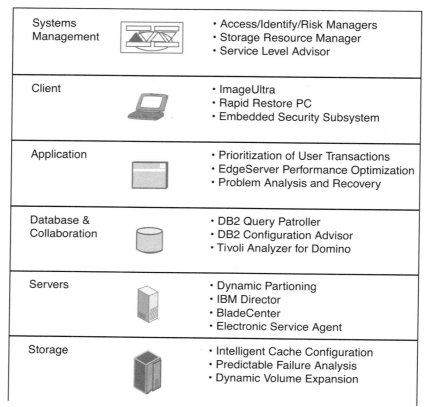

FIGURE 3.8 IBM Autonomic product capability examples.

It is helpful to employ metrics to guide and track progress in moving products forward toward Autonomic Computing goals. This will include tracking new autonomic features, adoption of core technologies, and measurements that reflect the reduced complexity achieved by each release in deploying and using the products.

5. *Autonomic Computing service offerings that can define the roadmap of required on demand business initiatives for any given business and deliver the technology and services as prescribed by those initiatives*

   Automation methodologies and tooling for systems integrators communicate the benefits of using new autonomic technologies as they become available in the delivery of customer solutions, providing the enterprise and its partners with Autonomic Computing service offerings. This will help solve specific customer problems as these new technologies become available, while providing insights as to how these technologies can best be utilized in various scenarios.

To this end, IBM has developed an Autonomic Computing Adoption Model assessment, which is a six- to eight-week holistic assessment of the autonomic maturity of an IT implementation. This provides the most practical set of initiatives required to advance the state of the on demand Operating Environment. The assessment encompasses not only the technology, but also the processes that surround the technology and the required adjustment of skills and organizational constructs. This Adoption Model is based on experiences from IBM customer engagements, and leverages best practices for managing systems while providing the opportunity to leverage the most recent Autonomic Computing core technologies from IBM and industry partners.

6. Customer/partner programs to co-create standards and technologies and validate the openness of initiatives to get the most relevant and complete self-managing systems to market sooner

Partnerships through customer and business partner programs provide the opportunity to get more complete and relevant self-managing systems to market sooner. The partner ecosystem of Autonomic Computing runs the gamut, from element vendors (e.g., servers, storage, middleware, and network services and equipment) to systems integrators, to system management partners, development information systems vendors (ISVs), and channel partners. IBM has well-established partner programs that are being leveraged to engage partners through joint collaboration, thus actively engaging the industry as a whole through participation in standards-based work.

An obvious element of the IBM strategic perspectives is to build on the industry-wide on demand business vision to rally participation and endorsement from other enterprises in global industries. Autonomic Computing covers a broad spectrum of partners, including:

- Vendors that provide technology at all levels of the customer solution stack on concepts, architecture, and standards, and how to best leverage each product and solution.
- Systems integrators that provide services and implement network designs, including hardware and software, which incorporate Autonomic Computing specifications and provide technical support post-installation.
- Development ISVs that license Autonomic Computing technology for embedding within broader customer solutions. (In some cases, IBM's technology is directly embedded; however, in other cases, joint development and integration are needed to accelerate time to market.)

The adoption rate of the Autonomic Computing architecture and related core technologies depends on the IT industry's software developers having relevant open standards. It also depends on open access to architectures and integrating technologies in the form of toolkits, or Software Development Kits (SDKs), and partner programs that make it easy to incorporate them into their products. Partner enablement programs have been established using IBM Solution Partnership Centers to enable ISVs to code, port, and test their relevant Autonomic Computing functions/features for use in autonomic solution offerings. IBM also provides individual technical assistance for partners, and wherever possible, develops references and joint marketing materials that promote the concepts of Autonomic Computing and on demand business.

While engaged in the provisioning of partner enablement technologies, it is important to supply SDKs to make it easier for industry partners operating in the Autonomic Computing arena to align with Autonomic Computing interfaces and information architectures. Autonomic Computing toolkits are utilized to build autonomic managers that implement Autonomic Computing specifications; allowing vendors to easily adapt their products according to the specifications also utilize them. IBM on demand business design centers have also proven to be highly effective in communicating Autonomic Computing messages to prospective customers, and engaging with customers in design workshops, solution assessments, and developing "proofs of concept" for their Autonomic Computing applications. This includes technologies and architectures in the solution set of any customer-defined problem. In addition to individual or group-facilitated meetings, continuous education, distance learning, and collaboration of the Autonomic Computing business partner community are facilitated via events such as PartnerWorld and developerWorks.

It is important to IBM and other enterprises to facilitate the creation of customer and partner references that serve as "lighthouses" to demonstrate what is possible to achieve with Autonomic Computing and on demand business. These activities serve as catalysts for further advancements. It is important within IBM to drive early adoption of Autonomic Computing and Grid Computing-based technologies and solutions. IBM does this by virtue of the delivery of Autonomic Computing and Grid Computing services/solutions to IBM employees to demonstrate commitment and leadership by example. IBM is (and has been for some time) currently engaged in extensive projects to do just that.

The good news is that many of the inventions required to build autonomic technologies are available and need only to be formalized and assimilated through the mechanism of standards-based collaboration across the industry. For those technologies that have yet to be developed, IBM is making significant investments in research, joint university programs, and industry collaboration projects. IBM will drive the rapid development of on demand business through such leadership initiatives; the Autonomic Computing Organization, a highly focused, cross-corporate business unit within IBM, is one organization deeply involved in these types of activities.

Another facilitator in the industry is Web-accessible content through various existing IBM "e-venues," such as developerWorks and alphaWorks, as well as via IBM's Autonomic Computing Web site, which can be found at *www.ibm.com/autonomic*.

The on demand business is focused on realizing efficiencies in business operations and improving ROI. This is true whether one is the provider of on demand business products and services or the recipient of on demand business products and services. In the early 1990s, the Internet linked scientists across academia, government, and research. The Internet then evolved to provide e-mail, and then the World Wide Web, which was excellent for communicating marketing messages, but did not have the demonstrative capabilities of the values we are noting today with on demand Operating Environments.

IT has matured, enabling a plurality of global electronic commerce solutions and advanced Web services across the Internet. This is presented in thought-provoking detail in the book *Exploring e-Commerce: Global e-Business and e-Societies* [Fellenstein01]. In the late 1990s, with e-Commerce still driving new business processes and seemingly brilliant new start-up companies, the world encountered a tremendous stock market boom. However, this new horizon and capability came with a cost; it has experienced some serious investments in technology. The on demand business has since evolved and has now been instantiated with new advanced forms of Web services. Conducting on demand business is a new technological and business operations frontier, yielding advanced new capabilities and services provider models, which help to strengthen service level delivery capabilities and QoS.

When any business invests in IT, it clearly expects to derive credible benefits from its investments. Issues surrounding the needs for ongoing cost reductions will never disappear, nor will issues surrounding technology advancements, integration, and automation. To become an on demand business is not a revolution, nor a short-term strategic endeavor; it is, simply stated, a very precise strategic *evolution*. Implementing on demand business practices, including Autonomic Computing, is not a short-lived business trend; these advanced topics represent the combined manifestation of incredibly powerful business leadership and carefully planned strategic investments. This notion of becoming an on demand Operating Environment is a new way of operating that embodies transformed business processes, utilizing advanced services, while implementing very effective operational environment capabilities.

The Autonomic Computing Vision

The following discussion surrounding the Autonomic Computing vision explores a broad set of ideas and approaches to Autonomic Computing, as well as some first steps in what we at IBM see as a journey for our customers to create more self-managing computing system environments.

Autonomic Computing represents a collection and integration of technologies and business processes, which in turn enhances the creation of an IT computing infrastructure in many beneficial ways.

ONE MAJOR GOAL OF AUTONOMIC COMPUTING

A fundamental goal of Autonomic Computing is to provide *self-configuration* capabilities for the entire IT infrastructure, not just individual servers, software, and storage devices.

This conceptual goal extends far beyond traditional approaches found in many of today's steady-state environments.

As published in the *IBM Systems Journal* dedicated to Autonomic Computing [IBMSYSJ], the article entitled "The dawning of the Autonomic Computing era," [Ganek02] very eloquently describes the new, innovative discipline of Autonomic Computing. In this section, we will provide the reprint of this innovative work, which outlines some key Autonomic Computing perspectives.

In the following discussion, we will explore this new and innovative era of computing with an overview of IBM's Autonomic Computing efforts. These concepts will explore some of IBM's strategic thoughts surrounding the subject of on demand business. Many of the key strategic thinkers in this area, such as Ric Telford, John Sweitzer, and Jim Crosskey of IBM, have contributed in very significant ways to these On Demand perspectives.

This article will explore the industry and marketplace drivers, the fundamental characteristics of autonomic systems, a framework for how systems will evolve to achieve a more self-managing state, and the key roles of the open industry standards necessary to support autonomic behavior in heterogeneous system environments.

The Next Section Is an Early Reprint of [Ganek02].

This *IBM Systems Journal* article, "*The dawning of the Autonomic Computing era*," exactly as published by Alan Ganek and Thomas A. Corbi, sets a precedent for one of the two key underpinnings of On Demand computing. This particular strategic underpinning, Autonomic Computing, will be further referenced throughout this book.

The Dawning of the Autonomic Computing Era

On March 8, 2001, Paul Horn, IBM Senior Vice President and Director of Research, presented the theme and importance of Autonomic Computing to the National Academy of Engineering at Harvard University. (1)

One month later, Irving Wladawsky-Berger, Vice President of Strategy and Technology for the IBM Server Group, introduced the Server Group's Autonomic Computing project (then named eLiza*)(2) with the goal of providing self-managing systems to address those concerns. Thus began IBM's commitment to deliver "Autonomic Computing"—a new company-wide and, it is to be hoped, industry-wide, initiative targeted at coping with the rapidly growing complexity of operating, managing, and integrating computing systems.

Information Technology and the Impossible Problem

"The information technology industry loves to prove the impossible possible. We obliterate barriers and set records with astonishing regularity. But now we face a problem springing from the very core of our success—and too few of us are focused on solving it. More than any other I/T problem, this one—if it remains unsolved—will actually prevent us from moving to the next era of computing. The obstacle is complexity. Dealing with it is the single most important challenge facing the I/T industry."

—Paul Horn, IBM Senior Vice President and Director of Research, presented this theme and the importance of Autonomic Computing to the National Academy of Engineering at Harvard University on March 8, 2001.

We do not see a change in Moore's law (3) that would slow development as the main obstacle to further progress in the information technology (IT) industry. Rather, it is the IT industry's exploitation of the technologies in accordance with Moore's law that has led to the verge of a complexity crisis. Software developers have fully exploited a four- to six-orders-of-magnitude increase in computational power—producing ever more sophisticated software applications and environments. There has been exponential growth in the number and variety of systems and components. The value of database technology and the Internet has fueled significant growth in storage subsystems to hold petabytes[4] of structured and unstructured information. Networks have interconnected the distributed, heterogeneous systems of the IT industry. Our information society creates unpredictable and highly variable workloads on those networked systems. And today, those increasingly valuable, complex systems require more and more skilled IT professionals to install, configure, operate, tune, and maintain them.

IBM is using the phrase "Autonomic Computing"[5] to represent the vision of how IBM, the rest of the IT industry, academia, and the national laboratories can address this new challenge. By choosing the word "autonomic," IBM is making an analogy with the autonomic nervous system. The autonomic nervous system frees our conscious brain from the burden of having to deal with vital but lower-level functions. Autonomic computing will free system administrators from many of today's routine management and operational tasks. Corporations will be able to devote more of their IT skills toward fulfilling the needs of their core businesses, instead of having to spend an increasing amount of time dealing with the complexity of computing systems.

Need for Autonomic Computing As Frederick P. Brooks, Jr., one of the architects of the IBM System/360*, observed, "Complexity is the business we are in, and complexity is what limits us."[6] The computer industry has spent decades creating systems of marvelous and ever-increasing complexity. But today, complexity itself is the problem.

The spiraling cost of managing the increasing complexity of computing systems is becoming a significant inhibitor that threatens to undermine the future growth and societal benefits of information technology. Simply stated, managing complex systems has grown too costly and prone to error. Administering a myriad of system management details is too labor-intensive. People under such pressure make mistakes, increasing the potential of system outages with a concurrent impact on business. And, testing and tuning complex systems is becoming more difficult. Consider:

- It is now estimated that one-third to one-half of a company's total IT budget is spent preventing or recovering from crashes.[7] Nick Tabellion, CTO of Fujitsu Softek, said: "The commonly used number is: For every dollar to purchase storage, you spend $9 to have someone manage it."[8]

- Aberdeen Group studies show that administrative cost can account for 60 to 75 percent of the overall cost of database ownership (this includes administrative tools, installation, upgrade and deployment, training, administrator salaries, and service and support from database suppliers).[9]
- When you examine data on the root cause of computer system outages, you find that about 40 percent are caused by operator error,[10] and the reason is not because operators are not well-trained or do not have the right capabilities. Rather, it is because the complexities of today's computer systems are too difficult to understand, and IT operators and managers are under pressure to make decisions about problems in seconds.[11]
- A Yankee Group report[12] estimated that downtime caused by security incidents cost as much as $4,500,000 per hour for brokerages and $2,600,000 for banking firms.
- David J. Clancy, chief of the Computational Sciences Division at the NASA Ames Research Center, underscored the problem of the increasing systems complexity issues: "Forty percent of the group's software work is devoted to test," he said, and added, "As the range of behavior of a system grows, the test problem grows exponentially." [13]
- A recent Meta Group study looked at the impact of downtime by industry sector as shown in Figure 1.

Although estimated, cost data such as shown in Figure 1 are indicative of the economic impact of system failures and downtime. According to a recent IT resource survey by the Merit Project of Computer Associates International, 1867 respondents grouped the most common causes of outages into four areas of data center operations: systems, networks, database, and applications.[14] Most frequently cited outages included:

- For systems: operational error, user error, third-party software error, internally developed software problem, inadequate change control, lack of automated processes
- For networks: performance overload, peak load problems, insufficient bandwidth
- For database: out of disk space, log file full, performance overload
- For applications: application error, inadequate change control, operational error, non-automated application exceptions

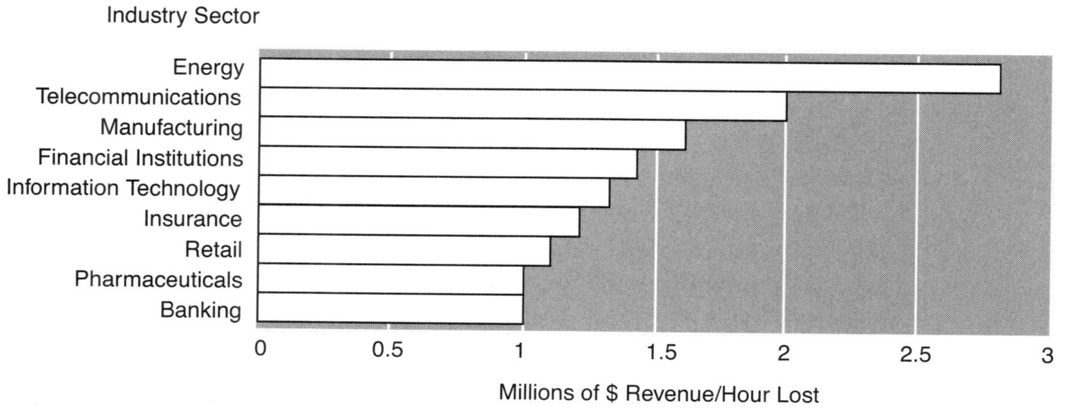

Figure 1 The illustration depicts data from *IT Performance Engineering and Measurement Strategies: Quantifying Performance Loss*, Meta Group, Stamford, Connecticut, USA (October 2000).

Well-engineered autonomic functions targeted at improving and automating systems operations, installation, dependency management, and performance management can address many causes of these "most frequent" outages and reduce outages and downtime.

Confluences of marketplace forces are driving the industry toward Autonomic Computing. Complex heterogeneous infrastructures composed of dozens of applications, hundreds of system components, and thousands of tuning parameters are a reality. New business models depend on the IT infrastructure being available 24 hours a day, 7 days a week. In the face of an economic downturn, there is an increasing management focus on "return on investment" and operational cost controls—while staffing costs exceed the costs of technology. To compound matters further, there continues to be a scarcity of highly skilled IT professionals to install, configure, optimize, and maintain these complex, heterogeneous systems.

To respond, system design objectives must shift from the "pure" price/performance requirements to issues of robustness and manageability in the total-cost-of-ownership equation. As a profession, we must strive to simplify and automate the management of systems. Today's systems must evolve to become much more self-managing, that is: self-configuring, self-healing, self-optimizing, and self-protecting.

Dr. Irving Wladawsky-Berger outlined the solution at the Kennedy Consulting Summit in November 2001: "There is only one answer: The technology needs to manage itself. Now, we do not mean any far out AI {*Artificial Intelligence*} project; what we mean is that we need to develop the right software, the right architecture, the right mechanisms . . . So that instead of the technology behaving in its usual pedantic way and requiring a human being to do everything for it, it starts behaving more like the 'intelligent' computer we all expect it to be, and starts taking care of its own needs. If it does not feel well, it does something. If someone is attacking it, the system recognizes it and deals with the attack. If it needs more computing power, it just goes and gets it, and it does not keep looking for human beings to step in."[15]

What is Autonomic Computing? Automating the management of computing resources is not a new problem for computer scientists. For decades system components and software have been evolving to deal with the increased complexity of system control, resource sharing, and operational management. Autonomic computing is just the next logical evolution of these past trends to address the increasingly complex and distributed computing environments of today. So why then is this something new? Why a call to arms to the industry for heightened focus and new approaches? The answer lies in the radical changes in the information technology environment in just the few short years since the mid-1990s, with the use of the Internet and Business extending environments to a dramatically larger scale, broader reach, and a more mission-critical fundamental requirement for business. In that time the norm for a large on-line system has escalated from applications such as networks consisting of tens of thousands of fixed-function automated teller machines connected over private networks to rich suites of financial services applications that can be accessed via a wide range of devices (personal computer, notebook, handheld device, smart phone, smart card, etc.) by tens of millions of people worldwide over the Internet.

IBM's Autonomic Computing initiative has been outlined broadly. Paul Horn[1] described this "grand challenge" and called for industry-wide collaboration toward developing Autonomic Computing systems that have characteristics as follows:

- To be autonomic, a system needs to "know itself"—and consist of components that also possess a system identity.
- An autonomic system must configure and reconfigure itself under varying and unpredictable conditions.
- An autonomic system never settles for the status quo—it always looks for ways to optimize its workings.
- An autonomic system must perform something akin to healing—it must be able to recover from routine and extraordinary events that might cause some parts to malfunction.

- A virtual world is no less dangerous than the physical one, so an Autonomic Computing system must be an expert in self-protection.
- An Autonomic Computing system knows its environment and the context surrounding its activity, and acts accordingly.
- An autonomic system cannot exist in a hermetic environment (and must adhere to open standards).
- Perhaps most critical for the user, an Autonomic Computing system will anticipate the optimized resources needed to meet a user's information needs while keeping its complexity hidden.

Fundamentals of Autonomic Computing. In order to incorporate these characteristics in "*self-managing*" systems, future Autonomic Computing systems will have four fundamental features. Various aspects of these four fundamental "self" properties are further explored in the "Autonomic Computing" *IBM Systems Journal*, Volume 42, Number 1, 2003.

Self-configuring. Systems adapt automatically to dynamically changing environments. When hardware and software systems have the ability to define themselves "on-the fly," they are self-configuring. This aspect of self-managing systems means that new features, software, and servers can be dynamically added to the enterprise infrastructure with no disruption of services. Systems must be designed to provide this aspect at a feature level with capabilities such as plug and play devices, configuration setup wizards, and wireless server management. These features will allow functions to be added dynamically to the enterprise infrastructure with minimum human intervention. Self-configuring not only includes the ability for each individual system to configure itself in real-time, but also for systems within the enterprise to configure themselves into the on demand Operating Environment. The goal of Autonomic Computing is to provide self-configuration capabilities for the entire IT infrastructure, not just individual servers, software, and storage devices.

Self-healing. Systems discover, diagnose, and react to disruptions. For a system to be self-healing, it must be able to recover from a failed component by first detecting and isolating the failed component, taking it off line, fixing or isolating the failed component, and reintroducing the fixed or replacement component into service without any apparent application disruption. Systems will need to predict problems and take actions to prevent the failure from having an impact on applications. The self-healing objective must be to minimize all outages in order to keep enterprise applications up and available at all times. Developers of system components need to focus on maximizing the reliability and availability design of each hardware and software product toward continuous availability.

Self-optimizing. Systems monitor and tune resources automatically. Self-optimization requires hardware and software systems to efficiently maximize resource utilization to meet end-user needs without human intervention. IBM systems already include industry-leading technologies such as logical partitioning, dynamic workload management, and dynamic server clustering. These kinds of capabilities should be extended across multiple heterogeneous systems to provide a single collection of computing resources that could be managed by a "logical" workload manager across the enterprise. Resource allocation and workload management must allow dynamic redistribution of workloads to systems that have the necessary resources to meet workload requirements. Similarly, storage, databases, networks, and other resources must be continually tuned to enable efficient operations even in unpredictable environments. Features must be introduced to allow the enterprise to optimize resource usage across the collection of systems within their infrastructure, while also maintaining their flexibility to meet the ever-changing needs of the enterprise.

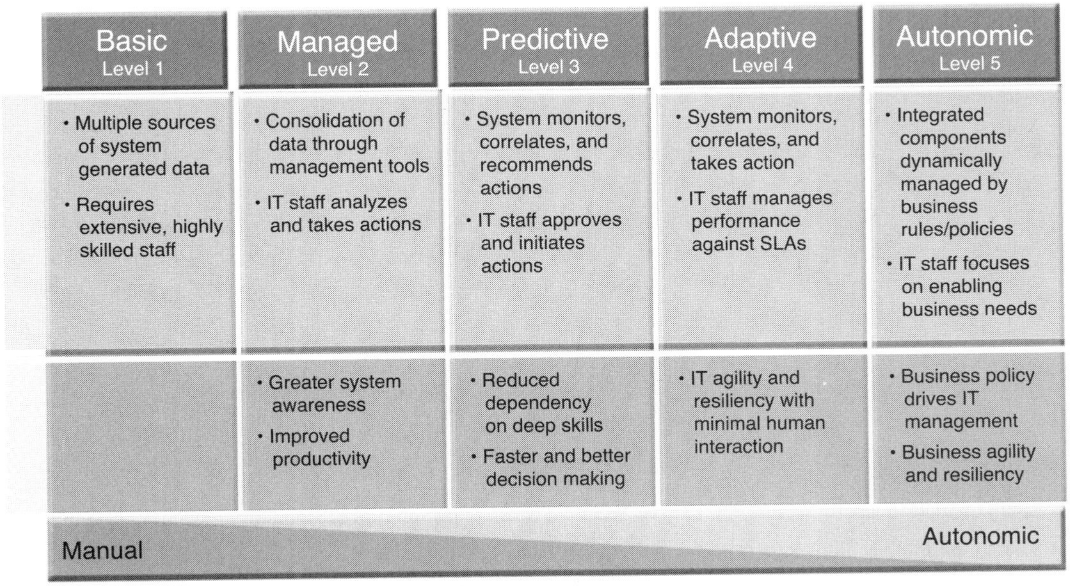
Basic
Level 1
• Multiple sources of system generated data
• Requires extensive, highly skilled staff
Managed
Level 2
• Consolidation of data through management tools
• IT staff analyzes and takes actions
• Greater system awareness
• Improved productivity
Predictive
Level 3
• System monitors, correlates, and recommends actions
• IT staff approves and initiates actions
• Reduced dependency on deep skills
• Faster and better decision making
Adaptive
Level 4
• System monitors, correlates, and takes action
• IT staff manages performance against SLAs
• IT agility and resiliency with minimal human interaction
Autonomic
Level 5
• Integrated components dynamically managed by business rules/policies
• IT staff focuses on enabling business needs
• Business policy drives IT management
• Business agility and resiliency
Manual
Autonomic

Figure 2 This illustration depicts the 5 transformation levels, from manual to autonomic, evolving to autonomic operations.

Self-protecting. Systems anticipate, detect, identify, and protect themselves from attacks from anywhere. Self-protecting systems must have the ability to define and manage user access to all computing resources within the enterprise, to protect against unauthorized resource access, to detect intrusions and report and prevent these activities as they occur, and to provide backup and recovery capabilities that are as secure as the original resource management systems. Systems will need to build on top of a number of core security technologies already available today, including LDAP (Lightweight Directory Access Protocol), Kerberos, hardware encryption, and SSL (Secure Socket Layer). Capabilities must be provided to more easily understand and handle user identities in various contexts, removing the burden from administrators.

An Evolution, not a Revolution To implement Autonomic Computing, the industry must take an evolutionary approach and deliver improvements to current systems that will provide significant self-managing value to customers without requiring them to completely replace their current IT environments. New open standards must be developed that will define the new mechanisms for interoperating heterogeneous systems. Figure 2 is a representation of those levels, starting from the basic level, through managed, predictive, and adaptive levels, and finally to the autonomic level.

As seen in Figure 2, the basic level represents the starting point where some IT systems are today. IT professionals who set it up, monitor it, and eventually replace it manage each system element independently. At the managed level, systems management technologies can be used to collect information from disparate systems onto fewer consoles, reducing the time it takes for the administrator to collect and synthesize information as the systems become more complex to operate. In the predictive level, as new technologies are introduced that provide correlation among several elements of the system, the system itself can begin to recognize patterns, predict the optimal configuration, and provide advice on what course of action the administrator should take.

As these technologies improve and as people become more comfortable with the advice and predictive power of these systems, we can progress to the adaptive level where the systems themselves can automatically take the correct actions based on the information that is available to them and the knowledge of what is happening in the systems. Service Level Agreements (SLAs)[16] guide operation of the system. Finally, at the fully autonomic level, the system operation is governed by business policies and objectives. Users interact with the system to monitor the business processes or alter the objectives.

Table 1 Aligning with the on demand business goals

Basic	Managed	Predictive	Adaptive	Autonomic
Process: Informal, reactive, manual	*Process*: Documented, improved over time, leverage of industry best practices; manual process to review IT performance	*Process*: Proactive, shorter approval cycle	*Process*: Automation of many resource mgmt best practices and transaction mgmt best practices, driven by service level agreements	*Process*: All IT service mgmt and IT resource mgmt best practices are automated
Tools: Local, platform and product specific	*Tools*: Consolidated resource mgmt consoles with problem mgmt system, automated software install, intrusion detection, load balancing	*Tools*: Role-based consoles with analysis and recommendations; product configuration advisors; real-time view of current & future IT performance; automation of some repetitive tasks; common knowledge base of inventory and dependency management	*Tools*: Policy management tools that drive dynamic change based on resource specific policies	*Tools*: Costing/ financial analysis tools, business and IT modeling tools, tradeoff analysis; automation of some on demand business mgmt roles
Skills: Platform-specific, geographically dispersed with technology	*Skills*: Multiple skill levels with centralized triage to prioritize and assign problems to skilled IT professionals	*Skills*: Cross-platform system knowledge, IT workload management skills, some bus process knowledge	*Skills*: Service objectives and delivery per resource, and analysis of impact on business objectives	*Skills*: on demand business cost & benefit analysis, performance modeling, advanced use of financial tools for IT context

As companies progress through the five levels of Autonomic Computing, the processes, tools, and benchmarks become increasingly sophisticated, and the skills requirement becomes more closely aligned with the business. The preceding Table illustrates this correlation.

The basic level represents the starting point for most IT organizations. If they are formally measured at all, they are typically measured on the time required to finish major tasks and fix major problems. The IT organization is viewed as a cost center, in which the variable costs associated with labor are preferred over an investment in centrally coordinated systems management tools and processes.

At the managed level, IT organizations are measured on the availability of their managed resources, their time to close trouble tickets in their problem management system, and their time to complete formally tracked work requests. To improve on these measurements, IT organizations document their processes and continually improve them through manual feedback loops and adoption of best practices. IT organizations gain efficiency through consolidation of management tools to a set of strategic platforms and through a hierarchical problem management triage organization.

In the predictive level, IT organizations are measured on the availability and performance of their business systems and their return on investment. To improve on these measurements, IT organizations measure, manage, and analyze transaction performance. The implications of the critical nature of the role of the IT organization in the success of the business are understood. Predictive tools are used to project future IT performance, and many tools make recommendations to improve future performance.

In the adaptive level, IT resources are automatically provisioned and tuned to optimize transaction performance. Business policies, business priorities, and service-level agreements guide the autonomic infrastructure behavior. IT organizations are measured on end-to-end business system response times (i.e., transaction performance), the degree of efficiency with which the IT infrastructure is utilized, and their ability to adapt to shifting workloads.

In the autonomic level, IT organizations are measured on their ability to make the business successful. To improve business measurements, IT tools understand the financial metrics associated with on demand business activities and supporting IT activities. Advanced modeling techniques are used to optimize on demand business performance and quickly deploy newly optimized on demand business solutions.

Today's software and hardware system components will evolve toward more autonomic behavior. For example:

- *Data management.* New database software tools can use statistics from the databases, analyze them, and learn from the historical system performance information. The tools can help an enhanced database system automatically detect potential bottlenecks, as they are about to occur and attempt to compensate for them by adjusting tuning parameters. Query optimizers can learn the optimal index and route to certain data and automatically seek out that path based on the historical access patterns and associated response times.
- *Web servers and software.* Web servers can provide real-time diagnostic "dashboard" information, enabling customers to more quickly become aware of resource problems, instead of relying on after-the-fact reports to identify problems. Once improved instrumentation is available, autonomic functions can be introduced that enable the Web server infrastructure to automatically monitor, analyze, and fix performance problems. As an example, suppose an application server is freezing-up intermittently, and no customer transactions are being processed for several seconds, thus losing thousands of dollars in business, as well as customer confidence and loyalty. Using real-time monitoring, predictive analysis, and auto-tuning, the freeze-up is anticipated before it happens. The autonomic function compares real-time data with historical problem data (i.e., suggesting that the cache sizes were set too low). The settings are reset automatically without service disruption, and a report is sent to the administrator that shows what action was taken.
- *Systems management.* Systems management software can contain improved problem determination and data collection features designed to help businesses better diagnose and prevent interruptions (e.g., breaches of security). Such systems management software must enable customers to take an "end-to-end" view of their computing environment across multiple, independently installed hardware and software elements. A bank transaction, for example, might "touch" a discrete database, another transaction, and Web application servers as it is processed across a network. If a problem occurs with processing on one of the individual components, lack of an integrated problem determination infrastructure makes it more difficult to determine what prevented that bank transaction from completing successfully. A consolidated view created by the system management software would enable the system and IT staffs to identify and quickly react to problems as they happen by providing an end-to-end view of the application. The end-to-end view of the environment allows companies to

understand problems and performance information in the context of their business goals.

- *Servers.* Computers can be built that need less human supervision. Computers can try to fix themselves in the event of a failure, protect themselves from hacker attacks, and configure themselves when adding new features. Servers can use software algorithms that learn patterns in Internet traffic or application usage, and provision resources in a way that gives the shortest response time to the task with the highest business priority. Server support for heterogeneous and enterprise workload management, dynamic clustering, dynamic partitioning, improved setup wizards, improved user authentication, directory integration, and other tools to protect access to network resources are all steps toward more autonomic functioning.

IBM hardware and software systems have already made significant progress in introducing Autonomic Computing functionality.[(2)] However, there is much more work ahead. The efforts to achieve cohesive system behavior must go beyond improvements in the individual components alone. These components must be federated, employing an integrating architecture that establishes the instrumentation, policy, and collaboration technologies so that groups of resources can work in concert, as for example, across systems in a grid. System management tools play a central role in coordinating the actions of system components, providing a simplified mechanism for system administration and for translating business objectives into executable policies to govern the actions of the IT resources available.

Industry Standards Are Needed to Support Autonomic Computing

Most IT infrastructures are composed of components supplied by different vendors. Open industry standards are the key to the construction of Autonomic Computing systems. Systems will need more standardization to introduce a uniform approach to instrumentation and data collection, dynamic configuration, and operation. Uniformity will allow the intersystem exchange of instrumentation and control information to create the basis for collaboration and autonomic behavior among heterogeneous systems.

For example, in storage systems, a standard that has been proposed for specifying data collection items is the Bluefin specification. Bluefin[(17)] defines a language and schema that allow users to reliably identify, classify, monitor, and control the physical and logical devices in storage area networking. The Storage Networking Industry Association (SNIA) has taken this standard to the Distributed Management Task Force (DMTF). SNIA is using Bluefin as the basis for its storage management initiative, the intent of which is to become the SNIA standard for management.

In the case of application instrumentation, the standard that has been proposed for obtaining the transaction rate, response time, failure rate, and topology data from applications is the Open Group Application Response Measurement (ARM)[18] application programming interfaces (APIs). The Application Response Measurement API defines the function calls that can be used to instrument an application or other software for transaction monitoring. It provides a way to monitor business transactions by embedding simple cells in the software that can be captured by an agent supporting the ARM API. The calls are used to capture data, allowing software to be monitored for availability, service levels, and capacity.

Other standards, such as the DMTF Common Information Model (CIM)[19] and Web Service Level Agreement (WSLA), provide languages and schemas for defining the available data. CIM is an object-oriented information model that provides a conceptual view of physical and logical system components. WSLA is a language to express SLA contracts, to support guaranteed performance, and to handle complex dynamic fluctuations in service demand. SLA-based system management would enable service providers to offer the same Web service at different performance levels, depending on contracts with their customers. WSLA is available through the IBM alphaWorks* Web Services Toolkit[20] that features a WSLA document approach based on Extensible Markup Language (XML) to define SLAs.

These standards are technologies that enable the building of "inter-communicating" autonomic system elements that are the foundation for cooperation in a federation of system components. Each individual autonomic "element" is responsible for managing itself, that is, for configuring itself internally, for healing internal failures when possible, for optimizing its own behavior, and for protecting itself from external probing and attack. Autonomic elements are the building blocks for making autonomic systems.

Autonomic elements continuously monitor system (or component) behavior through "sensors" and make adjustments through "effectors." By monitoring behavior through sensors, analyzing those data, then planning what action should be taken next (if any), and executing that action through effectors, a kind of "control loop"[21] is created (see the proceeding Figure 3).

Interconnecting autonomic elements requires distributed computing mechanisms to access resources across the network. "Grid computing" [22] encompasses the idea of an emerging infrastructure that is focused on networking together heterogeneous, multiple regional and national computing systems.

It has been called the next evolutionary step for the Internet. The term "grid" was chosen as an analogy with the electric power grid, which supplies pervasive access to power. Grids are persistent computing environments that enable software applications to integrate instruments, displays, and computational and information resources that are managed by diverse organizations in widespread locations.

In 2001, the Globus Project [23, 24] launched a research and development program aimed at creating a toolkit based on the Open Grid Service Architecture (OGSA) that defines standard mechanisms for creating, naming, and discovering services and specifies various protocols to support accessing services. Essentially, OGSA is a framework for distributed computing, based on Web services protocols. Although OGSA is a proposed standard that will be developed and defined in the Global Grid Forum (GGF)[25] it is applicable whether the environment consists of a multi-organization grid or simply distributed resources within an enterprise. IBM, Microsoft Corporation, and others have already announced support for the OGSA framework. Work efforts on grid and OGSA are creating important architectural models and new open industry standards that are enablers for the IT industry to make progress toward more self-managing systems. Since grid deployments can expand the domain of computing across many systems, in our view, a successful grid system will require autonomic functionality.

Individual autonomic elements can interact through OGSA mechanisms. For example, today there is no accepted "sensor and effector's" standard. But, the Globus Toolkit provides information services utilities to provide information about the status of grid resources. One of these utilities is the Monitoring and Discovery Service (MDS).[26] MDS 2.2 GRIS "Information Providers" that are essentially sensors or probes. The Globus Toolkit also provides a mechanism for authenticated access to MDS. Fault detection allows a client process to be monitored by a heartbeat monitor. Resource management APIs provide some job management capabilities.

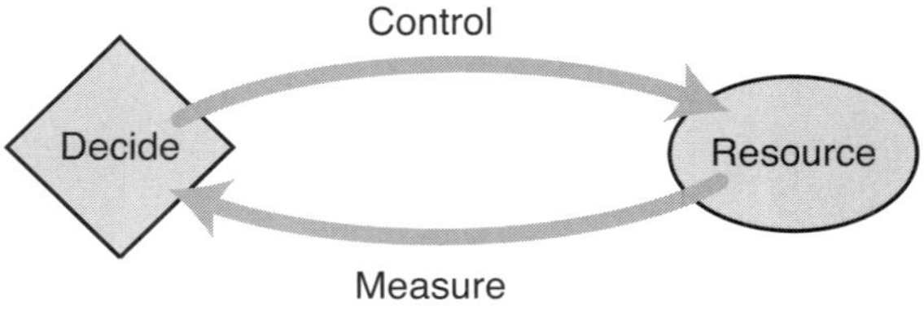

Figure 3 This Illustration depicts the *"Control Loop"* structure in Autonomic Computing. Source: Autonomic Computing Concepts IBM White Paper, 2001.

Thus we are seeing the emergence of some basic standard mechanisms needed for distributed "control loops" that in turn are needed for Autonomic Computing. When control loop standards are in place, the industry must address the more complex issues of specifying and automating policy management and service level agreements (SLAs).

A typical enterprise has a heterogeneous set of routers, firewalls, Web servers, databases, and workstations, all with different system management mechanisms. So again, industry standards will be needed in order to enable true policy management. We expect that policy specifications will be widely used in enterprises for defining quality of service management, storage backup, and system configuration, as well as security authorization and management.

A common approach to specifying and deploying policy would enable an enterprise to define and disseminate policies that reflect its overall IT service goals. A common, standard set of tools and techniques used throughout the enterprise could simplify analysis and reduce inconsistencies and conflicts in the policies deployed across the various components within the enterprise and also allow a policy exchange with external service providers.

Various standards bodies are working on specifying policies for network and systems management, security, and role-based access control (RBAC). The Internet Engineering Task Force (IETF)(27) and DMTF(28) have been concentrating on information models for management policies, protocols for transferring policies to network devices, and routing policies; the National Institute of Standards and Technology (NIST)(29) is working toward an RBAutonomic Computing standard; and the Oasis consortium (Organization for the Advancement of Structured Information Standards)(30) is working on an XML-based specification of access control policies and authentication information.

It will take some time for the current divergent standards policy-based solutions to come to embrace a common approach. Meanwhile, research on policy-based management approaches continues.(31,32) Advances in policy management are needed to enable enterprises to eventually specify the behaviors of IT services in terms of the business process objectives of the enterprises.

Exploratory Research and Development Presented in This Issue

{IBM Systems Journal, Volume 42, No. 1, 2003} Autonomic Computing represents an exciting new research direction in computing. IBM believes that meeting the grand challenge of Autonomic Computing systems will involve

researchers in a diverse array of fields, including systems management, distributed computing, networking, operations research, software development, storage, artificial intelligence, and control theory, as well as others.

The challenge of Autonomic Computing requires more than the re-engineering of today's systems. Autonomic Computing also requires new ideas, new insights, and new approaches. This issue of the IBM Systems Journal provides just a glimpse into an array of research and development efforts underway for Autonomic Computing. Below we present the topics in the issue.

D. C. Verma, S. Sahu, S. Calo, A. Shaikh, I. Chang, and A. Acharya in their paper, "SRIRAM: A Scalable Resilient Autonomic Mesh,"(33) propose a method that facilitates instantiating mirroring and replication of services in a network of servers.

The ability to redistribute hardware resources dynamically is essential to both the self-configuring and self-optimizing goals of Autonomic Computing. J. Jann, L. M. Browning, and R. S. Burugula describe this new server capability in "Dynamic Reconfiguration: Basic Building Blocks for Autonomic Computing on IBM pSeries Servers."(34)

In the first of two invited papers, D. A. Norman and A. Ortony from Northwestern University, along with D. M. Russell of IBM, discuss in "Affect and Machine Design: Lessons for the Development of Autonomous Machines"(35) how studying the human characteristics of cognition and affect will help designers in developing complex autonomic systems that will interact with unpredictable situations.

K. Whisnant, Z. T. Kalbarczyk, and R. K. Iyer examine the difficulties of dynamically reconfiguring application software in their paper, "A System Model for Dynamically Reconfigurable Software."(36) They believe that both static structure and runtime behaviors must be captured in order to define a workable reconfiguration model.

One technology to support self-healing and self-configuring is the ability to dynamically insert new pieces of software and remove other pieces of code, without shutting down the running system. This technology is being explored in the K42 research operating system and is presented in the paper by J. Appavoo, K. Hui, C. A. N. Soules, R. W. Wisniewski, D. M. Da Silva, O. Krieger, M. A. Auslander, D. J. Edelsohn, B. Gamsa, G. R. Ganger, P. McKenney, M. Ostrowski, B. Rosenburg, M. Stumm, and J. Xenidis, entitled "Enabling Autonomic Behavior in Systems Software with Hot Swapping."(37)

L. W. Russell, S. P. Morgan, and E. G. Chron introduce the idea of a predictive autonomic system in their paper entitled "Clockwork: A New Movement in Autonomic Systems."[38] They explore the idea of a system that anticipates workload needs based on statistical modeling, tracking, and forecasting.

Component-based development, where multiple distributed software components are composed to deliver a particular business function, is an emerging programming model in the Web services world. D. M. Yellin in his paper, "Competitive Algorithms for the Dynamic Selection of Component Implementations,"[39] proposes a strategy and framework for optimizing component performance based on switching between different component implementations.

In an example of "self-optimizing," V. Markl, G. M. Lohman, and V. Raman discuss improving query performance by comparing estimates with actual results toward self-validating query planning in "LEO: An Autonomic Query Optimizer for DB2."[40]

As noted, system and network security are fundamental to Autonomic Computing systems. In "Security in an Autonomic Computing Environment," [41] D. M. Chess, C. C. Palmer, and S. R. White outline a number of security and privacy issues in the design and development of autonomic systems.

G. Lanfranchi, P. Della Peruta, A. Perrone, and D. Calvanese describe what they see as a paradigm shift in system management needed for Autonomic Computing. In their paper, "Toward a New Landscape of Systems Management in an Autonomic Computing Environment,"[42] they introduce a knowledge-based resource model technology that extends across design, delivery, and run time.

In the second invited paper, "Comparing Autonomic and Proactive Computing,"[43] R. Want, T. Pering, and D. Tennenhouse of Intel Research present a high-level discussion of the similarities between proactive computing and Autonomic Computing with an emphasis on their research in proactive computing—an environment in which computers anticipate what users need and act accordingly.

Today, optimizing performance in multisystem e-commerce environments requires considerable skill and experience. In "Managing Web Server Performance with AutoTune Agents,"[44] Y. Diao, J. L. Hellerstein, S. Parekh, and J. P. Bigus describe intelligent agents that use control theory techniques to autonomically adjust an Apache** Web server to dynamic workloads.

The backbone of a grid or typical Autonomic Computing system is an intelligent, heterogeneous network infrastructure. Management issues related to topology, R. Haas, P. Droz, and B. Stiller in "Autonomic Service Deployment in Networks" explore service placement, cost and service metrics, as well as dynamic administration structure.[45]

Although much of the discussion on Autonomic Computing often focuses on servers, networks, databases, and storage management, we realize that personal computer users would also benefit greatly by the introduction of autonomic features. D. F. Bantz, C. Bisdikian, D. Challener, J. P. Karidis, S. Mastrianni, A. Mohindra, D. G. Shea, and M. Vanover explore these possibilities in their paper, "Autonomic Personal Computing."[46]

People will still need to interact with Autonomic Computing systems. D. M. Russell, P. P. Maglio, R. Dordick, and C. Neti in their paper entitled "Dealing with Ghosts: Managing the User Experience of Autonomic Computing"[47] argue that the lessons we have learned in human-computer interaction research must be applied to effectively expose and communicate the runtime behavior of these complex systems and to better define and structure the user system operation scenarios.

In the Technical Forum section, the complex challenges of life-cycle management and providing capacity On Demand are examined in a project as described by A. Abbondanzio, Y. Aridor, O. Biran, L. L. Fong, G. S. Goldszmidt, R. E. Harper, S. M. Krishnakumar, G. Pruett, and B.-A. Yassur in "Management of Application Complexes in Multitier Clustered Systems."[48]

Topic Conclusion In his keynote speech at the Almaden Institute 2002,[49] John Hennessy, President of Stanford University, presented his view of the autonomic challenge. While acknowledging the significant accomplishments in hardware architecture over the past 20 years, he urged industry and academia to look forward and to shift focus to a set of issues related to how services will be delivered over networks in the Internet/Web-centric "post-desktop" era: "As the business use of this environment grows and as people become more and more used to it, the flakiness that we've all accepted in the first generation of the Internet and the Web—will become unacceptable." Hennessy emphasized an increased research focus on availability, maintainability, scalability, cost, and performance—all fundamental aspects of Autonomic Computing.

Autonomic computing is a journey. Progress will be made in a series of evolutionary steps. This {*referenced*} issue of the *IBM Systems Journal* presents some of the technology signposts that can serve to guide the ongoing research in this new direction.

"The Dawning" Acknowledgments The authors of "The Dawning" express thanks to the many authors who submitted their work for publication. Special thanks go to Lorraine Herger, Kazuo Iwano, Pratap Pattnaik, Connie Marconi, and Felicia C. Medley, who organized the call for papers, managed the process to select the topics and the papers, and worked with the authors and *IBM Systems Journal* staff to produce this issue. We also thank the numerous referees whose comments helped all the authors in improving their papers. We particularly express our appreciation to Paul Horn, IBM Senior Vice President and Director of Research, for launching the Autonomic Computing grand challenge in his 2001 presentation on Autonomic Computing and to Irving Wladawsky-Berger who led the launch of the Autonomic Computing project in the IBM Server Group. We acknowledge the efforts of Robert Morris, Anant Jhingran, Tushar Chandra, and K. M. Moiduddin, who organized the Almaden Autonomic Computing Institute held at San Jose, California in April 2002. We also acknowledge Sam S. Adams, Lisa F. Spainhower, William H. Tetzlaff, William H. Chung, Steve R. White, and Kazuo Iwano who organized the Autonomic Computing Conference sponsored by the IBM Academy of Technology and held in Yorktown Heights, New York in May 2002. Our thanks go to Christopher W. Luongo for his article, "Server to IT: Thanks, But I Fixed It Myself." The authors would like to thank Ric Telford, John Sweitzer, and Jim Crosskey of the Autonomic Computing department for their contributions to this manuscript. The authors also wish to acknowledge the IBM Systems Journal staff for their many helpful suggestions during the creation of this issue.

Cited References and Notes

1. P. Horn, *Autonomic Computing: IBM's Perspective on the State of Information Technology*, IBM Corporation (October 15, 2001); available at http://www.research.ibm.com/autonomic/manifesto/autonomic_computing.pdf.
2. IBM Server Group, *eLiza: Building an Intelligent Infrastructure for E-business—Technology for a Self-Managing Server Environment*, G520-9592-00, IBM Corporation (2001); also at http://www-1.ibm.com/servers/eserver/introducing/eliza/eliza_final.pdf.

3. For more than 30 years, Moore's Law has forecasted progress in the computing industry like an immutable force of nature. In 1965, Gordon E. Moore, then the director of research and development at Fairchild Semiconductor Corporation, made the casual observation that processing power will double every 18 to 24 months, suggesting healthy growth ahead over the next decade for the then-nascent silicon chip industry. Five years later, Moore co-founded Intel Corporation, and "Moore's law" was well on its way to becoming a self-fulfilling prophecy among researchers and developers in the industry.
4. A petabyte (PB) is 1000 terabytes, and a terabyte is 1000 gigabytes. CERN, the European Center for Nuclear Research located just outside of Geneva, Switzerland, has started building a 100-PB archive to house data from the particle accelerator of the research center. See "From Kilobytes to Petabytes in 50 Years," *Science and Technology Review,* UCRL-52000-02-4, Lawrence Livermore National Laboratory, Livermore, CA (April 15, 2002), http://www.llnl.gov/str/March02/March50th.html.
5. The term "Autonomic Computing" comes from an analogy to the autonomic central nervous system in the human body, which adjusts to many situations without any external help. "Autonomic" is defined as: (a) Of, relating to, or controlled by the autonomic nervous system; (b) Acting, or occurring involuntarily; automatic; an autonomic reflex.
6. F. P. Brooks, Jr., *The Mythical Man-Month: Essays on Software Engineering,* Twentieth Anniversary Edition, Addison-Wesley Publishing Co., Reading, MA (1995), p. 226. See also, F. P. Brooks, Jr., "No Silver Bullet: Essence and Accidents of Software Engineering," Computer 20, No. 4, 1019 (1987).
7. D. A. Patterson, A. Brown, P. Broadwell, G. Candea, M. Chen, J. Cutler, P. Enriquez, A. Fox, E. Kiciman, M. Merzbacher, D. Oppenheimer, N. Sastry, W. Tetzlaff, J. Traupman, N. Treuhaft, *Recovery-Oriented Computing (ROC): Motivation, Definition, Techniques, and Case Studies,* U.C. Berkeley Computer Science Technical Report, UCB//CSD-02-1175, University of California, Berkeley (March 15, 2002).
8. K. Evans-Correia, "Simplifying Storage Management Starts with More Efficient System Utilization," Interview with N. Tabellion, *searchStorage* (August 29, 2001), see http://searchstorage. techtarget.com/qna/0, 289202,sid5_gci764063,00.html.
9. *IBM Data Management Tools: New Opportunities for Cost-Effective Administration,* Profile Report, Aberdeen Group, Inc., Boston (April 2002), p. 3.

10. D. Patterson, "Availability and Maintainability Performance: New Focus for a New Century," *USENIX Conference on File and Storage Technologies* (FAST '02), Keynote Address, Monterey, CA (January 29, 2002).

11. A. Brown and D. A. Patterson, "To Err Is Human," *Proceedings of the First Workshop on Evaluating and Architecting System dependability* (EASY '01), Goeteborg, Sweden (July 2001).

12. "How Much Is an Hour of Downtime Worth to You?" from *Must-Know Business Continuity Strategies*, Yankee Group, Boston (July 31, 2002).

13. D. J. Clancy, "NASA Challenges in Autonomic Computing," *Almaden Institute 2002*, IBM Almaden Research Center, San Jose, CA (April 10, 2002).

14. Merit Project, Computer Associates International, http://www.merit-project.com/it_survey_results.htm.

15. I. Wladawsky-Berger, "Advancing E-business into the Future: The Grid," *Kennedy Consulting Summit 2001*, New York (November 29, 2001).

16. In this context, a Service Level Agreement (SLA) is a compact between a customer or consumer and a provider of an IT service that specifies the levels of availability, serviceability, performance (and tracking/reporting), problem management, security, operation, or other attributes of the service, often established via negotiation. Typically, an SLA identifies and defines the customer's needs, provides a framework for discussion and understanding, attempts to simplify complex requirements, outlines methods for resolving disputes, and helps eliminate unrealistic expectations.

17. " 'Bluefin' A Common Interface for SAN Management," White Paper, Storage Networking Industry Association (August 13, 2002), http://www.snia.org/tech_activities/SMI/bluefin/Bluefin_White_Paper_v081302.pdf from http://www.snia.org/tech_activities/SMI/bluefin/.

18. *Application Response Measurement Issue 3.0 Java Binding*, Open Group Technical Standard CO14, The Open Group (October 2001), at http://www.opengroup.org/products/publications/catalog/c014.htm.

19. *Common Information Model (CIM) Specification Version 2.2*, DSP0004, Distributed Management Task Force (June 14, 1999), at http://www.dmtf.org/standards/standard_cim.php.

20. Web Services Toolkit, alphaWorks, IBM Corporation (July 26, 2000), http://www.alphaworks.ibm.com/tech/webservices toolkit.

21. The control loop is the essence of automation. By measuring or sensing some activity in a process to be controlled, a controller component

decides what needs to be done next and executes the required operations through a set of actuators. The controller then re-measures the process to determine whether the actions of the actuator had the desired effect. The whole routine is then repeated in a continuous loop of measure, decide, actuate, and repeat.

22. *The Grid: Blueprint for a New Computing Infrastructure*, I. Foster and C. Kesselman, Editors, Morgan Kaufmann Publishers, Inc., San Francisco, CA (1999).

23. See http://www.globus.org, the home of the Globus Project, the Globus Toolkit (GT3), and work related to the Open Grid Services Architecture (OGSA).

24. I. Foster, C. Kesselman, and S. Tuecke, "The Anatomy of the Grid: Enabling Scalable Virtual Organizations," *International Journal of High Performance Computing* 15, No. 3, 200222 (2001); see also, http://www.globus.org/research/papers/anatomy.pdf.

25. Global Grid Forum, http://www.gridforum.org/.

26. *MDS 2.2 User's Guide*, The Globus Project, available at www.globus.org/mds/mdsusersguide.pdf.

27. Internet Engineering Task Force, http://www.dmtf.org.

28. Distributed Management Task Force, http://www.dmtf.org.

29. National Institute of Standards and Technology, http://www.nist.org.

30. Organization for the Advancement of Structured Information Standards, http://www.oasis-open.org.

31. L. Lymberopoulos, E. Lupu, and M. Sloman, "An Adaptive Policy Based Management Framework for Differentiated Services Networks," *Proceedings of the 3rd IEEE Workshop on Policies for Distributed Systems and Networks (Policy 2002)*, Monterey, CA (June 2002), pp. 147158.

32. V. Sander, W. A. Adamson, I. Foster, and A. Roy, "End-to-End Provision of Policy Information for Network QoS," *Proceedings of the Tenth IEEE Symposium on High Performance Distributed Computing* (HPDC-10), IEEE Press (August 2001).

33. D. C. Verma, S. Sahu, S. Calo, A. Shaikh, I. Chang, and A. Acharya, "SRIRAM: A Scalable Resilient Autonomic Mesh," *IBM Systems Journal* **42**, No. 1, 1928 (2003, this issue).

34. J. Jann, L. A. Browning, and R. S. Burugula, "Dynamic Reconfiguration: Basic Building Blocks for Autonomic Computing on IBM pSeries Servers," *IBM Systems Journal* **42**, No. 1, 2937 (2003, this issue).

35. D. A. Norman, A. Ortony, and D. M. Russell, "Affect and Machine Design: Lessons for the Development of Autonomous Machines," *IBM Systems Journal* **42**, No. 1, 3844 (2003, this issue).

36. K. Whisnant, Z. T. Kalbarczyk, and R. K. Iyer, "A System Model for Dynamically Reconfigurable Software," *IBM Systems Journal* **42**, No. 1, 4559 (2003, this issue).

37. J. Appavoo, K. Hui, C. A. N. Soules, R. W. Wisniewski, D. M. Da Silva, O. Krieger, M. A. Auslander, D. J. Edelsohn, B. Gamsa, G. R. Ganger, P. McKenney, M. Ostrowski, B. Rosenburg, M. Stumm, and J. Xenidis, "Enabling Autonomic Behavior in Systems Software with Hot Swapping," *IBM Systems Journal* **42**, No. 1, 6076 (2003, this issue).

38. L. W. Russell, S. P. Morgan, and E. G. Chron, "Clockwork: A New Movement in Autonomic Systems," *IBM Systems Journal* **42**, No. 1, 7784 (2003, this issue).

39. D. M. Yellin, "Competitive Algorithms for the Dynamic Selection of Component Implementations," *IBM Systems Journal* **42**, No. 1, 8597 (2003, this issue).

40. V. Markl, G. M. Lohman, and V. Raman, "LEO: An Autonomic Query Optimizer for DB2," *IBM Systems Journal* **42**, No. 1, 98106 (2003, this issue).

41. D. M. Chess, C. C. Palmer, and S. R. White, "Security in an Autonomic Computing Environment," *IBM Systems Journal* **42**, No. 1, 107118 (2003, this issue).

42. G. Lanfranchi, P. Della Peruta, A. Perrone, and D. Calvanese, "Toward a New Landscape of Systems Management in an Autonomic Computing Environment," *IBM Systems Journal* **42**, No. 1, 119128 (2003, this issue).

43. R. Want, T. Pering, and D. Tennenhouse, "Comparing Autonomic and Proactive Computing," *IBM Systems Journal* **42**, No. 1, 129135 (2003, this issue).

44. Y. Diao, J. L. Hellerstein, S. Parekh, and J. P. Bigus, "Managing Web Server Performance with AutoTune Agents," *IBM Systems Journal* **42**, No. 1, 136149 (2003, this issue).

45. R. Haas, P. Droz, and B. Stiller, "Autonomic Service Deployment in Networks," *IBM Systems Journal* **42**, No. 1, 150164 (2003, this issue).

46. D. F. Bantz, C. Bisdikian, C. Challener, J. P. Karidis, S. Mastrianni, A. Mohindra, D. G. Shea, and M. Vanover, "Autonomic Personal Computing," *IBM Systems Journal* **42**, No. 1, 165176 (2003, this issue).

47. D. M. Russell, P. P. Maglio, R. Dordick, and C. Neti, "Dealing with Ghosts: Managing the User Experience of Autonomic Computing," *IBM Systems Journal* **42**, No. 1, 177188 (2003, this issue).

48. A. Abbondanzio, Y. Aridor, O. Biran, L. L. Fong, G. S. Goldszmidt, R. E. Harper, S. M. Krishnakumar, G. Pruett, and B.-A. Yassur, "Management of Application Complexes in Multitier Clustered Systems," Technical Forum, *IBM Systems Journal* **42**, No. 1, 189195 (2003, this issue).

49. J. Hennessy, "Back to the Future: Time to Return to Some Long-Standing Problems in Computer Science," *Almaden Institute 2002*, IBM Almaden Research Center, San Jose, CA (April 10, 2002).

An Architectural Blueprint for Autonomic Computing

This section provides full treatment of the IBM strategic perspectives in Autonomic Computing. The discussions presented in this section will describe the strategic *Autonomic Computing Blueprint*. This Autonomic Computing Blueprint [ACBLUEPRINT] is developed, published, and maintained by the IBM Autonomic Computing Group located in the IBM Thomas J. Watson Research Center in Hawthorne, New York. This architectural Autonomic Computing Blueprint is one of the key instruments to understanding the on demand business strategy perspectives.

WHAT IS *AUTONOMIC COMPUTING*?

The term *"Autonomic Computing"* is derived and paralleled from the body's autonomic nervous system. The same way we take for granted the human body's capabilities for the management of breathing, digestion, and fending off viruses, companies will one day take for granted the network's ability to manage, repair, and protect itself from an enterprise perspective.

In this stage of on demand business, realizing autonomic capabilities will bring a totally new kind of transformation—or, more specifically, new levels of integration: of processes and applications inside the business; of suppliers and distributors at either end of the business; of customers outside the enterprise and of employees inside it. Until now, companies have been "on the 'Net." The on demand business transformation will now place companies in such a way that they will become an integrated part of the 'Net.

The high-tech industry has spent decades creating computer systems with ever-mounting degrees of complexity to solve a wide variety of business problems. Ironically, complexity itself has become part of the problem. As networks and distributed systems grow and change, they can become increasingly hampered by system deployment failures and hardware and software issues, not to mention human error. Such scenarios in turn require further human intervention to enhance the performance and capacity of IT components. This drives up the overall IT costs—even though technology component costs continue to decline. As a result, many IT professionals seek ways to improve the ROI in their IT infrastructure, by reducing the total cost of ownership (TCO) of their environments while improving the QoS for users.

We do not see a slowdown in Moore's law as the main obstacle to further progress in the IT industry. Rather, it is our industry's exploitation of the technologies that has arisen in the wake of Moore's law that has led us to the verge of a complexity crisis. Software developers have fully exploited a four to six order-of-magnitude increase in computational power—producing ever more sophisticated software applications and environments. There has been exponential growth in the number and variety of systems and components. The value of database technology and the Internet has fueled significant growth in storage subsystems, which are now capable of holding petabytes of structured and unstructured information. Networks have interconnected our distributed, heterogeneous systems. Our information society creates unpredictable and highly variable workloads on those networked systems. And today, those increasingly valuable, complex systems require more and more skilled IT professionals to install, configure, operate, tune, and maintain them.

Autonomic Computing helps address these complexity issues by using technology to manage technology. The idea is not new—many of the major players in the industry have developed and delivered products based on this concept.

The term "autonomic" is derived from human biology. The autonomic nervous system monitors your heartbeat, checks your blood sugar level, and keeps your body temperature close to 98.6°F without any conscious effort on your part. In much the same way, Autonomic Computing components anticipate computer system needs and resolve problems—with minimal human intervention. Figure 3.9 shows this computer intersection.

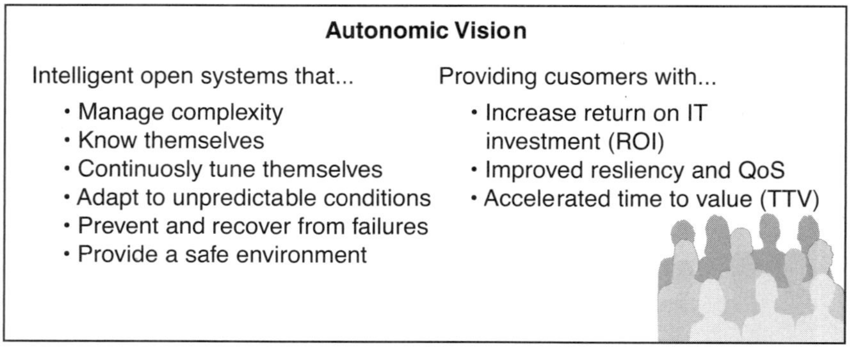
Autonomic Vision
Intelligent open systems that...
• Manage complexity
• Know themselves
• Continuosly tune themselves
• Adapt to unpredictable conditions
• Prevent and recover from failures
• Provide a safe environment
Providing cusomers with...
• Increase return on IT investment (ROI)
• Improved resliency and QoS
• Accelerated time to value (TTV)

FIGURE 3.9 The vision for Autonomic Computing incorporates "intelligent" open systems and important customer values.

However, there is an important distinction between autonomic activity in the human body and autonomic responses in computer systems. Many of the decisions made by autonomic elements in the body are involuntary, whereas autonomic elements in computer systems make decisions based on tasks you choose to delegate to the technology. In other words, adaptable policy—rather than rigid hard-coding—determines the types of decisions and actions autonomic elements make in computer systems.

Autonomic Computing can result in significant improvements in system management efficiency when the disparate technologies that manage the environment work together to deliver performance results system-wide. For this to be possible in a multi-vendor infrastructure, however, IBM and other vendors must agree on a common approach to architecting autonomic systems.

The Customer Value of Autonomic Computing

An on demand business enterprise is one whose business processes—integrated end-to-end across the company and with key partners, suppliers, and customers—can respond with agility and speed to any customer demand, market opportunity, or external threat.

To realize the benefits of on demand business, customers will need to embrace a new computing architecture that allows them to best leverage existing assets as well as those that lie outside traditional corporate boundaries. This on demand Operating Environment has four essential characteristics: It is integrated, open, virtualized, and autonomic.

Autonomic Computing was conceived as a way to help reduce the cost and complexity of owning and operating an IT infrastructure. In an autonomic environment, system components—from hardware such as desktop computers and mainframes to software such as operating systems and business applications—are self-configuring, self-healing, self-optimizing, and self-protecting.

These self-managing attributes are at the core of an Autonomic Computing environment. They suggest that the tasks involved in configuring, healing, optimizing, and protecting the IT system are initiated due to situations the technologies themselves detect, and that these tasks are performed by those same technologies. Collectively, these intuitive and collaborative characteristics enable enterprises to operate efficiently with fewer human resources, while decreasing costs and enhancing the organization's ability to react to change. For instance, in a self-managing system, a new resource is simply deployed and then optimization occurs. This is a notable shift from traditional implementations, in which a significant amount of analysis is required before deployment to ensure that the resource will run effectively.

Finally, it is important to be aware that the self-configuring, self-healing, self-optimizing, and self-protecting attributes are not independent of one another. Specifically, all four attributes allow the ability to make changes to any configuration of one or more aspects of the IT system. The motivation for the configuration change is different for each attribute.

The Autonomic Computing Blueprint

The architectural Autonomic Computing Blueprint (ACBLUEPRINT, hereinafter simply referred to as the "blueprint") is an overview of the basic strategic perspectives, architectural concepts, technological constructs, and behaviors for building autonomic capabilities into on demand Operating Environments.

The blueprint also describes, in a concise and hard-hitting manner, the initial set of core capabilities for enabling Autonomic Computing, and it discusses the technologies that support these core capabilities. The blueprint also discusses industry standards, emerging standards, and new areas for standardization that will deliver Autonomic Computing open system architectures.

Autonomic Computing Architectural Concepts

The architectural concepts presented in this section begin the process of developing a common approach and terminology to architecting Autonomic Computing systems. The Autonomic Computing architecture concepts provide a mechanism for discussing, comparing, and contrasting the approaches different vendors use to deliver self-managing attributes in an Autonomic Computing system. The Autonomic Computing architecture starts from the premise that implementing self-managing attributes involves an intelligent control loop. This loop collects information from the system, makes decisions, and then adjusts the system as necessary. An intelligent control loop can enable the system to do things such as:

- Self-configure, by installing software when it detects that software is missing
- Self-heal, by restarting a failed element
- Self-optimize, by adjusting the current workload when it observes an increase in capacity
- Self-protect, by taking resources offline if it detects an intrusion attempt

Figure 3.10 illustrates that these control loops can be delivered in two different ways:

- Various combinations of management tools or products can implement a loop. In Figure 3.10, the three examples are the configuration manager, workload manager, and risk manager. These tools use the instrumentation interfaces (for example, a Simple Network Management Protocol Management Information Base [SNMP MIB]) provided by IT system components to make the control loop manageable. This interface is referred to as the "manageability interface" in the figure.
- A control loop, which embeds a loop in the runtime environment for a particular resource, can be provided by a resource provider. In this case, the control loop is configured through the manageability interface provided for that resource (for example, a hard drive). In some cases, the control loop may be hard-wired or hard-coded so it is not visible through the manageability interface.

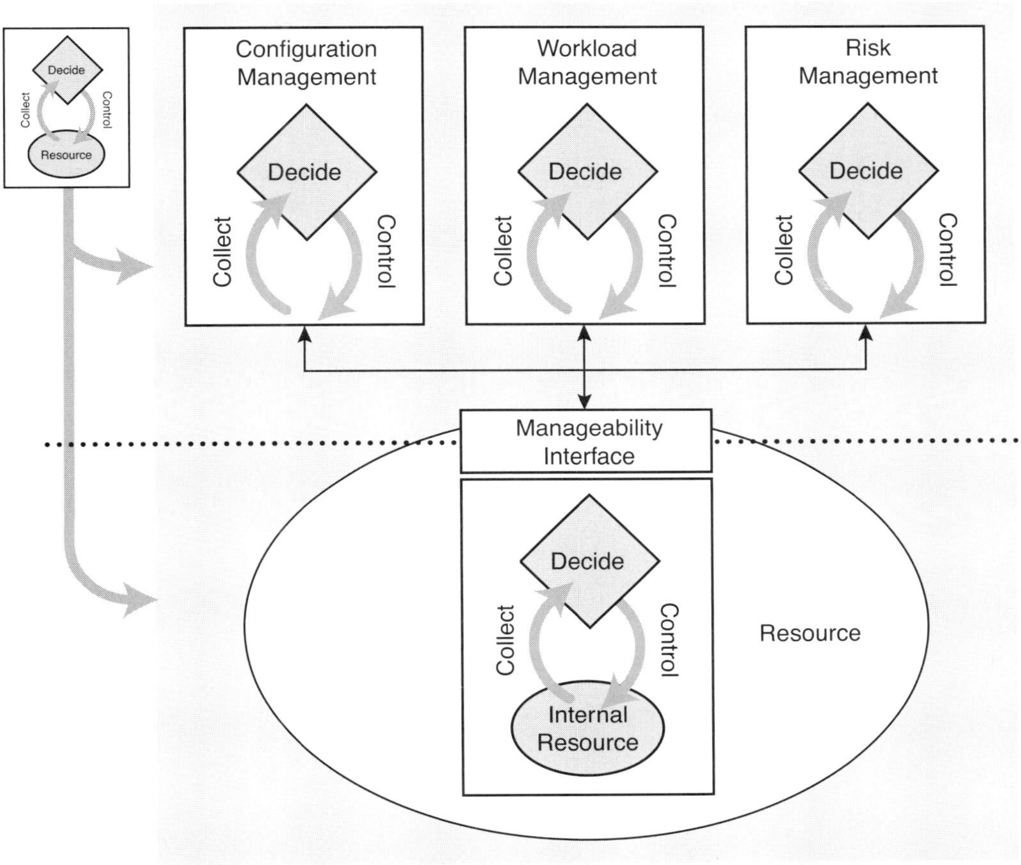

FIGURE 3.10 The flow of a control loop. A control loop can be delivered in two ways: in management tools and in embedded system resources.

Decision-Making Contexts

In the previous figure, three management functions were shown: configuration management, workload management, and risk management. Each of these management functions implements a different control loop, but they can potentially interact with the same resource. Thus, it is possible to have multiple control loops managing the same resource. In general, a robust IT system can have thousands of active control loops at any point in time.

To provide some order to this situation, the architecture for Autonomic Computing defines three different layers of management. Each layer involves implementing control loops to enable self-management in different decision-making contexts, or scopes:

- The resource element context is the most basic because its elements—networks, servers, storage devices, applications, middleware, and personal computers—manage themselves in an autonomic environment. The resource element layer is where autonomic function begins, by having intelligent control loops that configure, optimize, heal, and protect individual resources.
- Resource elements are grouped into a composite resources decision-making context. A pool of servers that work together to dynamically adjust workload and configuration to meet certain performance and availability thresholds can represent these groups; or, they can be represented by a combination of heterogeneous devices, such as databases, Web servers, and storage subsystems, which work together to achieve common performance and availability targets.

These different management levels define a set of decision-making contexts that are used to classify the purpose and role of a control loop within the Autonomic Computing architecture.

Control Loop Structure

In an autonomic environment, components work together, communicating with each other and with high-level management tools. They regulate themselves, and sometimes, each other. They can proactively manage the system, while hiding the inherent complexity of these activities from end-users and IT professionals.

Another aspect of the Autonomic Computing architecture is shown in the Figure 3.11. This portion of the architecture details the functions that can be provided for the control loops. The architecture organizes the control loops into two major elements: a managed element and an autonomic manager. A managed element is what the autonomic manager is controlling. An autonomic manager is a component that implements a particular control loop.

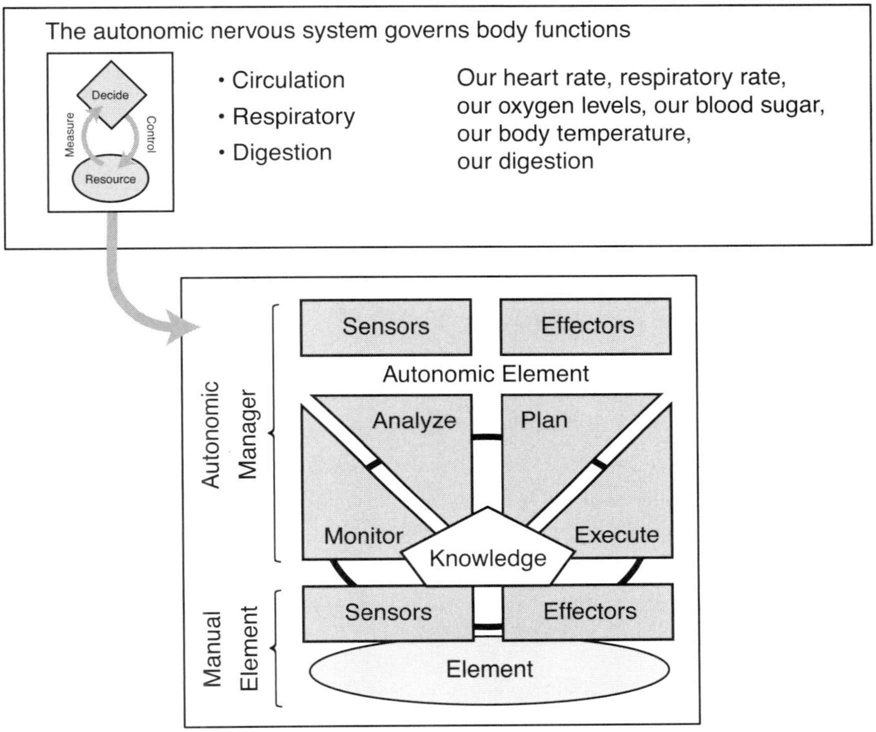

FIGURE 3.11 In an Autonomic Computing architectures, control loops facilitate system management.

Managed Elements

A *managed element* is a controlled system component. A managed element can be a single resource (a server, database server, or router) or a collection of resources (a pool of servers, cluster, or business application). A managed element is controlled through its *sensors* and *effectors*:

- Sensors provide mechanisms to collect information about the state and state transition of an element. To implement sensors, you can either use a set of "get" operations to retrieve information about the current state, or a set of management events (unsolicited, asynchronous messages or notifications) that flow when the state of the element changes in a significant way.
- Effectors are mechanisms that change the state (configuration) of an element. In other words, effectors are a collection of "set" commands or application programming interfaces (APIs) that change the configuration of the managed resource in some important way.

The combination of sensors and effectors forms the manageability interface that is available to an autonomic manager. As depicted by the black lines in the figure that connect the elements on the sensors and effectors, the architecture encourages the idea that sensors and effectors are linked together. For example, a configuration change that occurs through effectors should be reflected as a configuration change notification through the sensor interface.

Autonomic Manager

The *autonomic manager* is the component that implements the control loop. The architecture dissects the loop into four parts that share knowledge:

- The *monitor part* provides the mechanisms that collect, aggregate, filter, manage, and report details (metrics and topologies) collected from an element.
- The *analyze part* provides the mechanisms to correlate and model complex situations (time-series forecasting and queuing models, for example). These mechanisms allow the autonomic manager to learn about the IT environment and help predict future situations.
- The *plan part* provides the mechanisms to structure the action needed to achieve goals and objectives. The planning mechanism uses policy information to guide its work.
- The *execute part* provides the mechanisms that control the execution of a plan with considerations for on-the-fly updates.

Some autonomic elements will have as their main task the management of an IT resource, such as a DB2 information management system from IBM, a Linux Web server, an IBM TotalStorage Enterprise Storage Server storage array, or a network router or load balancer. The autonomic element's autonomic manager will make every effort to carry out the task as efficiently as possible, based on high-level policies that govern the apportionment of the resource, or specify who is to have access to it or place constraints on how the resource is to be made available. The autonomic manager relies on techniques such as feedback control optimization based on forecasting models.

This architecture does not prescribe a particular management protocol or instrumentation technology since the architecture needs to work with the various computing technologies and standards that exist in the industry today, such as SNMP, JavaManagement Extensions (JMX), Distributed Management Task Force, Inc. (DMTF), Common Information Model (CIM), vendor-specific APIs or commands, as well as any new technologies that emerge in the future. Given the diversity of the approaches that already exist in the IT industry, this architecture endorses Web services techniques for sensors and effectors. These techniques encourage implementers to leverage existing approaches and support multiple binding techniques as well as multiple marshalling techniques.

Figure 3.12 provides a more detailed view of these four parts by highlighting some of the functions each part uses:

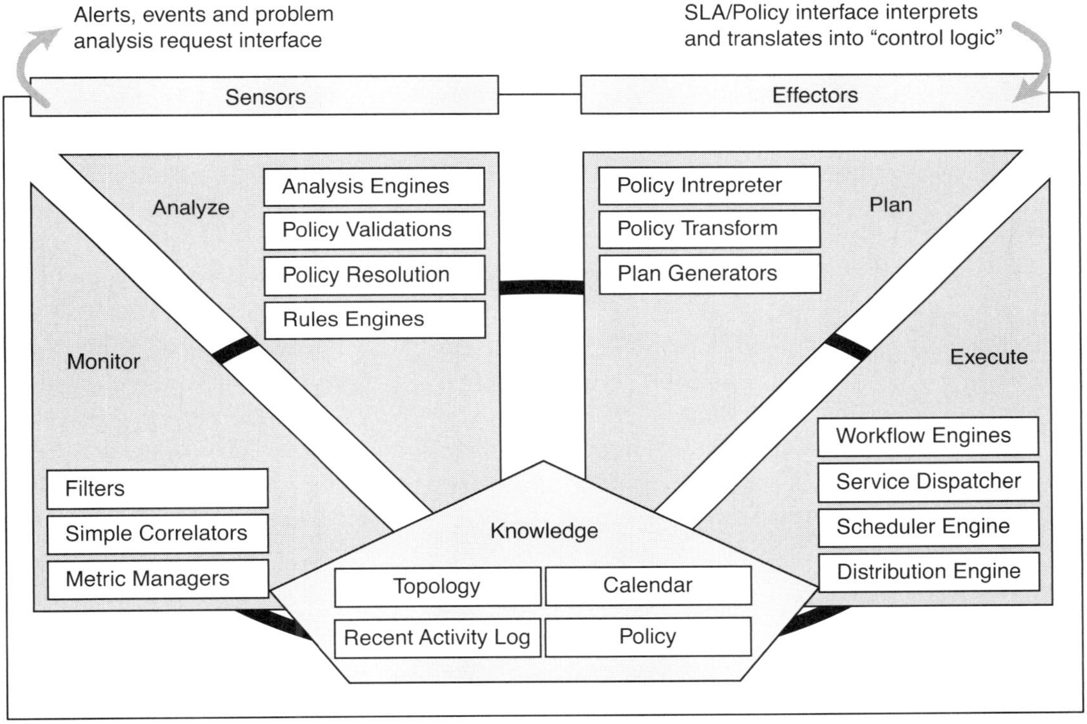

FIGURE 3.12 The functional details of an autonomic manager.

These four parts work together to provide control loop functionality. The diagram shows a structural arrangement of the parts—not a control flow. The bold line that connects the four parts should be thought of as a common messaging bus rather than a strict control flow. In other words, there can be situations where the plan part may ask the monitor part to collect more or less information. There could also be situations where the monitor part may trigger the plan part to create a new plan. The four parts collaborate using asynchronous communication techniques, like a messaging bus.

Autonomic Manager Collaboration

The numerous autonomic managers in a complex IT system must work together to deliver Autonomic Computing to achieve common goals. For example, a database system needs to work with the server, storage subsystem, storage management software, Web server, and other elements of the system for the IT infrastructure as a whole to become a self-managing system.

The sensors and effectors provided by the autonomic manager facilitate collaborative interaction with other autonomic managers. In addition, autonomic managers can communicate with each other in both P2P and hierarchical arrangements.

Figure 3.13 shows an example of a simple IT system that includes two business applications: a customer order application and a vendor relationship application. Separate teams manage these applications. Each of these applications depends on a set of IT resources—databases and servers—to deliver its functionality. Some of these resources—DB 3, DB 4, Server B, and Server C—are shared between the applications, which are managed separately.

There is a minimum of four management domains (decision-making contexts) in this example. Each of the applications (customer order and vendor relationship) has a domain that is focused on the business system it implements. In addition, there is a composite resource domain for managing the common issues across the databases and a composite resource domain for managing common issues for the servers.

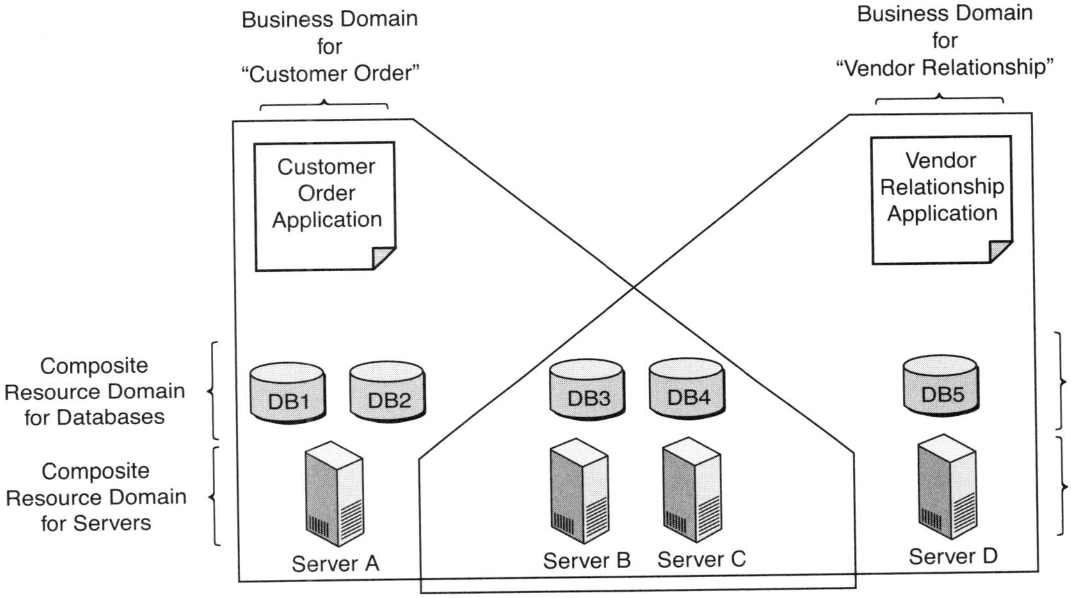

FIGURE 3.13 IT systems can share resources to increase efficiency.

Now, let us apply the Autonomic Computing architecture to this example, to see how autonomic managers would be used. Figure 3.14 illustrates some of the autonomic managers that either directly or indirectly manage DB 3 and some of the interaction between these autonomic managers. There are six autonomic managers in this illustration: one for each of the management domains, one embedded in the DB 3 resource, and one dedicated to the specific database resource.

Since the decision-making contexts for these autonomic managers are interdependent and self-optimizing, the autonomic managers for the various contexts will need to cooperate. This is accomplished through the sensors and effectors for the autonomic managers, using a "matrix management protocol." This protocol makes it possible to identify situations in which there are "multiple managers," and enables autonomic managers to electronically negotiate resolutions for domain conflicts based on a system-wide business and resource optimization policy.

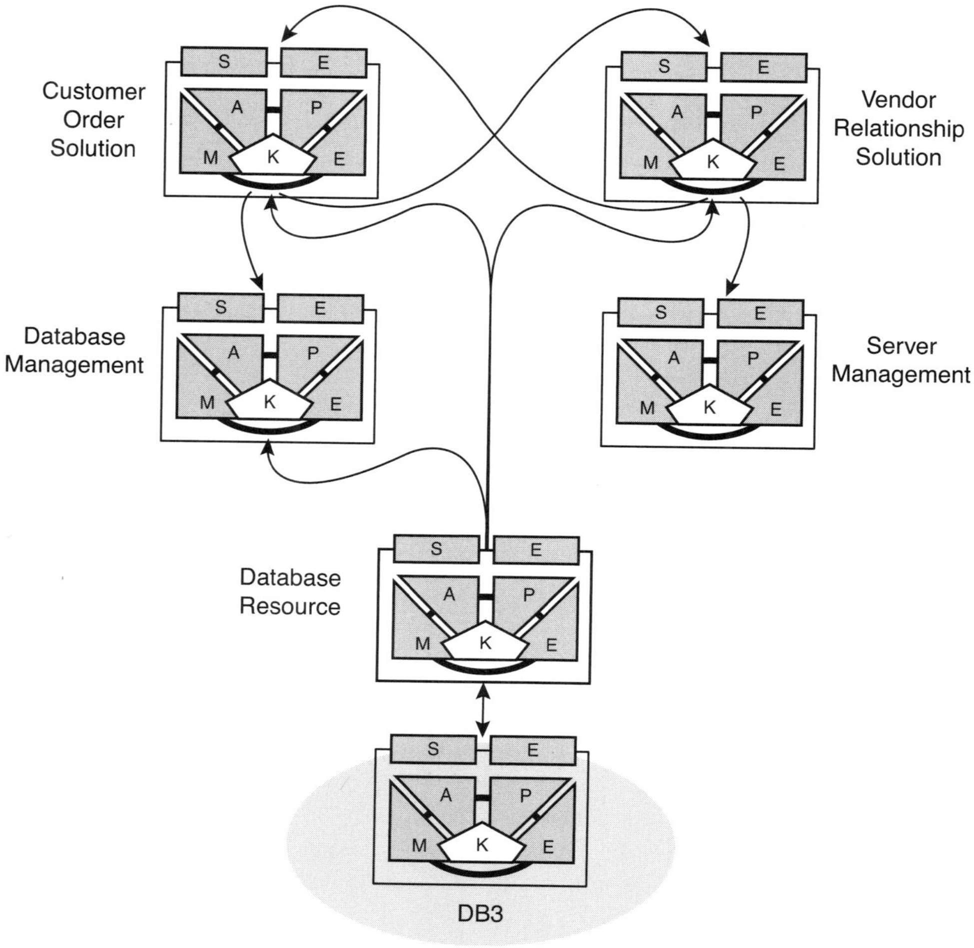

FIGURE 3.14 How six autonomic managers directly and indirectly manage the DB 3 resource.

Autonomic Manager Knowledge

Data used by the autonomic manager's four components is stored as shared knowledge. The shared knowledge includes things like topology information, system logs, performance metrics, and policies.

The knowledge used by a particular autonomic manager could be created by the monitor part, based on the information collected through sensors, or passed into the autonomic manager through its effectors. An example of the former occurs when the monitor part creates knowledge based on recent activities by logging the notification it receives from a managed element into a system log. An example of the latter is a policy. A policy consists of a set of behavioral constraints or preferences that influences the decisions made by an autonomic manager. Specifically, the plan part of an autonomic manager is responsible for interpreting and translating policy details. The analysis part is responsible for determining if the autonomic manager can abide by the policy, now and in the future.

Self-Managing Systems Change the IT Business

Ideally, the IT business operates through a collection of best practices and processes. Principles of the IT Infrastructure Library (from the Office of Government Commerce in the UK) and the IBM IT Process Model (developed by IBM Global Services) influence key IT best practices and processes. Figure 3.15 shows an example of a typical process flow for incident management, problem management, and change management. The actual mechanics of how these flows are implemented in a particular IT organization varies, but the functionality remains the same.

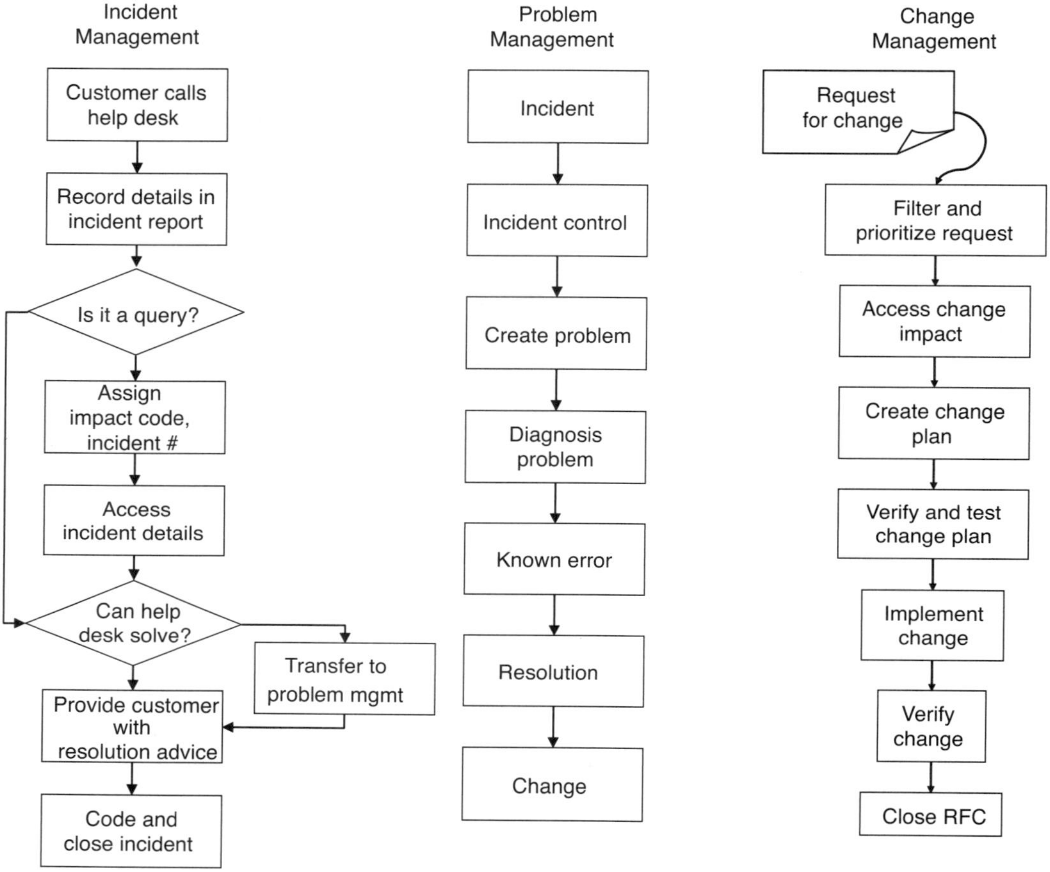

FIGURE 3.15 How typical IT processes can be represented as autonomic control loops.

The efficiency and effectiveness of these processes are measured using metrics such as elapsed time to complete a process, percentage executed correctly, and people and material costs to execute a process. Autonomic systems can positively affect these types of metrics, improving responsiveness, reducing TCO, and enhancing TTL through:

- *Quick process initiation*—Typically, implementing these processes requires an IT professional to initiate the process, create the request for change, spend time collecting incident details, and open a problem record. In a self-managing system, components can initiate the processes based on information derived directly from the system. This helps reduce the manual labor and time required to respond to critical situations, resulting in two immediate benefits: more timely initiation of the process and more accurate data from the system.

- *Reduced time and skill requirements*—Some tasks or activities in these processes usually stand out as skills-intensive, long-lasting, and difficult to complete correctly because of system complexity. In a change management process, such an activity is the "change impact analysis task."In problem management, such an activity is problem diagnosis. In self-managing systems, resources are built so that the expertise required to perform these tasks can be encoded or automated into the system. This helps reduce the amount of time and degree of skill needed to perform these tedious tasks, since technology rather than people can perform the tasks.

The mechanics and details of IT processes, such as change management and problem management, are different, but it is possible to categorize these into four common functions: Collect the details, analyze the details, create a plan of action, and execute the plan. These four functions correspond to the monitor, analyze, plan, and execute parts of the architecture. The approximate relationship between the activities in some IT processes and the parts of the autonomic manager are illustrated in Figure 3.16:

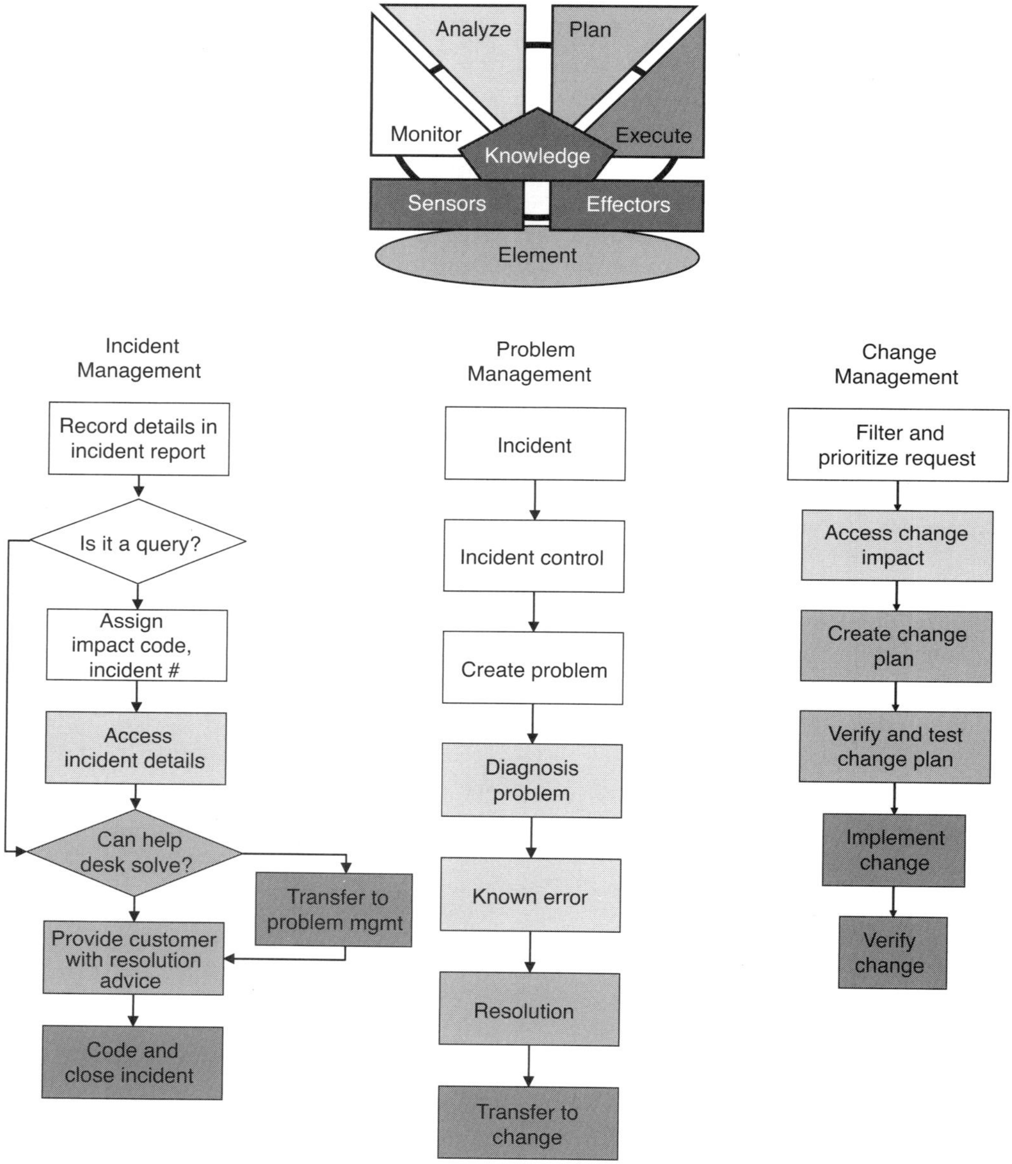

FIGURE 3.16 How Autonomic Computing affects IT processes.

The analyze and plan mechanisms are the essence of an Autonomic Computing system because they encode the "know-how" to help reduce the skill and time required of the IT professional.

An Evolution, Not a Revolution: Levels of Management Maturity and Sophistication

Incorporating autonomic capabilities into a computing environment is an evolutionary process enabled by technology. It is ultimately implemented by an enterprise through the adoption of these technologies, supporting processes, and skills. Throughout the evolution, the industry will continue delivering self-management tools to improve IT professionals' productivity.

To understand the level of sophistication of the tools and capabilities that are—and will be—delivered by the industry, consider the following five levels of autonomic maturity, which are also illustrated in Figure 3.17:

- At the *basic level*, IT professionals manage each infrastructure element independently and set it up, monitor it, and eventually replace it.
- At the *managed level*, systems management technologies can be used to collect information from disparate systems onto fewer consoles, helping to reduce the time it takes for the administrator to collect and synthesize information as the IT environment becomes more complex.
- At the *predictive level*, new technologies are introduced to provide correlation among several infrastructure elements. These elements can begin to recognize patterns, predict the optimal configuration, and provide advice on what course of action the administrator should take.
- At the *adaptive level*, the IT environment can automatically take action based on the available information and knowledge of what is happening in the environment. As these technologies improve and as people become more comfortable with the advice and predictive power of these systems, the technologies can progress to the autonomic level.
- At the *autonomic level*, business policies and objectives govern the IT infrastructure operation. Users interact with the autonomic technology tools to monitor business processes, alter the objectives, or both.

	Basic Level 1	Managed Level 2	Predictive Level 3	Adaptive Level 4	Autonomic Level 5
Characteristics	Rely on system reports, product documentation, and manual actions to configure, optimize, heal and protect individual IT components	Management software in place to provide consolidation, facilitation and automation of IT tasks	Individual IT components and systems able to monitor, correlate and analyze the environment and recommend actions	IT components, individually and collectively, able to monitor, correlate, analyze and take action with minimal human intervention	Integrated IT components are collectively and dynamically managed by business rules and policies
Skills	Requires extensive, highly skilled IT staff	IT staff analyzes and takes actions	IT staff approves and initiates actions	IT staff manages performance against SLAs	IT staff focuses on enabling business needs
Benefits	Basic requirements addressed	Greater system awareness Improved productivity	Reduced dependency on deep skills Faster/better decision making	Balanced human/system interaction IT agility and resiliency	Business policy drives IT management Business agility and resiliency

Manual → Autonomic

FIGURE 3.17 The Autonomic Computing evolution occurs gradually across five phases.

How Technology Must Evolve to Support Autonomic Computing

The earlier discussion about autonomic maturity levels demonstrated that self-managing capabilities would not be incorporated in one quick step. Rather, they constitute a concept that permeates all aspects of a system. Figure 3.18 reinforces this observation by showing a possible relationship between the maturity levels, the three decision-making contexts (resource element context, composite resource context, and business solution context), and the parts of the autonomic manager. This mapping results in two important observations:

- First, as the maturity levels increase, the decision-making context for the autonomic manager changes. The pyramid on the right-hand side summarizes the three different decision-making contexts in which autonomic managers can implement self-managing capabilities.
- Second, different parts of the autonomic manager are implemented at each maturity level. The monitor and execute parts of the autonomic manager are implemented at the basic and managed levels. So, at these two levels, IT professionals are responsible for performing the func-

tions of the analyze and plan parts. The analyze part of the autonomic manager is supplied at the predictive maturity level. At this level, the IT professional is responsible for the plan function. At the adaptive and autonomic levels, all the parts of the autonomic manager are working, so the IT professional can delegate work to the system. The difference between these two maturity levels is the decision-making context. The adaptive maturity level supports either the resource element or the composite element context, and the autonomic level supports the business solution context.

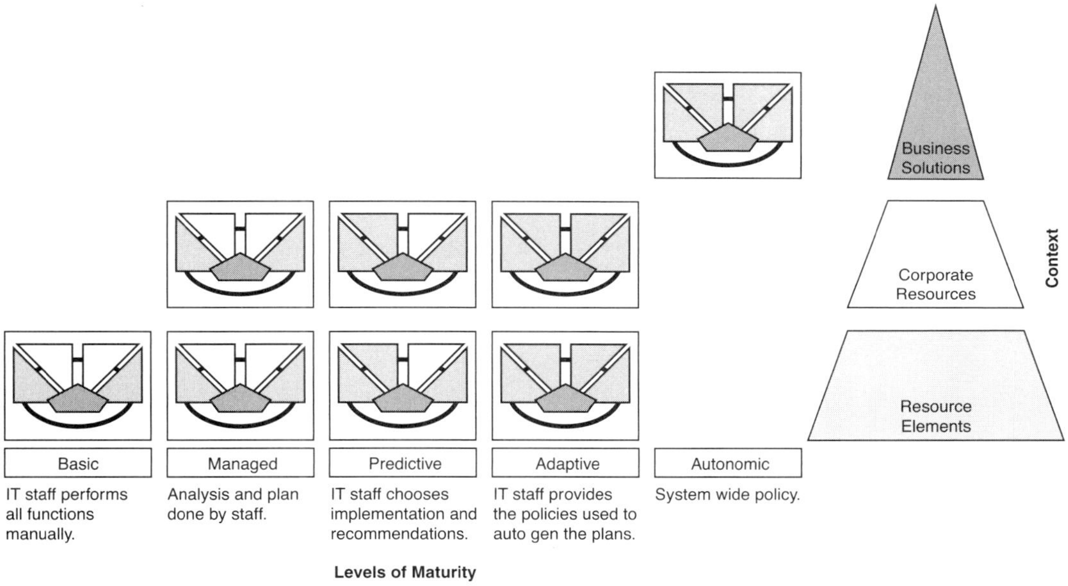

FIGURE 3.18 By progressing along the five autonomic maturity levels, businesses can evolve IT environments to fully autonomic levels.

As Figure 3.18 shows, the progressive implementation of the architecture occurs for each of the three contexts. This is because it is difficult to deliver a self-managing capability in a business system context if there is no self-managing capability in the lower contexts.

Core Autonomic Capabilities

For the autonomic managers and managed elements in an autonomic system to work together, the developers of these components need a common set of capabilities. This section describes an initial set of core capabilities that are needed to build autonomic managers. These core capabilities include: solution knowledge, common system administration, problem determination, autonomic monitoring, complex analysis, policy for autonomic managers, and transaction measurements. Technologies that deliver these capabilities will accelerate the delivery of autonomic managers that can collaborate in an autonomic system.

Solution Knowledge

Today, there are a myriad of installation, configuration, and maintenance mechanisms. The differences and idiosyncrasies of these many system administration tools and distribution packaging formats create significant problems in managing complex system environments. These problems are further compounded in a Web services environment, where application functionality can be composed dynamically. From an autonomic systems perspective, lack of solution knowledge inhibits important elements of self-configuring, self-healing, and self-optimizing.

A common solution knowledge capability eliminates the complexity introduced by many formats and installation tools. By capturing installation and configuration information in a consistent manner, it creates knowledge that autonomic managers can use in contexts beyond installation, such as problem determination or optimization. Solutions are combinations of platform capabilities (operating systems and middleware) and application elements (such as Enterprise JavaBeans [EJB], DB2 tables, hypertext markup language [HTML] pages, and flow definitions) that solve a particular customer problem.

The Autonomic Computing Blueprint defines a set of constructs for composing installable units and design patterns that make it possible to standardize solution knowledge. An *installable unit* is composed of a descriptor that describes the content of the installable unit and the actual artifact to be installed. The descriptor and artifact comprise the package—like a Java archive file. The target environment for the installable unit is called the *hosting environment*, or the container that will accept the artifact to be installed.

There are three categories of installable units that build on each other:

- *Smallest installable unit*—This unit contains one atomic artifact.
- *Container installable unit*—This unit aggregates a set of artifacts for a particular container type.
- *Solution module installable unit*—This unit contains multiple instances of container installable units.

The Autonomic Computing Blueprint identifies a number of enabler technology components for solution knowledge. These include:

- *A dependency checker*—This determines whether the dependency of an artifact is satisfied in the targeted hosting environment.
- *An installer*—The functionality that knows how to extract the artifacts in the installable units and invoke the appropriate operations on the target hosting environment.
- *An installable unit database*—A library for installable units.
- *Deploy logic*—This functionality knows how to distribute an installable unit to an installer component.
- *An installed unit "instances" database*—This database stores the configuration details about installable units and hosting environments.

The installable unit schema definitions and enabler components create the basis for coherent installation, configuration, and maintenance processes at the solution level versus different product-specific mechanisms.

Common System Administration

Autonomic systems require common console technology to create a consistent human-facing interface for the autonomic manager elements of the IT infrastructure. The common console capability provides a framework for reuse and consistent presentation for other autonomic core technologies.

The primary goal of a common console is to provide a single platform that can host all the administrative console functions in server, software, and storage products in a manner that allows users to manage solutions rather than managing individual systems or products. Administrative console functions range from setup and configuration to solution runtime monitoring and control.

The values to the customer in having a common administrative console are: reduced cost of ownership, attributable to more efficient administration, and reduced learning curves as new products and solutions are added to the autonomic system environment. The reduced learning curve results from using both standards and the familiar Web-based presentation style. By enabling increased consistency of presentation and behavior across administrative functions, the common console creates a familiar user interface that promotes reusing learned interaction skills versus learning new, different, product-unique interfaces.

The common console functionality could be a platform for IBM products with extensions for ISVs and business partners. Common console interfaces could also be made available outside IBM to enable development of new components for IBM products or to enable bundling of common console components in non-IBM products. Since the common console architecture is standards-based, IBM could propose it as a system administration user interface infrastructure standard.

A common console instance consists of a framework and a set of console-specific components provided by other product development groups. Administrative activities are executed as portlets. Consistency of presentation and behavior is key to improving Autonomic Computing system administrative efficiency, and will require ongoing effort and cooperation among many product communities. Console guidelines will take time to emerge, given the large number of human factors and design organizations involved.

Problem Determination

Whether healing, optimizing, configuring or protecting, autonomic managers take actions based on problems or situations they observe in their managed elements. Therefore, one of the most basic capabilities is being able to extract high-quality data to determine whether or not a problem exists in a managed element. In this context, a problem is a situation in which an autonomic manager needs to take action. A major cause of poor-quality information is the diversity in the format and content of the information provided by the managed element.

There is a relatively small, finite, canonical set of situations that is reported by components. This common set covers a large percentage of the situations that are reported by most system components. Currently, components use different terminology to report common situations. For example, one component may report the situation that a "component has started," where another component may report that the "component has begun execution." This variability in the description of the situation makes writing and maintaining autonomic systems difficult.

To address this diversity of the data collected, the Autonomic Computing Blueprint requires a common problem determination architecture that normalizes the data collected, in terms of format, content, organization, and sufficiency. To do this, it defines a base set of data that must be collected or created when a situation or event occurs. This definition includes information on both the kinds of data that must be collected as well as the format that must be used for each field collected. The problem determination architecture categorizes the collected data into a set of situations, such as component starts and stops.

The technologies used to collect autonomic data must be capable of accommodating legacy data sources (e.g., logs and traces) as well as data that is supplied using the standard format and categorization. To accommodate this legacy data, the architecture defines an adapter/agent infrastructure that provides the ability to plug in adapters to transform data from a component-specific format to the standard format as well as sensors to control data collection (e.g., filtering, aggregation, etc.).

Autonomic Monitoring

Autonomic monitoring is a capability that provides an extensible runtime environment for an autonomic manager to gather and filter data obtained through sensors. Autonomic managers can utilize this capability as a mechanism for representing, filtering, aggregating, and performing analyses on sensor data. This capability includes:

- A common way to capture the information that surfaces from managed elements through sensors. This should utilize the CIM, SNMP, Windows Management Instrumentation (WMI), and JMX industry standards.
- Built-in sensor data-filtering functions.
- A set of pre-defined resource models (and a mechanism for creating new models) that enables the combination of different pieces of sensor data to describe the state of a logical resource. Resource models describe business-relevant "logical objects" from the perspective of common problems that can affect those objects. Examples of frequently used resource models include "machine memory" and "machine connectivity."
- A way to incorporate policy knowledge.
- A way to plug in analysis engines that can provide basic event isolation, basic root cause analysis, and server-level correlation across multiple IT systems, and automate initiation of corrective actions.

An autonomic manager using this autonomic monitoring functionality can help manage certain applications or resources more effectively through:

- *Multiple source data capture*—Allows processing of data from industry-standard APIs, and from any custom data interfaces that a particular application uses
- *Local persistence checking*—Links corrective actions or responses to the repeated occurrence of a problem condition so that a single point-in-time threshold exception does not immediately trigger a costly and unnecessary troubleshooting response

- *Local intelligent correlation*—Recognizes a number of metrics in aggregate as a "problem signature," enabling root cause identification and responses to problems rather than symptoms
- *Local data store and reporting*—Provides a real-time "heart monitor" that determines whether the application environment and individual applications are functioning properly

The reference model component of autonomic monitoring should provide built-in intelligence, a set of embedded best-practices data that:

- Interprets the quality of a logical object against a defined baseline
- Logs performance data related to the business object
- Proactively manages the application through a pre-defined collection of problem signatures

This resource management model demonstrates a plurality of capabilities, and can be exploited in a number of ways to:

- Manage systems and resources reactively, allowing escalation of the status change of a resource
- Adopt a proactive management strategy, using resource models to automatically diagnose and fix problems at the local level
- Use predictive management tasks, allowing the utilization of data mining tools to analyze performance metrics and predict abnormal behavior
- Use adaptive management, automatically tuning model baselines based on historical trends of performance data collected by the model itself

Complex Analysis

Autonomic managers need to have the capability to perform complex data analysis and reasoning on the information provided through sensors. The analysis will be influenced by stored knowledge data. The Autonomic Computing Blueprint defines complex analysis technology building blocks that autonomic managers can use to represent knowledge, perform analysis, and do planning.

Complex analysis technology components and tools provide the power and flexibility required to build practical autonomic managers. Autonomic managers must collect and process large amounts of data from sensors and managed resources. This data includes information about resource configuration, status, offered workload, and throughput. Some of the data is static or changes slowly, while other data is dynamic, changing continuously through time. An autonomic manager's ability to quickly analyze and make sense of this data is crucial to its successful operation. Common data analysis tasks include classification, clustering of data to characterize complex states and detect similar situations, prediction of anticipated workload and throughput based on past experience, and reasoning for causal analysis, problem determination, and optimization of resource configurations.

The complex analysis technology uses a rule-based language that supports reasoning through procedural and declarative rule-based processing of managed resource data. The underlying rule engines can be used to analyze data with scripting as well as forward and backward inference methods using if-then rules, predicates, and fuzzy logic. Application classes can be imported directly into rule sets so that data can be accessed (using sensors) and control actions can be invoked directly from rules (using effectors). The rule-based language features both Java programming language-like text and XML source rule set representations, enhancing productivity for rule authors familiar with Java syntax and allowing portable knowledge interchange. Rule sets can include multiple rule blocks so that a mix of procedural and inference methods can be used to analyze data and define autonomic manager behavior.

Other complex analysis technology components can be used to augment the basic rule capabilities. These include Java Beans to access data from flat files and relational databases, to filter, transform, and scale data using templates, and to write data to flat files and databases. Complex analysis technology also includes machine learning beans and agents to perform classification, clustering, and time-series prediction using neural networks, decision trees, and Bayesian classifiers, and to perform statistical data analysis and optimization using genetic algorithms.

As a core Autonomic Computing technology, complex analysis components can enhance productivity when used as building blocks to implement specific plan, analyze, or knowledge functionality.

Policy for Autonomic Managers

An Autonomic Computing system requires a uniform method for defining the policies that govern the decision-making for autonomic managers. A *policy* specifies the criteria that an autonomic manager uses to accomplish a definite goal or course of action. As shown in Figure 3.19, policies are a key part of the knowledge used by autonomic managers to make decisions, essentially controlling the planning portion of the autonomic manager. By defining policies in a standard way, they can be shared across autonomic managers to enable entire systems to be managed by a common set of policies.

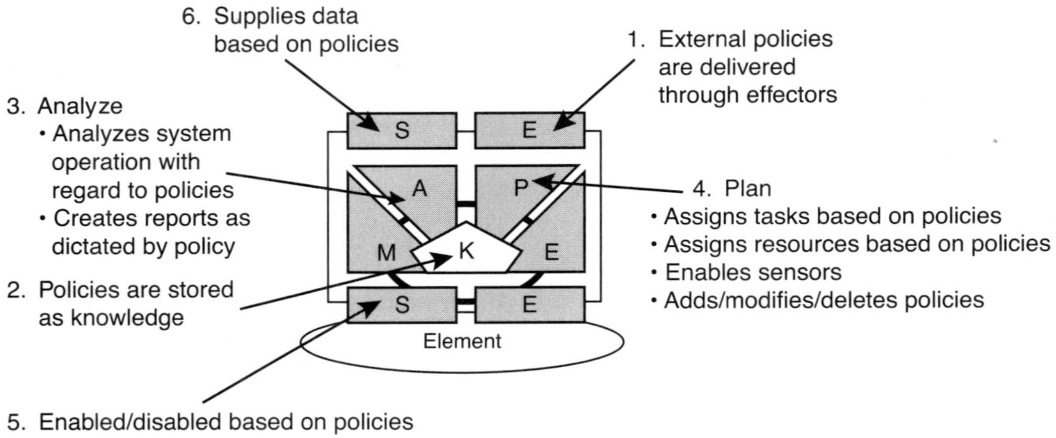

FIGURE 3.19 Policy-based management in autonomic managers.

Today, the term "policy" is used in various contexts to mean seemingly different things, and exists in various forms and formats. Table 3.1 contains some examples:

TABLE 3.1 Forms of Autonomic Policies

Typical Domain	Examples
IT resource policies	If a packet is gold, queue it with high priority.
Business process policies	If it's a frequent customer, apply a 3% discount.
Interaction policies	Require Kerberos authentication.
SLA policies	If 2-second response time is not delivered, refund 30%.

Despite the apparent differences, these examples exhibit substantial commonality and must be specified consistently for an autonomic system to behave cohesively. The Autonomic Computing Blueprint is currently defining the specifications and capabilities for policy-based autonomic managers. This definition includes:

- Specification of canonical configuration parameters for management elements
- Format and schema to specify user requirements or criteria
- Mechanisms, including wire formats, for sharing and distributing policies
- Schema to specify and share policy among autonomic managers

One of the key functions that an autonomic system must perform is to share policies among autonomic managers, so this capability will leverage and extend policy standards.

Transaction Measurements

Autonomic managers need a transaction measurement capability that spans system boundaries to understand how the resources of heterogeneous systems combine into a distributed transaction execution environment. By monitoring these measurements, the autonomic manager can analyze and plan to change resource allocations to optimize performance across these multiple systems according to policies, as well as determine potential bottlenecks in the system.

Tuning servers individually cannot ensure the overall performance of applications that span a mix of platforms. Systems that appear to be functioning well on their own may not, in fact, be contributing to optimal end-to-end processing. Inefficiencies created by infrastructure complexity and the growing number of servers are outstripping the increasing productivity provided to system administrators by powerful management tools. While hardware and software costs decline, people costs rise and larger staffs are needed. Additionally, when hundreds or even thousands of different servers are involved, it may not be possible with current technology for any number of administrators to discover failing systems in time to isolate or repair them before any damage is done.

Furthermore, the average utilization of most distributed systems is very low today. Many on demand business applications must be capable of handling large spikes in volume, so companies typically buy hardware to meet the needs of those spikes. However, when the original application is not fully utilizing computing resources, there is no easy way to divert the excess capacity to lower priority work. Therefore, customers must buy and maintain a separate infrastructure for each application to meet that application's most demanding computing needs.

Instituting an end-to-end transaction measurement infrastructure enables a distributed workload management capability, and addresses these problems of rising administrative costs and low hardware utilization. The general philosophy behind distributed workload management is one of policy-based, goal-oriented management. The philosophy requires both a policy definition infrastructure (like that mentioned above) and an end-to-end transaction measurement infrastructure. The policy contains simple definitions of classes of service—broad categories of "work"—and an associated performance goal for each class of service. The goals are stated in terms such as "need to complete 90 percent in less than 1 second," or "an average response time of 2 seconds." In addition to a goal declaration, each class of service is accompanied by a business importance level, which indicates how important the achievement of the goal is to the business that owns the computing resources. The relationships are then quite simple: satisfy the goals of the most important workloads and then worry about the rest.

Once the service classes are defined and prioritized, the next step is to understand what systems are used to process the service classes, and to instrument these systems appropriately. An administrator already understands what servers are in place, and what applications are on each; but what the administrator typically does not know is the exact nature of the relationships between application environments, and the relationships between the various servers. Therefore, by applying instrumentation across the application environments uniformly, an administrator can determine these relationships and the flow of transactions through the system. The application response measurement (ARM) API is key to providing this uniform instrumentation.

The key capability needed for a distributed workload management system is the ability to understand the transaction topology and map the service classes to this topology. Autonomic managers involved in workload management need a transaction measurement capability to understand how the systems involved commit their resources to execute the workload, and how changes in allocation affect performance over time.

The combination of prioritized service classes and an understanding of the systems involved in delivering a particular class of service enables distributed workload management. This creates a general workload management infrastructure that can be used for many purposes because of the ability to prioritize different service classes according to the needs of the business. The distributed workload management system can optimize work across the distributed infrastructure in an attempt to meet all the goals associated with each service class. If all the goals cannot be achieved, then changes can be made to ensure that the most important applications meet their goals first. Overall, this enables a single infrastructure to, in an autonomic manner, "self-optimize" while meeting the needs of the business.

Standards for Autonomic Computing

The fundamental nature of Autonomic Computing systems precludes any one company from delivering a total autonomic solution. Enterprises have heterogeneous IT infrastructures and must deal with heterogeneous environments outside the enterprise. A proprietary implementation would be like a heart that maintains a regular, steady heartbeat but is not able to adjust to the needs of the rest of the body when under stress.

Autonomic Computing systems require deployment of autonomic managers throughout the IT infrastructure, managing resources that include other autonomic managers from a diverse range of suppliers. These systems, therefore, must be based on open industry standards.

The Autonomic Computing Blueprint identifies relevant existing computing industry standards. New open standards will continue to be developed and shared with the industry that will define new mechanisms for interoperating in a heterogeneous system environment.

Summary

Self-management is about shifting the burden of managing systems from people to technologies. When the self-management capabilities delivered by IBM and other vendors are able to collaborate, it will be possible to deliver Autonomic Computing capabilities for the entire IT infrastructure. In these environments, the elements of a complex IT system will manage themselves based on a shared view of system-wide policy and objectives.

Future of Computing

IBM has named its vision for the future of computing "Autonomic Computing." This new paradigm shifts the fundamental definition of the technology age from one of computing, to one defined by data. Access to data from multiple, distributed sources, in addition to traditional centralized storage devices, will allow users to transparently access information when and where they need it.

At the same time, this new view of computing will necessitate changing the industry's focus on processing speed and storage to one of developing distributed networks that are largely self-managing, self-diagnostic, and transparent to the user.

This chapter presented a high-level blueprint to assist in delivering Autonomic Computing in phases. The architecture reinforces the fact that self-managing implies intelligent control loop implementations that will execute in one of three decision-making contexts to monitor, analyze, plan, and execute using knowledge of the environment. In addition, the loops can be embedded in resource runtime environments or delivered in management tools. These control loops collaborate using a matrix management protocol.

The journey to a fully autonomic IT infrastructure is an evolution. The stages of this evolution were illustrated by showing which aspects of the architecture need to be addressed at the five management maturity levels. This model was then applied to the IT infrastructure using the three decision-making contexts.

Enterprises want and need to reduce their IT costs, simplify the management of their IT resources, realize a faster return on their IT investment, and ensure the highest possible levels of system availability, performance, security, and asset utilization. Autonomic Computing addresses these issues—not just through new technology, but also through a fundamental, evolutionary shift in the way IT systems are managed. Moreover, such systems will free the IT staff from detailed, mundane tasks, allowing them to focus on managing their business processes. True Autonomic Computing will be accomplished through a combination of process changes, skills evolution, new technologies, architecture, and open industry standards.

For readers wishing to further their understanding of the IBM Corporation's Autonomic Computing initiative, please refer to the book entitled *Autonomic Computing* by Richard Murch.

In the next chapter, we will explore the complementary side to Autonomic Computing, which together with autonomic disciplines enables very powerful enterprise-wide solutions for integrating into on demand business solutions. This second, complementary computing discipline is Grid Computing.

Glossary of Autonomic Computing Terms

This section contains definitions of some Autonomic Computing terms that were utilized in this chapter.

Analyze: The function of an autonomic manager that models complex situations.

Autonomic: Being accomplished without overt thought or action. Example: the human autonomic nervous system that monitors and regulates temperature, pupil dilation, respiration, heart rate, digestion, etc.

Autonomic Computing: An approach to self-managed computing systems with a minimum of human interference.

Autonomic manager: A part of an autonomic element that manages a managed element within the same autonomic element.

Data collection: Definitions for standard situational event formats in the Autonomic Computing architecture (also called logging). This notion is one of the Autonomic Computing core technologies.

Domain: A collection of resources that have been explicitly or implicitly grouped together for management purposes.

Effector: A way to change the state of a managed element.

Execute: The function of an autonomic manager that is responsible for interpreting plans and interacting with element effectors to insure that the appropriate actions occur.

Install (Installation): Definitions for standard methods to describe software deployment and installation. This notion is one of the Autonomic Computing core technologies.

Knowledge: The common information that the monitor, analyze, plan, and execute functions require to work in a coordinated manner.

Maturity index: A graduated scale that expresses the level of maturity of Autonomic Computing, where Level 1 is basic (completely manual), Level 2 is managed, Level 3 is predictive, Level 4 is adaptive, and Level 5 is completely autonomic.

Open Grid Services Architecture (OGSA): A Grid Computing system architecture based on an integration of Grid Computing and Web services concepts and technologies.

Plan: The function of an autonomic manager that provides a way to coordinate interrelated actions over time.

Policy: A definite goal, course, or method of action to guide and determine future decisions. Policies are implemented or executed within a particular context. This is a set of behavioral constraints and preferences that influence decisions made by an autonomic manager. This notion of policy utilization is one of the Autonomic Computing core technologies.

Policy-based management: A method of managing system behavior or resources by setting policies that the system interprets.

Self-configuring: Setting an element up for operation.

Self-healing: Repairing damage to an element regarding its own operational integrity.

Self-managing: Directing and controlling an element. This is most often regarding self-configuring, self-optimizing, self-protecting, and self-healing operations.

Self-optimizing: Tuning or improving an element's own performance.

Self-protecting: Maintaining an element's own operational integrity.

Sensor: A way to get information about a managed element.

Situations: Events that Autonomic Computing components report to the outside world. Situations vary in granularity and complexity, ranging from simple situations like the start of a component to more complex situations like the failure of a disk subsystem.

Grid Computing

4

Grid Computing is defined as controlled and coordinated resource sharing and problem-solving in dynamic, multi-institutional virtual organizations (VOs). These sharing resources, ranging from simple file transfers to complex and collaborative problem-solving, are accomplished within controlled and well-defined conditions and policies. The dynamic groupings of individuals, multiple groups, or organizations that define the conditions and rules for sharing are called *virtual organizations*, or *VOs*.

WHAT IS GRID COMPUTING?

Grid Computing enables the virtualization of distributed computing and data resources such as processing, network bandwidth, and storage capacity to create a single system image, granting users and applications seamless access to vast IT capabilities. Grid Computing utilizes untapped resources of a computing device, without interruption to whatever the end user may be doing at that moment in time on the computer. Just as an Internet user views a unified instance of content via the Web, a Grid Computing user essentially sees a single, large, virtual computer. At its core, Grid Computing is based on an open set of standards and protocols—for instance, OGSA—that enables communications across heterogeneous, geographically dispersed environments. With Grid Computing, organizations can optimize computing and data resources, pool them for large-capacity workloads, share them across networks, and enable collaboration.

Grid Computing is important to on demand business; however, Grid Computing is not always required to create on demand business solutions. That being said, Grid Computing is one way of delivering very powerful on demand business solutions. Grid Computing, in fact, shares many of the same technologies with Autonomic Computing; for example, dynamic provisioning of resources. It is important to note, however, as we explore the interesting and slightly diverse subjects in this chapter, that one can deliver on demand business solutions without developing and/or integrating Grid Computing applications.

In the previous chapter, we discussed Autonomic Computing strategic perspectives. In this chapter, we will introduce and define the "Grid Computing problem," discuss the core concepts of a *virtual organization* (VO), and define the Grid Protocol Architecture that will solve the Grid Computing problem. This chapter will present some very innovative work [Foster01] that defines the Grid Computing anatomy. In addition, this chapter will examine Grid Computing in relation to other distributed technologies such as the Web, framework(s) providing services, server clusters, and P2P (peer-to-peer) computing.

THE VO CONCEPT IN GRID COMPUTING IS KEY

One of the significant operational concepts in Grid Computing is the implementation of the *VO*. This involves the dynamic computational-oriented task of defining groupings of individuals, such as multiple groups or organizations. Although this is perhaps simple to understand in theory, it remains complex across several dimensions. The complexities involved in this dynamic assembly revolve around identifying and bringing together the humans who initially defined the conditions to instantiate the grid. This includes automated functionalities of the rules, the policies, and the specific conditions affecting operations in the Grid that are at the core surrounding processing and sharing of information with those individuals in a VO.

The simplest way to think about this advanced Grid Computing concept is captured in the term "VO." This type of grouping serves as the basis for identifying and managing the Grid Computing groups that are associated with any particular community of Grid Computing end-users.

The Grid Computing Problem

Grid Computing has evolved as an important field in the computer industry by differentiating itself from distributed computing with an increased focus on resource sharing, coordination, and high-performance orientation. Grid Computing tries to solve the problems associated with resource sharing among a set of individuals or groups.

Grid Computing resources include computing power, data storage, hardware instruments, on demand business software, and applications. Problems associated with resource sharing among a set of individuals or groups are resource discovery, event correlation, authentication, authorization, and access mechanisms. These problems become proportionately more complicated when the Grid Computing solution is introduced as a solution for Utility Computing. Utility Computer in a grid is where end-users in the grid VO share industrial applications and resources.

This commercial Utility Computing concept spanning across Grid Computing services has introduced a number of challenging problems to the already complicated Grid Computing problem domains. These challenging problems include service level management features, complex accounting, utilization metering, flexible pricing, federated security, scalability, open-ended integration, and a multitude of very difficult networking services to sustain. It is key to understand that networking services can no longer be taken for granted, as these very important services now become the central nervous system for the enablement of all worldwide Grid Computing environments.

The Concept of Virtual Organizations (VOs)

Understanding the concept of a VO is the key to understanding Grid Computing. A VO is a dynamic set of individuals and/or institutions defined around a set of resource sharing rules and conditions [Foster02]. All VOs share some commonality, including common concerns and requirements, but they may vary in size, scope, duration, sociology, and structure.

The members of any VO negotiate resource sharing based on defined rules and conditions. This allows the members of the VO to share the resources from the automatically constructed grid resource pool. Establishing the VO with users, resources, and organizations from different domains across multiple, worldwide geographics is one of the fundamental technical challenges in Grid Computing. This complexity includes the definitions of the resource discovery mechanism, resource sharing methods, rules and conditions by which this sharing can be achieved, security federation and/or delegation, and access controls among the participants of the VO. As you can see, this challenge is complicated across several dimensions.

Let us explore two examples of VOs to better understand their common characteristics:

1. Thousands of physicists from different laboratories join together to create, design, and analyze the products of a major detector at CERN, the European high-energy physics laboratory. This group forms a "data

grid," with resource sharing among intensive computing, storage, and network services to analyze petabytes of data created by the detector at CERN.

2. A company wants to do financial modeling for a customer based on the data collected from various data sources, both internal and external to the company. This specific VO customer may need a financial forecasting capability and advisory capability on its investment portfolio, which is based on actual historic and current real-time financial market data. This financial institution customer can then be responsive by forming a dynamic VO within the enterprise for achieving more benefit from advanced and massive forms of computational power (i.e., application service provider [ASP]) and for data (i.e., data access and integration provider). This dynamic, financially-oriented VO can now reduce undesirable customer wait time while increasing reliability on forecasting by using real-time data and financial modeling techniques.

By closely examining the VOs, we can infer the following: the number and type of participants, the resources being shared, the duration, the scale, and the interaction pattern between the participants. All these attributes vary between any one single VO and another. At the same time, we can also infer that there are common characteristics among competing and sometimes distrustful participants that contribute to their VO formation. They may include [Foster02] some of the following items for consideration:

1. *Common concerns and requirements on resource sharing*—A VO is a well-defined collection of individuals and/or institutions that shares a common set of concerns and requirements. For example, a VO created to provide financial forecast modeling shares the same concerns on security, data usage, computing requirements, resource usage, and interaction patterns.
2. *Conditional, time-bound, and rules-driven resource sharing*—Resource sharing is conditional, and each resource owner has full control on making the resource available to the sharable resource pool. These conditions are defined based on mutually understandable policies and access control requirements (authentication and authorization). The number of resources involved in the sharing may dynamically vary over time based on the policies defined.
3. *Dynamic collection of individuals and/or institutions*—Over its period of time, a VO should allow individuals and/or groups into and out of the collection, provided they all share the same concerns and requirements on resource sharing.

4. *Sharing relationship among participants is P2P in nature*—The sharing relationship among the participants in a VO is P2P, which emphasizes that a resource provider can become a consumer to another resource. This introduces a number of security challenges, including mutual authentication, federation, and delegation of credentials among participants.
5. *Resource sharing is based on an open and well-defined set of interaction and access rules*—Open definition and access information must exist for each sharable resource for better interoperability among the participants.

The above characteristics and non-functional requirements of a VO lead to the definition of an architecture for the establishment, management, and sharing of resources among participants. As we will see in the next section, the focus of the Grid Protocol Architecture is to define an interoperable and extensible solution for resource sharing within a VO.

The Grid Protocol Architecture

A new architecture model and technology were developed for the establishment, management, and cross-organizational resource sharing within a VO. This new architecture, called the *Grid Protocol Architecture,* identifies the basic components of a Grid Computing system, defines the purpose and functions of such components, and indicates how each of these components interacts with the others [Foster02]. The main focus of the architecture is on the interoperability among the resource providers and users to establish the sharing relationships. This interoperability means common protocols at each layer of the architecture model, which leads to the definition of a Grid Protocol Architecture, as shown in Figure 4.1. This Grid Protocol Architecture defines common mechanisms, interfaces, schema, and protocols at each layer, by which users and resources can negotiate, establish, manage, and share resources.

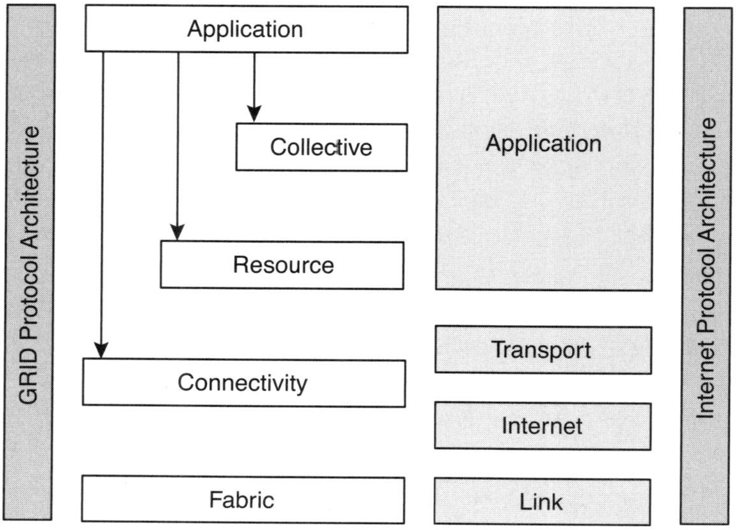

FIGURE 4.1 The layered Grid Protocol Architecture and its relationship to the IP architecture [Foster02].

Figure 4.1 illustrates the component layers of the architecture with specific capabilities at each layer. Each layer shares the behavior of the component layers described next. As we can see in this illustration, each of these component layers is compared with its corresponding IP layer(s) to further clarify its capabilities.

Fabric Layer: Interfaces to Local Resources This layer defines the resources that can be shared. These could include computational resources, data storage, networks, catalogs, and other system resources. These resources can be physical resources or logical resources by nature.

Typical examples of logical resources found in a Grid Computing environment are distributed file systems, computer clusters, distributed computer pools, software applications, and advanced forms of networking services. Logical resources are implemented by their own internal protocol (e.g., Network File System [NFS] for distributed file systems and Logical File System [LFS] for clusters). These resources then comprise their own network of physical resources.

Although there are no specific requirements for a particular resource that relate to integrating itself as part of any Grid Computing system, it is recommended to have two basic capabilities associated with the integration of resources; these basic capabilities should be considered "best practices" of Grid Computing disciplines:

1. Provide an "inquiry" mechanism that allows for the discovery of the resource against its own capabilities, structure, and state of operations. These are value-added features for resource discovery and monitoring.
2. Provide appropriate "resource management" capabilities to control the QoS (Quality of Service) that the Grid Computing solution promises, or has been contracted to deliver. This enables the service provider to control a resource for optimal manageability, such as (but not limited to) the following: starting and stopping activation, resolving problems, configuration management, load balancing, workflow, complex event correlation, and scheduling.

Connectivity Layer: Manages Communications The connectivity layer defines the core communication and authentication protocols required for Grid Computing-specific networking services transactions. Communication protocols, which include aspects of networking transport, routing, and naming, assist in the exchange of data between fabric layers of respective resources. The authentication protocol builds on top of the networking communication services to provide secure authentication and data exchange between users and respective resources.

The communication protocol can work with any of the networking layer protocols that provide the transport, routing, and naming capabilities in networking services solutions. The most commonly used network layer protocol is the Transmission Control Protocol/Internet Protocol (TCP/IP) stack; however, this discussion is not limited to that protocol. The authentication solution for VO environments requires significantly more complex characteristics. The following describes these characteristics:

1. *Single Sign-On (SSO)*—This allows any of multiple entities in the Grid Protocol Architecture fabric to be authenticated once, so the user can then access any available resources in the fabric layer without further user authentication intervention.
2. *Delegation*—This provides the ability to access a resource under the current user's permissions set, and the resource should be able to relay the same user credentials (or a subset of the credentials) to other resources respective to the chain of access.
3. *Integration with local resource-specific security solutions*—Each resource and hosting environment has specific security requirements and solutions that match the local environment. This may include (for example) Kerberos security methods, Windows security methods, Linux security methods, or UNIX security methods. Therefore, to provide proper secu-

rity in the Grid Protocol Architecture fabric model, all Grid Computing solutions must provide integration with the local environment and respective resources specifically engaged by the security solution mechanisms.

4. *User-based trust relationships*—In Grid Computing, establishing an absolute trust relationship among users and multiple service providers is very critical.
5. *Data security*—Data security provides both data integrity and confidentiality. The data passing through the Grid Computing solution, no matter what complications may exist, should be made secure using various cryptographic and data encryption mechanisms. These mechanisms are well-known in the world of technology, across all global industries.

Resource Layer: Shares a Single Resource This layer utilizes the communication and security protocols defined by the networking communications layer to control the secure negotiation, initiation, monitoring, metering, accounting, and payment involved in the sharing of operations across *individual resources*. The way this works is the resource layer calls the fabric layer's functions to access and control the local resources. This layer only handles the individual resources, and hence ignores the global state and atomic actions across the other resources, which in the operational context becomes the responsibility of the collective layer.

There are two primary classes of resource layer protocols; these protocols are key to the operation and integrity of any single resource:

- Information protocols—These protocols are used to get information about the structure and operational state of a single resource, including configuration, usage policies, SLAs, and the state of the resource. In most situations, this information is used to monitor the resource capabilities and availability constraints.
- Management protocols—The important functionalities provided by the management protocols are:
  - Negotiating access to a shared resource is paramount. These negotiations can include the requirements for QoS, advanced reservation, scheduling, and other key operational factors.
  - Performing operation(s) on the resource, such as process creation, is critical. Data access is also a very important operational factor.
  - Acting as the service/resource policy enforcement point for policy validation between a user and resource is critical to the integrity for the operations.

- Providing accounting and payment management functions on resource sharing is mandatory.
- Monitoring the status of an operation and controlling the operation, including terminating the operation and providing asynchronous notifications on operational status, is extremely critical to the operational state of integrity.

It is recommended that these resource level protocols be minimal from a functional overhead point of view, and they should focus on the functionality each provides from a utility aspect.

Collective Layer: Coordinates Multiple Resources The collective layer is responsible for all global resource management and interaction with any resource collection. This protocol layer implements a wide variety of sharing behaviors utilizing a small number of resource layer and connectivity layer protocols.

Some key examples of the common, more visible collective services in a Grid Computing system are:

- *Discovery services*—These services enable VO participants to discover the existence and/or properties of the specific available VO's resources.
- *Co-allocation, scheduling, and brokering services*—These services allow VO participants to request the allocation of one or more resources for a specific task, for a specific period of time, and to schedule those tasks on the appropriate resources.
- *Monitoring and diagnostic services*—These services provide the VO's resource failure recovery capabilities, monitoring of the networking and device services, and diagnostic services that include common event logging and intrusion detection. Another important aspect of this topic relates to the partial failure of any portion of a Grid Computing environment. It is critical to understand any and all *business impacts* related to any partial failure immediately, as the failure begins to occur—all the way through its corrective healing stages.
- *Data replication services*—These services support the management aspects of the VO's storage resources; they maximize data access performance with respect to response time, reliability, and cost.
- *Grid-enabled programming systems*—These systems allow familiar programming models to be utilized in Grid Computing environments while sustaining various Grid Computing networking services. These networking services are integral to the environment; they address

resource discovery, resource allocation, problem resolution, event correlation, network provisioning, and other very critical operational concerns related to Grid Computing networks.

- *Workload management systems and collaborative frameworks*—These provide multi-step, asynchronous, multi-component workflow management. This is a complex topic across several dimensions, yet it is a fundamental area of concern for enabling optimal performance and functional integrity.
- *Software discovery services*—These services provide the mechanisms to discover and select the best software implementation(s) available in the Grid Computing environment, and those available to the platform based on the problem being solved.
- *Community authorization servers*—These servers control resource access by enforcing community utilization policies and providing respective access capabilities by acting as policy enforcement agents.
- *Community accounting and payment services*—These services provide resource utilization metrics; they also generate payment requirements for members of any community.

As we can observe based on the previous discussion, the capabilities and efficiencies of these collective layer services are based on the underlying layers of the protocol stack. These collective networking services can be defined as general-purpose Grid Computing solutions that have been narrowed down to domain and application-specific solutions. As an example, one such service is accounting and payment, which is most often very specific to a domain or application. Other notable and very specialized collective layer services include schedulers, resource brokers, and workload managers.

Applications Layer: User-Defined Grid Computing Applications

These are user applications that are constructed by utilizing the services defined at each lower layer. Such an application can either access a resource directly or through the Collective Service API.

Each layer in the Grid Protocol Architecture provides a set of APIs and SDKs for the higher layers of integration. It is up to the application developers whether to use the collective services for general-purpose discovery and other high-level services across a set of resources or to start working directly with exposed resources. These user-defined Grid Computing applications are (in most cases) domain-specific and provide specific solutions.

The Grid Protocol Architecture and Its Relationship to Other Distributed Technologies

It is a known fact in technology that there are numerous well-defined and well-established technologies and standards developed for distributed computing. This foundation has been a huge success (to some extent) until we entered the domain of heterogeneous resource sharing and the formation of VOs.

Based on our previous discussions, the Grid Protocol Architecture is defined as coordinated and highly automated, dynamically sharing resources for a VO. It is appropriate that we turn our attention at this stage toward a discussion regarding how this architectural approach differs from distributed technologies, how the two approaches complement each other, and how we can leverage the best practices of both approaches.

Our discussion will now begin to explore notions of widely implemented distributed systems, including the following: World Wide Web environments, application and storage service providers, distributed computing systems, P2P computing systems, and clustering systems.

The World Wide Web

Numerous open and ubiquitous technologies are defined for the World Wide Web (TCP, Hypertext Transport Protocol [HTTP], SOAP, and XML) that in turn makes the Web a suitable candidate for the construction of VOs. However, as of now, the Web is defined as a browser/server messaging exchange model, and lacks the more complex interaction models required for a realistic VO.

Some areas of interest include: Delegation of authority, complex authentication mechanisms, and event correlation mechanisms. Once the browser-to-server interaction matures, the Web will be suitable for the construction of Grid Computing portals to support multiple VOs. This will be possible because the basic platforms and layers of technology will remain the same.

Distributed Computing Systems

The major distributed technologies, including CORBA, J2EE, and the Distributed Communication Object Model (DCOM) are well-suited for distributed computing applications; however, these technologies do not provide a suitable platform for sharing resources among the members of a VO. Some of the notable drawbacks include resource discovery across virtual participants, collaborative and declarative security, dynamic construction of the VO, and the scale factor involved in potential resource sharing environments.

Another major drawback in distributed computing systems involves the lack of interoperability among the technology protocols. However, even with these perceived drawbacks, some distributed technologies have attracted considerable Grid Computing research attention toward the construction of Grid Computing systems, the most notable of which is Java JINI.[1] The JINI system is focused on a platform-independent infrastructure that delivers services and mobile code to enable easier interaction with clients through service discovery, negotiation, and leasing.

Application and Storage Service Providers Application and storage service providers normally outsource their business and scientific applications and services, as well as very high-speed storage solutions to customers outside their organizations. Customers negotiate with these highly effective service providers on QoS requirements (i.e., hardware, software, and network combinations) and pricing (i.e., utility-based, fixed, or other pricing options).

Normally speaking, these types of advanced service arrangements are executed over some type of a VPN (Virtual Private Network) or dedicated line by narrowing the domain of security and event interactions. This is oftentimes somewhat limited in scope because the VPN or private line is very static in nature. This, in turn, reduces the visibility of the service provider to a lower and fixed scale, with the lack of complex resource sharing among heterogeneous systems and inter-domain networking service interactions.

That being said, the introduction of the Grid Computing principles related to resource sharing across VOs, along with the construction of VOs yielding inter-domain participation, will alter this situation. Specifically, this will enhance the utility model of Application Service Providers/Storage Service Providers (ASPs/SSPs) to a more flexible and mature value proposition.

Peer-to-Peer (P2P) Computing Systems Similar to Grid Computing, P2P computing is a relatively new computing discipline in the realm of distributed computing. Both P2P and distributed computing are focused on resource sharing, and are now widely utilized throughout the world by the home, commercial, and scientific markets. Some of the major P2P systems are SETI@home[2] and file-sharing system environments (e.g., Napster, Kazaa, Morpheus, and Gnutella).

---

1. For more details on JINI and related projects, visit *www.jini.org*
2. Details on the SET1@home project can be found at *http://setiathome.ssl.berkeley.edu/*

The major differences between Grid Computing and P2P computing are centered on the following notable points:

1. They differ in their target communities. Grid Computing communities can be small with regard to the number of users, yet will yield more focused applications with higher levels of security requirements and application integrity. On the other hand, P2P systems define collaboration among a larger number of individuals and/or organizations, with a limited set of security requirements and a less complex resource sharing topology.
2. Grid Computing systems deal with more complex, more powerful, more diverse, and more highly interconnected sets of resources than P2P environments.

The convergence of these areas toward Grid Computing is highly probable since each discipline deals with the same problem of resource sharing among the participants in a VO. There has been some work done to date in the Global Grid Forum (GGF), which is focused on the merger of these complementary technologies for the interests of integrating a larger audience.

Cluster Computing Clusters are local to a domain and are constructed to solve inadequate computing power. Clustering is related to the pooling of computational resources to provide more computing power by parallel execution of the workload. Clusters are limited in scope with dedicated functionality and local to the domain; they are not suitable for resource sharing among participants from different domains. The nodes in a cluster are centrally controlled, and the state of the node is aware of the cluster manager. This forms only a subset of the Grid Computing principles of more widely available intra-/inter-domain communication and resource sharing.

Summary

This chapter introduced the *Grid Computing problem* in the context of a *VO*, combined with the proposed *Grid Protocol Architecture*. The overall Grid Protocol Architecture described in this chapter is effectively designed for controlled resource sharing with better interoperability among participants. This is, in practice, one commonly accepted and globally suggested solution to resolve the Grid Computing problem. This introduction to Grid Computing helps us to better understand aspects of the Grid Protocol Architecture's design, while understanding the Grid Computing protocols necessary for resource sharing with maximum interoperability.

In addition, this chapter discussed the relationships between Grid Computing systems and distributed systems by emphasizing exactly how these varying disciplines complement each other, while at the same time differ with respect to each other.

The next chapter will explore the future of Grid Computing, outlining more of the tactical perspectives of the technology, as well as the strategic perspectives.

The Future of Grid Computing 5

The last decade has noted a substantial change in the ways global industries, businesses, and home users apply computing devices, including a wide variety of ubiquitous computing resources and advanced Web services. Initially, the focus was on localized computing resources and respective services; however, the capabilities have changed over time and we are now in an environment consisting of sophisticated, virtualized, and widely distributed on demand business services. This chapter will outline both tactical and strategic perspectives in the Grid Computing discipline, which can be a key ingredient in on demand business services.

This chapter will explore the current and prominent technology initiatives that are affecting the recent Grid Computing revolution. Some of the prominent technology initiatives acting as catalysts to this revolution are: on demand Operating Environments, Autonomic Computing, Service-Oriented Architecture (SOA), and Semantic Grids.

Major global organizations, solution providers, service providers, and technology innovators, which we discussed in earlier chapters (and even those we have not yet discussed), have absolutely contributed to this technology evolution. In the previous chapter, we explored several Grid Protocol Architecture points, and the relationships between Grid Computing and other distributed computing technologies. As we can see from the previous discussions and Figure 5.1, the evolution of Grid Computing is progressing at a very rapid rate. This computing evolution is tightly aligned with the incredible and very rapid evolution of the Internet and other open standards architectures.

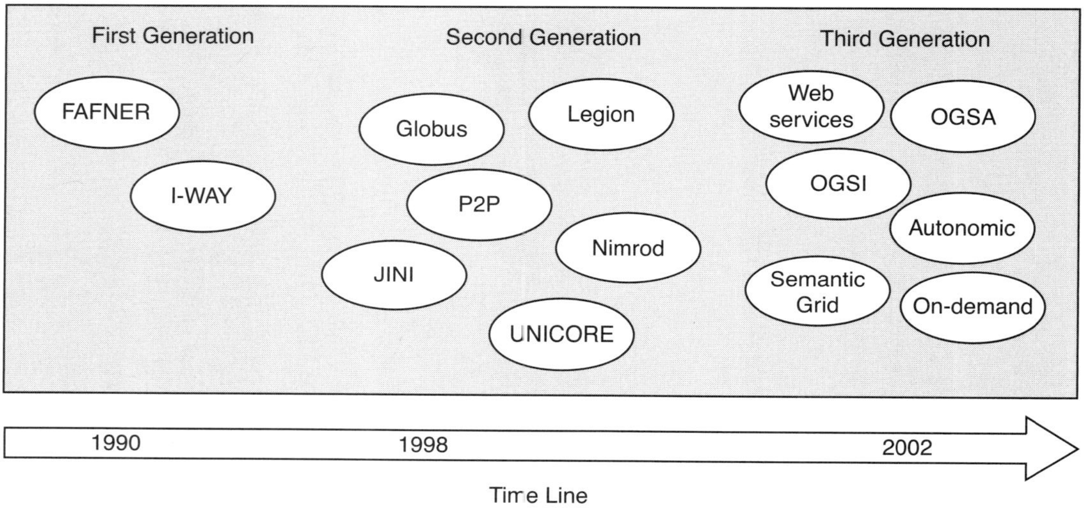

FIGURE 5.1 The future of Grid Computing technology in terms of generations over time.

As shown in Figure 5.1, the evolution of Grid Computing is broadly classified into three generations. This Grid Computing evolution is discussed in detail by [Roure01].

In previous chapters, we explored some of the major projects and organizational initiatives that contributed to the first and second generations of Grid Computing. The first two focus areas concentrate on large-scale resource discovery, utilization, and sharing within VO boundaries. The major drawback of these two phases was the lack of transparency in the middleware, which contributed to monolithic and non-interoperable solutions for each Grid Computing environment.

This difference in the first two stages results in a vertical tower of solutions and applications for resource sharing among organizations. Today, we are in the third generation of Grid Computing, where applications and solutions are focused on open technology-based, service-oriented, and horizontally oriented solutions that are aligned with other global industry efforts. Grid

Computing infrastructures are clearly transitioning from being *information-aware* to being *knowledge-centric* frameworks.

We will now begin to explore the third generation of technologies, the respective Grid Protocol Architecture, and the future of the next generation of Grid Computing technology initiatives.

The next generations of Grid Computing technologies that are channeling this third generation of Grid Computing initiatives are:

- Autonomic Computing
- Virtualization, and on demand business grids
- Service-Oriented Architecture (SOA)
- Semantic Grids

Autonomic Computing

As we discussed in previous chapters, the term "autonomic" comes from an analogy to the autonomic central nervous system in the human body, which adjusts to many situations automatically, without any external help. With the increasing complexity in dealing with distributed systems, solutions, and shared resources in Grid Computing environments, we require a significant amount of autonomic functions to manage a grid.

As previously discussed in this book, Figure 5.2 shows how basic Autonomic Computing systems must follow four basic principles [IBM01]:

- Self-configuring (able to adapt to changes in the system)
- Self-optimizing (able to improve performance)
- Self-healing (able to recover from mistakes)
- Self-protecting (able to anticipate and cure intrusions)

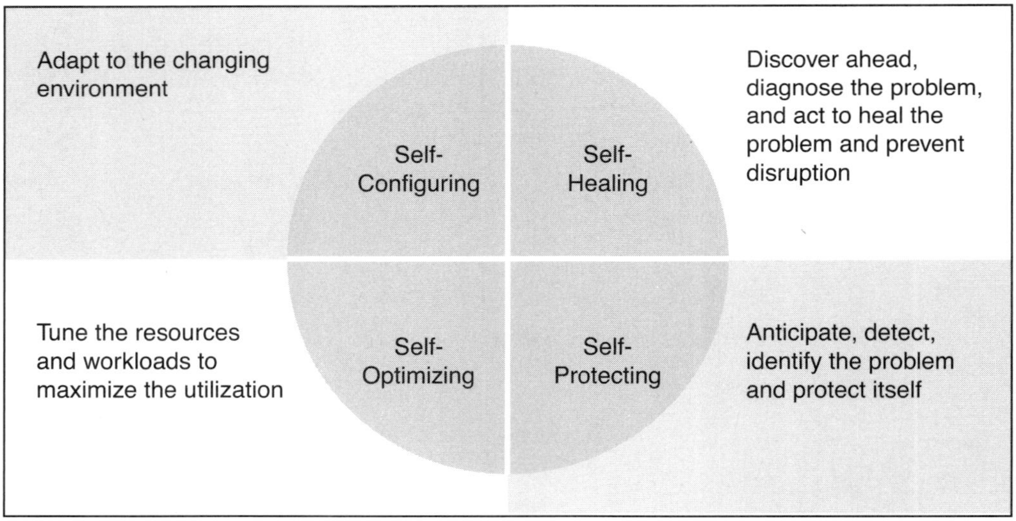

FIGURE 5.2 The Autonomic Computing vision [IBM01] is robust across several complementary dimensions.

Orchestrating complex connected problems on heterogeneous, distributed systems is a complicated job and requires a number of autonomic features for the infrastructure and resource management. Thus, it is important that systems be as self-healing and self-configuring as possible, to meet the requirements of resource sharing and to handle failure conditions. These autonomic enhancements to the existing Grid Computing framework at the application and middleware framework levels provide a scalable and dependable grid infrastructure.

The IBM Corporation, as the pioneer in worldwide Autonomic Computing initiatives, has already implemented a number of projects around the world in this strategic area, while keeping in mind the synergies needed to create complementary global Grid Computing solutions. These global Grid Computing solutions are continuously being enhanced to include Autonomic Computing capabilities. The Grid Computing and Autonomic Computing disciplines will continue to work closely together to develop highly reliable, efficient, self-managing grids.

On Demand Business and Infrastructure Virtualization

Utility Computing is one of the key technology resources that help businesses to develop advanced Web Services capabilities. Utility computing deals with the implementation of many types of programmatic utilities, an underlying fabric of a sort to support many services, which deal with a variety of functions; for example, functions such as billing, network bandwidth management, application provisioning, network provisioning, event correlation, monitoring, and other key functional areas. These utilities are designed for specific functions and reused throughout an on demand business infrastructure. Many companies, aside from IBM, are addressing Utility Computing approaches, all with a variety of "utility" definitions. However, on demand business is not just about Utility Computing; it has a much broader set of ideas about the transformation of business practices, process transformation, and technology implementation.

Companies striving to achieve the on demand business operational model will have the capacity to sense and respond to fluctuating market conditions in real time, while providing products and services to customers in a on demand business operational model.

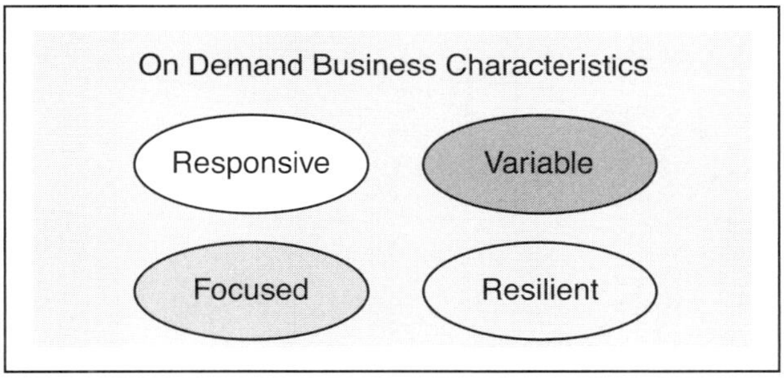

FIGURE 5.3 On demand business characteristics are shared across four distinct areas.

In general, on demand business has four essential characteristics, as depicted in Figure 5.3 [IBM02]:

1. *Responsive*—An on demand business has to be responsive to dynamic, unpredictable changes in demand, supply, pricing, labor, and competition.
2. *Variable*—An on demand business has to be flexible in adapting to variable cost structures and processes associated with productivity, capital, and finance.
3. *Focused*—An on demand business has to focus on its core competencies, its differentiating tasks, and its assets, along with closer integration with its partners.
4. *Resilient*—An on demand business must be capable of managing changes and competitive threats with consistent availability and security.

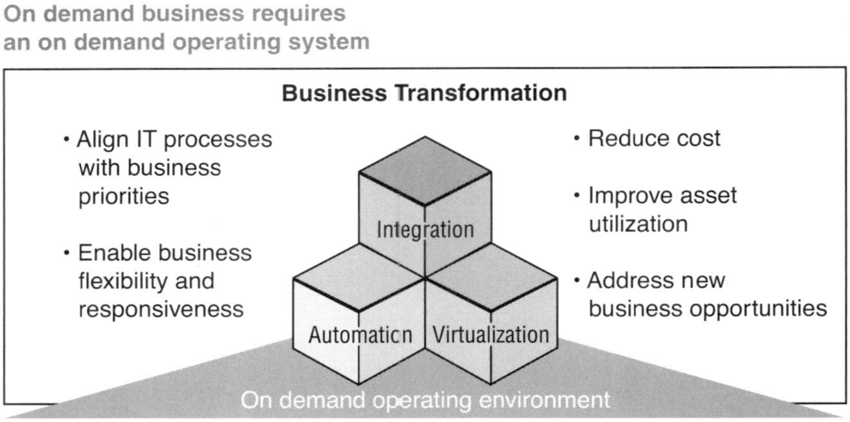

FIGURE 5.4 The on demand Operating Environment is rich with Autonomic Computing process enablement, sometimes combined with advanced concepts in Grid Computing, like virtualization.

To achieve the core capabilities of an on demand Operating Environment, as shown in Figure 5.4, must exhibit the following essential capabilities (which we will again review):

1. *Integration*—Integrated systems enable seamless linkage across the enterprise and across its entire range of customers, partners, and suppliers.
2. *Virtualization*—Resource virtualization enables the best use of resources and minimizes complexity for users. We can achieve the virtualization of resources through a number of existing and emerging technologies, including clusters, LPARs (logical partitions), server blades, and Grid Computing. Based on our earlier discussions, we know that Grid Computing provides the best use of virtualized resources for virtually organized customers within the constraints of SLAs and policies.
3. *Automation*—As we discussed in previous sections, *autonomic* capabilities provide a dependable technology framework for an on demand Operating Environment.
4. *Open standards*—An open, integrate-able technology allows resource sharing to be more modular. Some of the most notable open standards are XML, Web services, and the OGSA.

We have now seen that infrastructure virtualization transformation can be achieved by using the Grid Computing infrastructure. These sharable, highly virtualized resources indeed form the backbone for an on demand Operating Environment. In addition to this need for virtualization of hardware resources, the virtualization of data and software applications must also occur. This enables access to resources as a single entity, and allows applications to be enabled to respond quickly to the dynamic needs of the on demand business. This virtualization provides computational and/or data grids for highly computation-intensive throughput, as well as a uniform data access mechanism.

Service-Oriented Architecture (SOA) and Grid Computing

A distributed system consists of a set of software agents that all work together to implement some intended functionality. Furthermore, the agents in a distributed system do not operate in the same processing environment, so they must communicate by hardware/software protocol stacks that are intrinsically less reliable than direct code invocation and shared memory. This has important architectural implications because distributed systems require that developers of infrastructure and applications consider the unpredictable latency of remote access, and take into account issues of concurrency and the possibility of an unplanned partial failure [Kendall].

A *Service-Oriented Architecture (SOA)* is a specific type of distributed system in which the agents are "software services" that perform some well-defined operations (i.e., they provide a service like billing on demand, or provisioning); this type of architecture can be invoked outside the context of a larger application. By this, we can infer a service is acting as a user-facing software component of a larger application. This separation of functionality helps the users of the larger application to be concerned only with the interface description of the service.

In addition, an SOA stresses that all services have to be a network-addressable interface that communicates via standard protocols and data formats, simply called *messages*. The major functionality of an SOA is to define the messages (i.e., their format, content, and exchange policies) that are exchanged between users and services. The Web architecture [WEB] and Web Services Architecture [WSA] are instances of an SOA.

Grid Computing is a distributed system for the sharing of resources among participants. As noted in earlier discussions, Grid Computing, being a distributed architecture, has to deal with problems of the distributed computing grid, including latency, concurrency, and partial failures. We also stated that the current Grid Protocol Architecture is built on several existing distributed technologies, with a major focus on resource sharing, interoperability, and virtual organizational security. An SOA is a distributed architecture with more focus on service interoperability, easier integration, and extensible and secure access.

The WSA is gaining the most attention in the industry as an open, standards-based architecture with the main focus on interoperability. The World Wide Web Consortium (W3C) is leading this initiative. The core components of the WSA are shown in Figure 5.5:

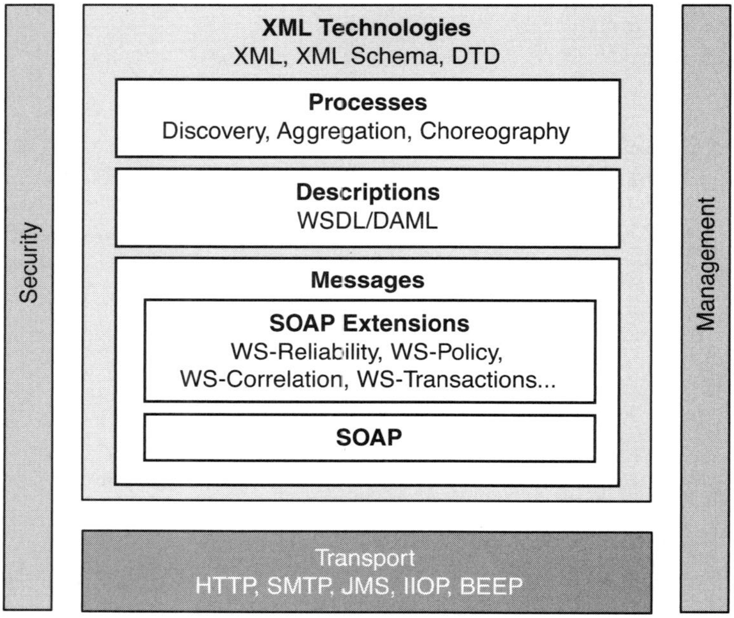

FIGURE 5.5 The WSA is a key enabler in the overall computing discipline of Grid Computing.

On closer inspection of Figure 5.5, one can infer that XML and related technologies (e.g., DTD, XML schema) form the base technologies of the Web services. Web services are invoked and combinational results are provided via messages that must be exchanged over some communications medium. This process is often accomplished where a communication medium can be a low-level networking services transport protocol (e.g., TCP), and/or a high-level communication protocol (e.g., HTTP), and/or a combination of both.

The message format can be specified through SOAP and its extensions, but this capability is not just limited to SOAP. SOAP functionality provides a robust and standard way for exchanging messages.

Interoperability across heterogenous systems requires a mechanism to define the precise structure and data types of the messages that have to be exchanged between a message producer and consumer. The Web Service Description Language (WSDL) is another desirable choice for the message and exchange pattern.

The SOAP specification provides the definition of the XML-based information that can be used for exchanging structured and typed information between peers in a decentralized, distributed environment. SOAP is fundamentally a stateless, one-way message exchange paradigm, but applications can create more complex interaction patterns (including request/response, request/multiple responses). SOAP is silent on the semantics of any application-specific data it conveys. At the same time, SOAP provides a framework (SOAP header) by which application-specific information may be conveyed in an extensible manner. Also, SOAP provides a full description of the required actions taken by a SOAP node on receiving a SOAP message. In short, a SOAP message is a SOAP envelope with a SOAP header and a SOAP body, where the header contains semantic and meta-data information about the contents of the SOAP body, which forms the message. Most of the Web service vendors today use SOAP as their message payload container.

The WSDL provides a model and an XML format for describing Web services. WSDL enables one to separate the description of the abstract functionality offered by a service from the concrete details of a service description.

Grid Computing is all about resource sharing by integrating services across distributed, heterogeneous, dynamic VOs formed from disparate sources within a single institution and/or external organization. This integration cannot be achieved without a global, open, extensible architecture agreed upon by the participants of the VO.

The OGSA achieves these integration requirements by providing an open service-oriented model for establishing Grid Computing architectures. The OGSA is described in detail in the "physiology" paper [Foster04]. The OGSA is aligned with the SOA as defined by the W3C and utilizes a Web service as its framework and message exchange architecture. Thanks to the valuable and innovative concepts of the OGSA, and the open nature of the standard, the GGF formed an architecture work area to discuss the OGSA and its programming model.

The basic approach the OGSA has taken is to integrate itself with the WSA and define a programming model using this emerging architecture. The OGSI uses WSDL as its service description mechanism and the Web service infrastructure for message exchange. We will further explore the details surrounding the OGSA and the core components that constitute this architecture in a later section of the book.

Semantic Grids

The W3C initiated a "meta-data activity," which defines a standard for a meta-data definition surrounding a Semantic Web. A *Semantic Web* is defined as the next generation of Web services.

A Semantic Web Defines the Next Generation of Web Services

A Semantic Web is an extension of the current Web, in which information is given well-defined meaning, better enabling computers and people to work in cooperation. It's the ideas of having data on the Web defined and linked in a way that Webs can be used for more effective discovery, automation, integration and reuse across various applications.

The Web can reach its full potential if it becomes a place where data can be shared and processed by automated tools as well as enhancing critical people skills [W3C-SEM].

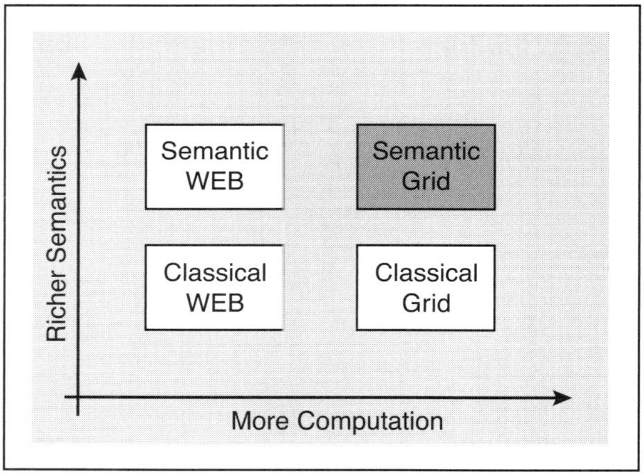

FIGURE 5.6 The Semantic Web evolution is straightforward across several dimensions.

Figure 5.6 shows the Semantic Web evolution. We will start exploring this evolution with the understanding of the Semantic Web.[1]

The two most important technologies for building Semantic Webs are XML and the Resource Description Framework (RDF). XML allows users to add arbitrary structure to documents through markup tags, but says nothing about the meaning of the structures. Meaning is expressed by the RDF,

1. For more information on the Semantic Web, visit *www.w3.org/2001/sw/*

which encodes itself in sets of triples, wherein each triple represents the subject, object, and predicate of an elementary sentence. These triplets can be written using XML tags. The subject and object are each identified by a universal resource identifier (URI).

An RDF document makes assertions that particular things, for instance, people and Web pages, have properties (e.g., "is a sister of" or "is the author of") with certain values (e.g., another person, another Web page). To further clarify, consider the following example: "Jeff is the author of the book *ABC*." In this example, the subject is "Jeff," the predicate is "is the author of," and the object is "the book *ABC*." This structure turns out to be a natural semantic means to describe the vast majority of the data processed by machines. The RDF has therefore evolved as a data model for logic processing.

Another important aspect of the Semantic Web is the use of "ontology" to describe collections of information like concepts and relationships that can exist in many semantic situations. This semantic taxonomy defines classes of objects and relations among them.

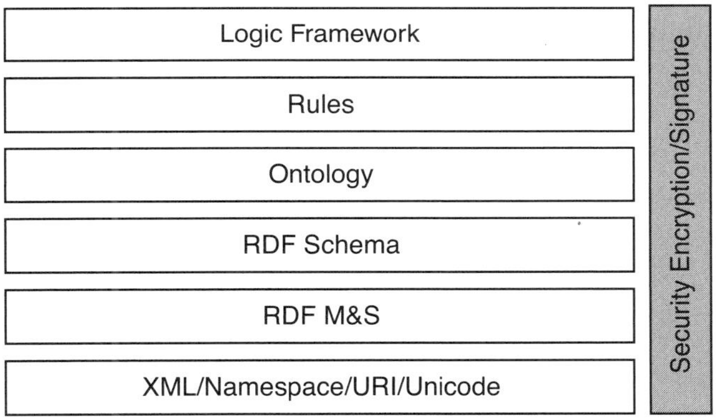

FIGURE 5.7 The Semantic Web architecture, with the necessary security and encryption required across all levels.

Figure 5.7 illustrates the Semantic Web architecture layers. The real power of the Semantic Web will be realized when people create many programs that collect data from diverse sources, process the information, and exchange the results with other programs.

The *Semantic Grid*[2] is a natural evolution of Grid Computing toward a knowledge-centric and meta-data-driven computing paradigm. The Semantic Grid is an effort to utilize Semantic Web technologies in Grid Computing

2. For more information on the Semantic Grid, visit *www.semanticgrid.org*

development efforts, from the Grid Computing infrastructure to the delivery of Grid Computing applications.

These concepts will enhance more automated resources and knowledge/information-based resource sharing. Knowing the importance of the evolution of a Semantic Web and its usability in Grid Computing, the GGF has created a research group for the Semantic Grid under the Grid Protocol Architecture area. We can find a number of Semantic Grid projects here, including the corporate ontology grid,[3] grid-enabled combinatorial chemistry,[4] collaborative advanced knowledge technologies[5] in the grid, and many other significant initiatives.

Summary

The future of Grid Computing technology leads to the natural evolution of distributed computing with more emphasis on open architecture models, knowledge-centric solutions, and simpler forms of integration. This Grid Computing evolution will enable easier resource discovery, virtualization, and autonomic enhancements for Grid Computing solutions.

At this stage of the book, we have introduced several strategic perspectives and technologies and numerous very key on demand business elements. We have discussed Autonomic Computing, its role, and the strategies of the discipline. We have discussed Grid Computing and its role, plus we have started to examine the strategies of this discipline. Additionally, we have discussed the natural synergies of each of these computing disciplines as they apply to an on demand Operating Environment.

The next chapter will detail the Grid Computing strategy in reference to the on demand business intersections unveiled in previous discussions of this book.

3. For more information, visit *www.cogproject.org/*
4. More information is available at *www.combechem.org/*
5. For more information, visit *www.aktors.org/coakting/*

Grid Computing Strategy Perspectives

6

This chapter treats specific strategy perspectives of Grid Computing. Strategic architecture topics and some low-level infrastructure and service topics for Grid Computing are also discussed in this chapter. Many of the IBM Corporation's global customer solutions in Grid Computing will also be discussed, providing an exploration of various implementations and detailed case study information related to Grid Computing.

As previously mentioned, Grid Computing[1] enables the virtualization of distributed computing resources such as processing, network bandwidth, and storage capacity to create a single system image, granting users and applications seamless access to vast IT resources. The following analogy may help you to better understand aspects of the Grid Computing topic: Just as an Internet user views a unified, single-server instance of content and other media via the Web, a Grid Computing user essentially sees a single, very large virtual computer being provided from many computing

1. For detailed information on Grid Computing, refer to the book entitled *Grid Computing* (J. Joseph and C. Fellenstein), Prentice Hall Professional Technical Reference, Upper Saddle River, NJ). [ISBN 0-13-145660-1]

resources and services across a large geographical environment. We will depart from previous discussions to take a deep look at this.

The vision of IBM related to Grid Computing is clear and powerful. The vision (and practice) of Grid Computing embodies massive numbers of computing device endpoints as a prime objective, applied to Grid Computing application-intensive problems. The IBM Corporation's Grid Computing General Manager, Tom Hawk, clearly states this vision as follows:

IBM's ultimate vision for Grid is a utility model over the Internet, where clients draw on computer power much as they do now with electricity. With more than 60 percent of IT budgets dedicated to maintenance and integration—a percentage that continues to rise—the need to reduce complexity and management demands is a pressing one.[2]

Although IBM is a world-class leader in the area of Grid Computing, there are a number of other contributing organizations in Grid Computing. For example, there is the Grid Computing Info Center,[3] which promotes the development and advancement of technologies that provide seamless and scalable access to wide area, distributed resources. Computational grids[4] enable the sharing, selection, and aggregation of a wide variety of geographically distributed computational resources. Supercomputers utilize computer clusters,[5] or Grid Computing storage[6] system solutions, data sources, instruments, and people. Grid Computing is capable of considering all these types of entities and presenting them as a single, unified resource for solving large-scale computation- and data-intensive applications. Examples of these intensive computing applications are: molecular modeling for drug design,[7] brain activity analysis,[8] and high-energy physics.[9]

This seemingly new idea of Grid Computing is not necessarily a new concept; it is simply that Grid Computing has now been applied to end-user computing machines. Grid Computing is analogous to an electric "utility" power network,[10] where power generators are distributed in a grid-like fash-

---

2. Tom Hawk, IBM Grid Computing General Manager, Grid Computing Planet Conference and Expo, San Jose, June 17, 2002.
3. For detailed information, refer to *www.gridcomputing.com*
4. For detailed information, refer to *www.mkp.com/grids*
5. For detailed information, refer to *www.buyya.com/cluster*
6. For detailed information, refer to *www.buyya.com/superstorage*
7. For detailed information, note the white paper "vlab-drug-design" in the CD backmatter of this book, or refer to *http://buyya.com/papers/vlab-drug-design.pdf*
8. For detailed information, note the white paper "neurogrid" in the CD backmatter of this book, or refer to *www.buyya.com/papers/neurogrid.pdf*
9. For detailed information, refer to *http://lcg.web.cern.ch/LCG/*

ion and the users are able to access the electric power without having to know anything about the source of the power and its locations.

One of the IBM Corporation's business leaders in this field, Chris Reech in the Global Services e-Technology Center, encourages the many global partnerships we share in Grid Computing to continue to forge ahead in the mission of developing and deploying on demand business solutions. He states that, "We encourage all of the vendors to embrace the OGSA, so that the individual strengths of each organization's product can be mixed and matched and interconnected via the open standards, in order to build the best-fit solution per customer."

Strategically speaking, the protocols within the OGSA, the Globus Toolkit's reference implementation, for example, are the primary choices as the technology backbone of any Grid Computing system. The OGSA provides the flexibility to extend a solution component, like the Globus Toolkit, with additional products to provide policy management, scheduling, file sharing, automated grid management, and other features as appropriate for any given customer situation. We will review the OGSA in greater detail in this chapter.

In the following discussion, we will explore a very important coalition of Grid Computing partners, referred to as the *Globus Project*. The Globus Project focuses on the combined achievements of many organizations instituting technologies and applications while utilizing Grid Computing *"utility"* practices. We will discuss both the OGSA and the OGSI. In conjunction with these two environments, we will later explore the Globus Grid Toolkit. First, let's take a look at the Globus Project itself.

The Globus Project

The *Globus Project* is assisting in the development of the fundamental technologies required to build worldwide computing grids. These grids are persistent computing environments that enable software applications to integrate instruments, displays, and computational and information resources. These types of grids are utilized and managed by diverse organizations in widespread geographic locations.

As previously discussed, Grid Computing is an important technology in the IBM on demand business strategy, and for this reason, IBM is a strong supporter of Grid Computing-based architectures. The IBM Corporation,

10. For detailed information, note the white paper "WeavingGrid" in the CD backmatter of this book, or refer to *www.cs.mu.oz.au/~raj/papers/WeavingGrid.pdf*

as a founding partner of the Globus Project, provides a world-class multi-institutional research and development effort tightly interlocked on the Grid Computing disciplines. The IBM Thomas J. Watson Research Laboratory along with IBM's Software Group, the e-TC, and IBM's Server Group, are all working together towards further industrial accomplishments in this field. The goal of the Globus Project, while working with IBM, is to address the technical and business challenges required to make Grid Computing a success across all sectors: government, public, and commercial.

Globus, founded by a team of technicians and researchers, has defined the OGSA and a corresponding set of tools to assist in many fascinating implementations of Grid Computing. IBM and Globus are sponsors of the "Grid Alliance" (formerly known as the Global Grid Forum, or GGF), which has accepted the mission of developing industry open standards for Grid Computing. Other members of the Alliance include major IT suppliers and representatives from a growing number of industries. In collaboration with the Globus Project, these IT organizations are dedicated to the development of Grid Computing technologies to support the specifications and open standards set forth by the Globus Project.

Open Grid Services Architecture (OGSA)

At the core of Grid Computing strategies, one will find "open"[11] sets of computing standards and networking services protocols; in addition, significant compartmentalized resources must also be present to enable a wide variety of Grid Computing application classes. This compartmentalization is declared within the OGSA, which enables communications across heterogeneous, geographically dispersed IT environments and networking services. The OGSA is focused on several significant areas, as shown in Figure 6.1:

11. The term *"open"* in this case refers to the ability of components (software and hardware) to interoperate with one another, in a seamless manner, using some form of a technology standard adopted by the one of the parties. This is commonly referred to as "open standards."

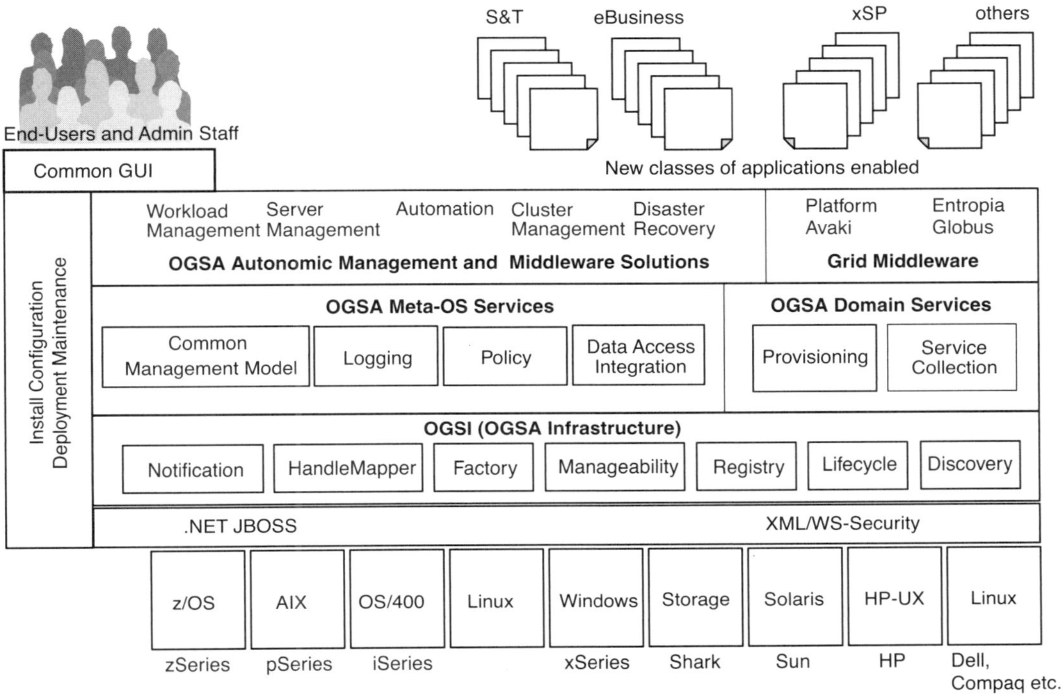

FIGURE 6.1 The topology of the OGSA is compartmentalized into sections, which contributes to many new classes of applications for Grid Computing.

Grid Computing strategies are defined as applying resources from many computers in a network—at the same time—to a single problem, usually a problem that requires a large number of processing cycles or access to large amounts of data. In subsequent sections of this chapter, you will find examples of several Grid Computing solutions.

It is interesting to note that in Grid Computing solutions, all the computing power can be rallied from within a company, or it can be a combination of the company and another third party, or it can be a combination of many companies and other private individuals around the world with excess computing power. The point here is that a grid can be established by anyone wishing to join the grid to take advantage of its tremendous problem-solving capabilities, should they desire and have a need for this type of information. The most powerful computing solutions, today, can be found in large global grids, VOs, with a common need for massive processing power to solve complex problems. The administrative functions of a grid are indifferent as to who joins, and from what country, as a grid end-user member.

Grid Computing strategies enable many devices—regardless of their operational characteristics—to be shared, managed, and accessed across an enterprise, industry, or workgroup. This *virtualization* of resources places all the necessary access, data, and processing power at the fingertips of those who need to rapidly solve complex business problems, conduct computation-intensive research and data analysis, and engage in real-time computational tasks.

Grid Computing strategies involve relationships with leading Grid Computing tool and application providers. For example, IBM is working closely with Globus,[12] an open source provider of Grid Computing middleware, to provide the OGSA reference implementation that is based on J2EE open standards. Globus provides a comprehensive set of software applications for security, information infrastructure development, resource management, data management, communications, fault detection analysis, and portability across a wide array of systems. This robust OGSA toolkit is available today across all IBM systems running within the AIX or Linux operating system environments.

The OGSA incorporates the proposed evolution of the current Globus Toolkit (OGSI, which will be discussed in a later section) toward the Grid Computing system architecture. The Globus Toolkit is based on the integration of Grid Computing initiatives, Web services concepts, and several other advanced technologies. The initial proposed OGSA technical specifications were developed by IBM and the Globus Project, and are being openly discussed at worldwide events such as the GGF. The fundamental purposes for these discussions are to enable strategy refinement and eventually standardize the technologies in the Grid Computing areas.

The Globus Project and IBM have published a set of specifications and standards that integrates and leverages the worlds of Web services and Grid Computing. The OGSA is the result of collaboration by some of the best researchers, architects, and deep technical minds in the industry. IBM is actively participating in the development of this standard.

The OGSA brings together standards such as XML, WSDL, Universal Description, Discovery, and Integration of Web Services (UDDI), and SOAP—all of which are very important to Web services—with the standards for Grid Computing established by the Globus Project. Through the OGSA, Globus and IBM have been able to create a specification for open standards, Grid Computing protocols, and Web services designed to enable large-scale cooperation and access to applications over global public and private networks.

---

12. For additional information on this and Globus, refer to *www.globus.org/ogsa/*

The IBM Corporation's Grid Computing strategic perspective focuses on (but is not limited to) the following best internal business practices related to transformation:

- Base all Grid Computing support on open industry standards
- Enable IBM hardware and software platforms for Grid Computing
- Build relationships with leading "open" Grid Computing toolkit providers
- Provide support for computing "grids," including computation grids and data grids
- Provide a full set of products, including on demand business services and computing infrastructure management utilities, to assist customers in implementing Grid Computing solutions to meet their own on demand business needs

Sample Use Cases for the OGSA

The global OGSA assembly, referred to as the "OGSA working group" (OGSA-WG), has defined a number of use cases from a wide variety of application scenarios, including those related to science and on demand business applications.

The main purposes of these use cases are:

- To identify and define core OGSA platform functionalities
- To define core platform components based on functionality requirements
- To define the high-level requirements of these core components and identify their interrelationships

These use cases are defined as part of the OGSA-WG charter, created by the Grid Alliance, which strives "to produce and document the use cases that drive the definition and prioritization of OGSA platform components, as well as document the rationale for our choices."

As we will soon see, some of these use cases defined from the on demand business and e-Science domains help to identify the general OGSA platform features, components, and their interrelationships. This will, in turn, pave the way for detailed discussions regarding the OGSA architecture platform components described later in this chapter.

Here are the use cases from the OGSA-WG, which we will use to illustrate elements of our discussion:

1. [*CASE 1*] Commercial Data Center (commercial grid)

2. [*CASE 2*] National Fusion "Collaboratory" (science grid)
3. [*CASE 3*] Online Media and Entertainment (commercial grid)

[*CASE 1*] Commercial Data Center (CDC)

Data centers (in the generalized context of the word), or On Demand centers, are common in most enterprises to consolidate the huge number of servers (e.g., blade servers[13]) and reduce the TCO. Data centers also play a key role in the outsourcing business, where major businesses outsource their IT resource management to concentrate on their core business competence and excellence. These data centers are required to manage a huge number of IT resources (e.g., servers, storages, and networks). Since these data centers are providing resource-sharing capabilities across a VO, Grid Computing forms the technology of choice for their resource management.

To support a CDC, the Grid Computing technology platform, middleware, and applications should possess a number of core functionalities. We identify and enlist these data center-oriented functionalities by defining the customers of a CDC and their usage scenarios.

The following discussion introduces the term referred to as "actor." An actor is a computer science term describing a programmatic agent, which delivers powerful forms of computational functions. Actors are programmatic code objects, with classes and sub-classes that have a position in a larger virtual grid world.

The simplest way to think about the concept of *actors* is to consider large online gaming environments, and what kinds of codes are required to make these environments become their own virtual gaming worlds. For example, in an online gaming environment[14] there are classes of actors that seek to alter weather conditions, resource scenarios that might involve items such as tanks, jeeps, rockets, and other gaming objects utilized in the collaborative environment.

13. With the recent launch of IBM eServer BladeCenter servers, IBM's mainframe-inspired technology now extends to the entire eServer product line. These new blade servers are so compact—just about an inch thick—and so easy to use that customers can increase the performance of their systems by simply sliding an additional BladeCenter server into their rack. It's just like putting books on a shelf. And, since BladeCenter servers share basic components such as power, system ports, and fans, they also lower power consumption and reduce system complexity. The blade server market is expected to reach $3.7 billion (U.S. dollars [USD]) by 2006.
14. For more detailed information on "actors," as utilized in the example of "online gaming," reference the types of actors and classes being discussed at http://wiki.beyondunreal.com/wiki/Actor.

Let us now explore how actors play a very key role in any type of grid computing environment, such as a CDC.

Customers/Providers The following topics describe the various roles and responsibilities of customer/provider "actors," conducting operations within a CDC.

- *Grid administrator*—This actor's goals are to ensure both the maximum utilization of the resources in the data center, and to enforce the management of these resources so they are appropriately controlled through the environments resource policies.
- *IT system integrator*—This actor's goal is to reduce the complexity of the distributed and heterogeneous systems. Also, this actor is responsible for the construction of the heterogeneous system and the management of all service changes.
- *IT business activity manager*—A business manager actor seeks a scalable and reliable platform, at a lower cost and an agreed-upon QoS (Quality of Service).

Scenarios The following describes some of the overall operations to be performed:

- *Multiple in-house systems support within the enterprise*—Consolidate all the in-house systems and make resources available. This reduces the cost of ownership and increases resource utilization. This scenario is suitable for human resource services, customer resource management, finance, and accounting systems.
- *Time-constrained commercial campaign*—Provide resources to run time-constrained campaigns and levy charges on the basis of usage. Examples of these campaigns include sales promotion campaigns and game ticket sales.
- *Disaster recovery*—Disaster recovery is an essential part of major IT systems today. Commercial Grid Computing systems are able to provide standard disaster recovery framework(s) across remote CDCs at low cost.
- *Global load balancing*—Geographically separated CDCs can share high workloads while providing very scalable and extensible systems. This is accomplished by balancing the sometimes-planned (or unplanned) loads of heavy networking services traffic, and computational operations in a dynamic real-time exchange.

Functional Requirements for the OGSA After a thorough and careful examination of the static and dynamic behaviors present in this use case, the following functional requirements of the Grid Protocol Architecture can be derived:

- Discovery of available resources
- Secure authentication, authorization, and auditing of resource utilization
- Resource brokerage services, to better utilize the resources, and to achieve the QoS requirements
- Scalable and manageable data sharing mechanisms
- Provisioning of resources based on need
- Scheduling of resources for specific tasks
- Advanced reservation facilities to achieve the scale of QoS requirements
- Enable metering and accounting to quantify resource usage amounts into pricing units
- Enable system capabilities for fault handling and partial failure detection and correction
- Use of static and dynamic policies is mandatory
- Manage transport and message levels, and end-to-end security
- Construct dynamic VOs with common functionalities and agreements
- Facilitate resource monitoring
- Enable facilities for disaster recovery in case of unplanned outages

Let us now explore another use case, where we will discuss a scientific research project with geographically distributed participants.

[*CASE 2*] National Fusion "Collaboratory"

The National Fusion project defines a VO that is devoted to fusion research, and provides the "codes (or programs)" developed by this community to the end-users (i.e., researchers). Earlier, this "code" software was installed in the end-user's machine. This became a complex and unmanageable process of software management, distribution, versioning, and upgrading.

Due to these change management and configuration problems, the National Fusion community decided to adopt the ASP model, known as the "*network services model*," where the "code" is maintained by the service provider and made accessible to remote clients.

This approach eliminates the burden on the end-user; however, it adds some QoS requirements to the service provider, including executing the "code" as efficiently as possible, executing it within a certain time boundary, and producing the results with extreme accuracy. As you can imagine, this is the best-case usage model for a computational grid.

Let us now explore the usage scenarios of this Grid Computing environment, and derive the functional requirements of the Grid Protocol Architecture.

Customers/Providers (Actors) This National Fusion project defines a VO that is devoted to fusion research, and provides the "codes" developed by this community to the end-users (i.e., researchers and/or scientists).

- *Scientists*—These are customers of the codes provided by the National Fusion service provider. Some of the customer requirements are, for example:
  - The ability to run the "codes" on remote resources based on the condition that end-to-end QoS can be sustained, with a guarantee of time boundary execution.
  - Availability of the resource (code execution) in the computational grid.
  - A policy-based management of resources, including what resources can run the codes, how many hardware resources are available, etc.
  - The ability to use community services by accreditation with the community, rather than individual service providers. This is a form of "dynamic account" creation and usage.

Scenarios

- A remote client (e.g., a scientist at a National Fusion facility) can run codes on a remote site within a time boundary. The service provider downloads the necessary data and executes a workflow script.
- A networking services event correlation-monitoring agent starts and watches the submitted job for operational anomalies. This helps to sustain SLA validation. This also helps the service provider to provision more resources, recover from unpredictable failure conditions, etc.

Functional Requirements for the OGSA After a thorough and careful examination of the static and dynamic behaviors present in this use case, the following functional requirements of the Grid Protocol Architecture can be identified:

- Discovery of available resources

- Workflow management for job distribution across resources
- Scheduling of service tasks
- Enable facilities for disaster recovery in case of outages
- Provisioning of resources based on need
- Resource brokering services, to better utilize the resources and achieve the QoS requirements
- Load balancing to manage workloads
- Network transport management
- Integration with legacy applications and their management
- Handling application- and network-level firewalls
- SLA and agreement-based interaction
- Provide end-to-end security and security authorization and use policies

In the next scenario, we will discuss an online media and entertainment project with some highly interactive content and data sharing among participants. This is an on demand business media and entertainment system, which can be thought of as a classic representation of the next generation of on demand business applications.

[*CASE 3*] Online Media and Entertainment

In this scenario, the entertainment and media content may consist of different forms (e.g., movies on demand, online gaming on demand, etc.), with different hosting capacity demands and lifecycle properties. One of the primary goals of this use case is the ability to dynamically manage a resource based on the workload demand and current system configurations.

Another observation with entertainment media is the frequent need for changing the content during its lifecycle, and changing the roles of the actors involved. The user involvement and responsiveness with the entertainment content drive this use case into two categories:

1. The consumption of the media content movie on demand with very limited user interaction
2. Frequent user interaction with the content as in online on demand games

A number of new commercial consumer experiences will emerge from the economic factors of content subscription with "digital rights management." Some examples of the commercial opportunities arising from this type of scenario are: usage-based pricing capabilities, content availability, and product inventories, plus differentiation services among competitors.

Most online entertainment media (e.g., on demand gaming and video on demand) are designed based on a narrow domain of solutions for each entertainment medium, and each solution tends to be managed separately from the others. This will become a cumbersome solution because of the lack of reusability and over-provisioning of the networking services resources.

The Grid Protocol Architecture will provide mechanisms for on demand provisioning, new business models (e.g., pricing models), and a plurality of resource sharing models.

Actors

1. A customer who consumes entertainment content
2. A service provider that hosts entertainment content
3. A publisher that offers entertainment content
4. A developer who consumes entertainment content

Scenarios

- A consumer, for example, a game player, accesses the game portal and authenticates with the game server and starts the selected game.
- Several providers working in concert provide the required services to the consumer. For example, the network service provider offers the consumer the required bandwidth; the hosting provider provides server and storage; and the ASP offers common services like the game engine, accounting and billing applications, and help.
- The content provider or media studio provides the content for the customer experience.

Each of the above activities actually presents itself as an interaction between actors.

Functional Requirements for the OGSA After a thorough and careful examination of the static and dynamic behaviors presented in this use case, the following functional requirements of the Grid Protocol Architecture can be derived:

- Discovery of resources
- Instantiate new service
- Service-level management to meet user expectations

- Enable metering and accounting to quantify resource usage into pricing units
- Monitoring resource usage and availability
- Manage service policies
- Provide service grouping and aggregation to provide better indexing and information
- Manage end-to-end security
- Service lifecycle and change management
- Failure management
- Provisioning management
- Workload management
- Load balancing to provide a scalable system

We can see that the requirements yielded in each of the use cases are complex in nature. To provide a solution to these complex requirements is a challenging task.

Use Case Conclusions

The use cases in this section introduced some of the core scientific and commercial usage patterns for Grid Computing. After going through these representative use cases, and the functional requirements exhibited by each of them, we can classify these use cases into four categories:

1. Basic functions
2. Security functions
3. Resource management functions
4. System properties

Further discussion of the details in these classifications will be covered when we discuss platform components. Based on the above functional requirements, the OGSA-WG started identifying platform services and component model definitions for each identified service.

Open Grid Services Infrastructure (OGSI)

This section introduces the OGSI specification. The key focus of this section will explore Grid Computing infrastructures, technologies, and strategies, including considerations for what Grid Computing developers must implement to deliver Grid Computing application environments. Specifically explored in this section will be the concepts from the article entitled *"A developer's overview of OGSI and OGSI-based GRID computing"* [Joshi01] including naming and technological referencing perspectives for Grid Computing services, the need for using interfaces and behaviors that are common to all Grid Computing services, the importance of specifying additional interfaces, and a set of Grid Computing behaviors and key extensions. Examples in this chapter will implement a simple Grid Computing service, hereinafter referred to as the *AcmeSearchEngine* service.[15]

As noted in this chapter, Grid Computing has attracted the attention of many technical communities worldwide with the evolution of on demand business capabilities. As mentioned earlier in this book, both Grid Computing and Autonomic Computing are important technologies in the on demand Operating Environment. As *The Anatomy of the Grid* explains (see "Grid Computing Resources" section), Grid Computing is a process of coordinated resource sharing and problem-solving in dynamic, multi-institutional VOs. In the context of on demand business and Autonomic Computing (i.e., self-regulation), Grid Computing is especially interesting regarding solving a new set of problems with a new set of technologies, not addressed in traditional distributed computing or centralized computing approaches. We will further explore these notions in this chapter.

In the plight to discover "just right" strategies and architectures for industrial deployments of Grid Computing solutions, prominent researchers have detailed strategic frameworks and roadmaps for current and future Grid Computing technologies. Significant contributors in this field of research and development assert that Grid Computing technologies are currently distinct from other major technology trends such as Internet-related, enterprise-related, distributed, and P2P computing. These other prior technological trends can sometimes significantly benefit from extending their innovative problem/solution approaches into the problem-solving space addressed by Grid Computing strategies and technologies. The service provider solution space is one example of an area that has already realized tremendous values in creating additional computing resource power through complex Grid Computing solution implementations (see "Grid Computing Solution Implementation Cases").

The OGSI technological specification assists in the management of required Grid Computing service behaviors, and is extensible to achieve maximum

15. The OGSI discussion is (at this time) based on the proposed final draft published by the GGF OGSI WG, dated March 13, 2003. Since this is a technology reference document, it is a working draft, and there may be changes to the OGSI specification when it reaches the final acceptance stage.

interoperability across service provider implementations. Several software frameworks available today are based on the OGSI specification. The Globus Toolkit version 3 (i.e., GT3, a Grid Computing development toolkit) is the most prominent and practical implementation of this specification. Many innovative solution frameworks for Grid Computing are also available today in several IBM products. These solutions are incorporated in many on demand business solutions IBM provides to customers worldwide. It should be noted that although the IBM Corporation is considered a leader in Grid Computing, IBM is not alone as the global provider of Grid Computing solutions. There are many key contributors working together to address the many complex Grid Computing technology and strategy perspectives.

Let us further explore the OGSI specification at deeper technology levels, which will clearly illustrate the intersections of technology and strategy, thereby providing a robust environment for Grid Computing solution deployments.

Grid Computing Services

Based on the OGSI specification (see "Grid Computing References" section), a Grid Computing service instance is a Web service that conforms to a set of conventions expressed by the WSDL as service interfaces, extensions, and behaviors. This technology specification defines the following:

- How Grid Computing service instances are named and referenced
- Interfaces and systematic behaviors common to all Grid Computing services
- How to specify additional interfaces, behaviors, and their extensions

The following discussion uses a simple service implementation of the AcmeSearchEngine service to demonstrate this technology. Note that industrial Grid Computing services are more complex in nature regarding resource sharing, provisioning, and management aspects. The AcmeSearchEngine service is a stateful Web service with Grid Computing service behaviors as indicated in Figure 6.2.

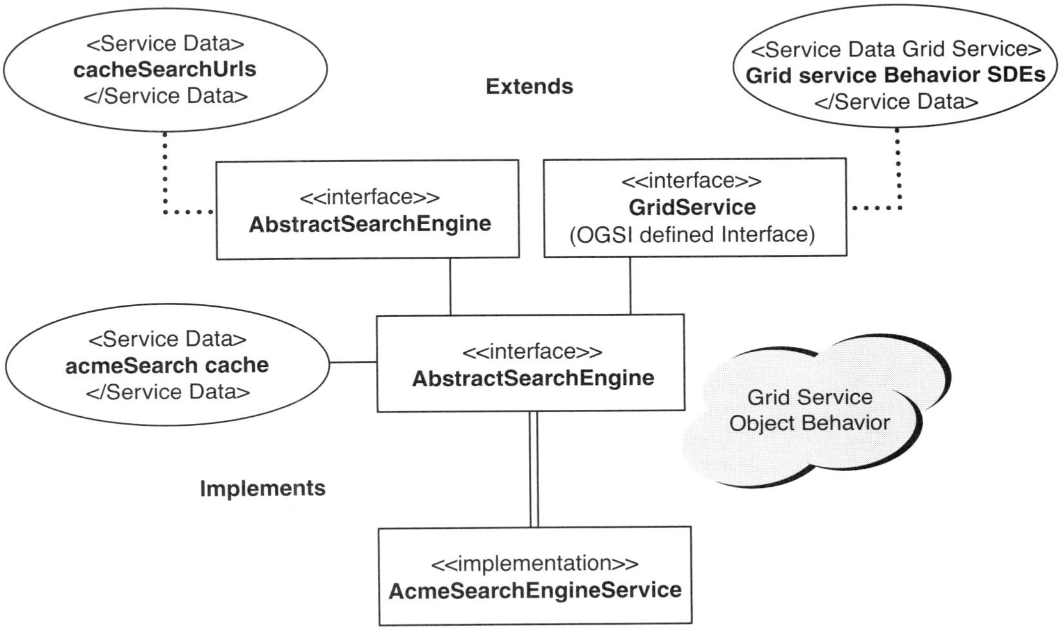

FIGURE 6.2 A simple Grid Computing service object diagram showing Grid Computing service behaviors.

The object diagram in Figure 6.2 introduces the following concepts of a Grid Computing service:

- Provides a stateful Web service implementation with the AcmeSearch-Engine public interface to access the service and its state.
- Supports interface inheritance. The AcmeSearchEngine service implements the AcmeSearchEngine interface, which in turn extends the GridService interface and AbstractSearchEngine interface.
- Specifies the common Grid Computing service behaviors (state data and operations) through the interface defined in the OGSI (GridService interface).
- Publishes the state information to the client through the findService-Data of the GridService interface.
- Provides a look at the state and meta-data of the service through the findServiceData of the GridService interface.

One of the requirements of services defined by the OGSI is the ability to describe the preceding concepts using an OGSI description model, which is a combination of Web service WSDL and OGSI WSDL (GWSDL).

The next section discusses the core concepts of the OGSI specification and shows how to use those concepts to describe and implement a Grid Computing service with the required behavior. Specific implementations, however, are reserved for a future discussion.

OGSI Utilization with WSDL

The OGSI technologies enable new extension models for WSDL called GWSDL. There are two core requirements for describing Web services based on the OGSI framework:

1. The ability to describe interface inheritance, which is a core concept with most distributed object systems.
2. The ability to describe additional information elements with interface definitions. One such information element is called "service data," and it is discussed in detail later in this chapter.

Like most Web services, OGSI services use WSDL as a service description mechanism, but the current WSDL 1.1 specification lacks these two abilities in its definition of *portType*. The WSDL 1.2 working group has agreed to support these features through portType inheritance and an open content model for portTypes. As an interim decision, the OGSI working group developed a new schema for the portType definition (extended from normal WSDL 1.1 schema portType Type) under a new namespace definition, GWSDL.

Another important aspect of OGSI is the naming convention adopted for the portType operations and the lack of support for operator overloading. In these cases, OGSI follows the same conventions as described in the suggested WSDL 1.2 specification. See the OGSI and WSDL 1.2 specifications for details on extension support and operation naming.

THE MERGING OF GWSDL AND WSDL 1.2

There is a consensus among OGSI working group members to follow the WSDL 1.2 specification in all respects. This agreement may eliminate the new schema and namespace (GWSDL) introduced by OGSI when WSDL 1.2 reaches the recommendation stage. Therefore, for current usage and backward compatibility, developers of tools and applications may want to use the GWSDL extensions.

In addition to the WSDL 1.2 recommendation, one should become familiar with the Web Services-Interoperability (WS-I) basic profile best practices guidelines (see "Grid Computing References" section) for interoperable Web services and Grid Computing services.

Service Data

A service data declaration is a mechanism used to publicly express the available state information of a service through a known schema. This concept is not limited to Grid Computing services. Any stateful Web service can declare its publicly available state information through *service data* concepts. The following Listing 6-1 is an example of a service data elements schema.

Listing 6-1

```
<wsdl:definitions xmlns:tns="abc"
targetNamespace="mynamespace">
<gwsdl:portType name="AbstractSearchEngine">
<wsdl:operation name="search" />
--------------------
<sd:serviceData name="cachedURL" type="tns: cachedURLType"
mutability="mutable" nilable="true", maxOccurs="1"
minOccurs="0"
modifiable="true"/>
</gwsdl:portType>
</wsdl:definitions>
```

A set of attributes defined for *service data elements (SDEs)* is described in Table 6.1:

TABLE 6.1 SDE Attributes

SDE Attributes	Description and Default Values
name	A required attribute with a uniquely identifiable name in the target namespace.
type	Defines the XML schema type of the service data value.
maxOccurs	Indicates the maximum number of SDE values that can appear in the service instance's SDE values or the portType staticServiceDataValues. Default value = 1
minOccurs	Indicates the minimum number of SDE values that can appear in the service instance's SDE values or the portType staticServiceDataValues. If minOccurs = 0, then this SDE is optional. Default value = 1

TABLE 6.1 SDE Attributes (Continued)

nilable	Indicates whether a service data value can be nil. You can declare this SDE value as <sd:cachedURL xsd:nil="true"/>. Default value = false
modifiable	A mechanism to specify a read-only or write-only SDE. If writable, you can use setServiceData to change its SDE value based on mutability, min, and max constraints. Default value = false (all SDEs are by default "read-only")
mutability	An indication of whether and how the values of an SDE can change. Possible values are "static" | "constant" | "extendable" | "mutable". Default value = "extendable"

This attribute set is extensible through the open attribute declaration of the schema's SDE. Therefore, the service can add more semantic information about service data through attribute extensibility. An example of this extensibility is presented later with lifecycle attributes. For now, remembering the default values of these attributes will help to understand and define a good state management framework for Grid Computing services. Table 6.1 showed that the only required attribute for an SDE is the *"name"* attribute.

A COMPARISON BETWEEN XSD:ELEMENT AND SERVICE DATA

The SDE declaration is similar to the xsd:element declaration, but the SDE declaration is a restriction of the XML schema's xsd:element, which uses only six properties of the xsd:element declaration (annotation, name, type, occurs(minOccurs, maxOccurs), nilable, and the open attribute model) and adds two new attributes, "mutability" and "modifiable," using the open attribute model.

This constitutes the xsd:element of the SDE declaration.

Types of Service Data

There are two types of SDEs:

1. *Static*—Declared as part of the service's interface definition (GWSDL portType definition).
2. *Dynamic*—Added to a service instance dynamically. This behavior is implementation-specific. The client may know the semantics (type and meaning) of the service data, or can acquire that information from somewhere else. For example, to process dynamic SDE values, you may need to get the schema for the SDE from a remote location.

To support both types of service data, the client must get the complete list of SDEs in a service during runtime. The client can query a service to get the current list of SDEs using the findServiceData method of the Grid Computing service (the service keeps a list of SDEs, both static and dynamic, in its serviceData SDE).

Service Data and WSDL portType Inheritance

A Grid Computing service can support the portType inheritance model as defined by the WSDL 1.2, but this inheritance model can affect the service data declarations associated with each portType in the inheritance chain. Listing 6–2 shows how the portType hierarchy diagram and WSDL declarations work:

Listing 6-2

```
<gwsdl:portType name="BasePT">
<sd:serviceData name="baseSD" type="xsd:string"
minOccurs="1" maxOccurs="1" mutability="static" />

<sd:serviceData name="myServiceData" type="xsd:string"
minOccurs="1" maxOccurs="1" mutability="static" />

<sd:staticServiceDataValues>
         <baseSD>base</baseSD>
</sd:staticServiceDataValues>
</gwsdl:portType>
<gwsdl:portType name="DerivedPT1" extends="BasePT">
<sd:serviceData name="derivedSD1" type="xsd:string"
minOccurs="1" maxOccurs="1" mutability="static" />

<sd:staticServiceDataValues>
         <derivedSD1>derived 1</ derivedSD1>
</sd:staticServiceDataValues>
</gwsdl:portType>

<gwsdl:portType name="DerivedPT2" extends="BasePT">
<sd:serviceData name="derivedSD2" type="xsd:string"
minOccurs="1" maxOccurs="1" mutability="static" />

<sd:staticServiceDataValues>
         <derivedSD2> derived 2</ derivedSD2>
</sd:staticServiceDataValues>
</gwsdl:portType>

<gwsdl:portType name="MostDerivedPT" extends=" DerivedPT1
DerivedPT2">
<sd:serviceData name="mostDerivedSD" type="xsd:string"
```

```
minOccurs="1" maxOccurs="1" mutability="static" />

<sd:serviceData name="myServiceData" type="xsd:string"
minOccurs="1" maxOccurs="1" mutability="static" />

<sd:staticServiceDataValues>
        <derivedSD>base</ derivedSD>
</sd:staticServiceDataValues>
</gwsdl:portType>
```

Following is a set of guidelines derived from the example inheritance definition of portTypes and service data in the previous listing:

- The service contains a union of all the service data defined in the portTypes it implements. Table 6.2 shows how service data aggregation is accomplished with inheritance:

TABLE 6.2 SDE Attributes

If a service implements:	Its service data set contains:
BasePT	BaseSD, myServiceData
DerivedPT1	derivedSD1, myServiceData
DerivedPT2	derivedSD2, myServiceData
MostDerivedPT	MostDerivedSD, myServiceData, derivedSD2, derivedSD1, baseSD

- This aggregation is based on the name of the SDE (QName), hence only one SDE with the same QName is present in the service. Listing 6–2 uses only one myServiceData element in the MostDerivedPT because BasePT's myServiceData and MostDerivedPT's myServiceData have the same local name and belong to the same target namespace.
- The cardinality requirements (minOccurs and maxOccurs) on SDEs must be preserved.

Refer to the OGSI specification for more examples and best practice guidelines on how these cardinality constraints can be utilized. The OGSI specification also contains information on how to define an abstract portType with an SDE, and how the implementation and tools will check for cardinality violations.

The SDE and Mutability Attributes

One of the most complex concepts of the SDE is its *mutability* attribute. Table 6.3 shows the possible mutability attributes and resulting SDE values. It also shows how to define and initialize those values.

TABLE 6.3 SDE Mutability Attributes

SDE Mutability Attribute Value	Description of SDE Value	How to Define and Initialize this SDE Value
static	Analogous to a language class member variable. All portType declarations carry this service data value.	Inside WSDL portType using staticServiceDataValues.
constant	This SDE value is constant and must not change.	This SDE value is assigned on creation of a Grid Computing service (runtime behavior).
extendable	Similar to the notion of appending values. Once added, these values remain with the SDE and new values are appended.	Programmatically can append new SDE values. The new values are appended while old ones remain.
mutable	The SDE values can be removed and others can be added.	Programmatically (setServiceData) can change the SDE values and add new ones.

The SDE and Lifetime Attributes

In addition to the expressed features of the service data discussed above, there is also a "hidden" concept in the specification with respect to the lifetime properties associated with SDEs. The concept is hidden because it is just a recommendation that one could possibly ignore. However, good designs, programs, and tools should be aware of this feature.

The SDE represents the real-time observations of the dynamic state of a service instance. This real-time observation forces the clients to understand the validity and availability of the state representation (certain SDEs, especially dynamic SDEs, may have a limited lifetime). If there is lifetime information associated with an SDE, it can help the client make decisions on whether the SDE is available and valid and when to revalidate the SDE (client-side service data cache revalidation).

Based on the preceding requirements, the specification provides three kinds of lifetime properties:

1. The time from which the contents of this element are valid (ogsi:goodFrom)
2. The time until which the contents of this element are valid (ogsi:goodUntil)
3. The time until which this element itself is available (ogsi:availableUntil)

The first two properties are related to the lifetime of the contents, while the *availableUntil* attribute defines the availability of the element itself.

According to the specification, these values are optional attributes of the SDE and its values, but it is recommended to include them in the XML schema design of the types for service data declarations. Listing 6–3 includes a service data declaration (myPortType), its type (myType), and the lifetime attributes through open content XML schema type attribute declarations (##any):

Listing 6–3

```
<wsdl:types>
     <xsd:complexType name="myType">
        <xsd:sequence>
          <xsd:element name="myElem" type="xsd:string"/>
</xsd: sequence >
<anyAttribute namespace="##any" />
</xsd:complexType>
</wsdl:types>

<gwsdl:portType name ="myPortType">
        <sd:serviceData name="mySDE" type="myType" />
        ........
</gwsdl:portType>
```

The service data declaration in Listing 6–3 can contain the following values:

```
<sd:serviceDataValues>
        < mySDE goodFrom="2003-04-01-27T10:20:00.000-06:00"
goodUntil="2003-05-20-27T10:20:00.000-06:00"
availableUntil="2004-05-01-27T10:20:00.000-06:00">
```

```
        <myElem goodUntil="2003-04-20-27T10:20:00.000-
06:00" >
test
</ myElem>
       </ mySDE>
</sd:serviceDataValues>
```

In the previous listing, the sub-elements in the SDE values (myElem) can override the goodUntil lifetime property to a new time earlier than that of the parent (mySDE). The attribute extension (anyAttribute) of the SDE schema and XML type schema of the SDE value make the override feature possible. See the OGSI specification for a more detailed discussion.

Following are several advantages of using this programming model:

- Provides an aggregated view of the service state rather than individual state values or properties
- Gives a document-centric view (XML documents) of the data, thereby avoiding specific methods for state access
- Shows flexibility in supporting dynamic state information and service state introspection
- Provides lifetime and subscription support for the state properties
- Enables dynamic construction of state data information elements and values

Grid Computing Service Instance Handles, References, and Usage Models

Grid Computing service instances incorporate handles, references, and usage models in the following ways:

- Developers must incorporate one or more Grid Computing service instance handles (GSHs). These handles uniquely identify a Grid Computing service instance that is valid for the entire life of a service. However, handles do not carry enough information to allow a client to communicate directly with the service instance. The service GSH is based on a URI scheme (for example, http://) and specific information (for example, abc.com/myInstance). A client must resolve the GSH information to a service-specific Grid Computing service reference (GSR)—discussed in the next section—in one of three ways: by itself, by using the mechanisms provided by the service provider (a HandleResolver service that implements a HandleResolver portType), or by delegating to a third-party resolver.

- Developers must incorporate one or more GSRs. A client can access a Grid Computing service through the use of a GSR, which can be treated as a pointer to a specific Grid Computing service instance. The format of a GSR is specific to the binding mechanism used by the client to communicate with the service. Examples of these binding formats are Interoperable Object Reference (IOR) for clients using the Remote Method Invocation/Internet Inter-ORB Protocol (RMI/IIOP), or WSDL for clients using SOAP. A Grid Computing service instance may have one or more GSRs available. The GSRs are associated with a lifecycle that is different from the service lifecycle. When a GSR is no longer valid, the client should get a reference to the service using an available GSH. Note that the specification recommends a WSDL encoding of a GSR for a service. This reference contains a WSDL definition element (with one service) as the child element.

As the developer of a Grid Computing service, one should be aware of another helpful construct defined by the specification called the *service locator*. This is a locator class with zero or more GSHs, zero or more GSRs, and zero or more portTypes (identified by portType QName) implemented by the service.

The hosting environment manages the lifecycle of a Grid Computing service (how and when a service is created, and how and when a service is destroyed). The OGSI specification does not dictate that behavior. For example, a Grid Computing service founded on an EJB has the same lifecycle properties, manageability interfaces, and policies as defined by the EJB specification. Following are several lifecycle events and service creation patterns defined by the OGSI specification to help the clients of Grid Computing services:

- A common Grid Computing service creation pattern is defined through the Grid Computing service factory interface (discussed below) and its createService operation. The service can decide whether or not to support this behavior based on the policies defined for the service.
- Two destruction mechanisms are defined: [1] calling an explicit destruction operation (destroy operation on GridService portType); and [2] using a service-supported soft-state mechanism based on the termination time attribute of a service. See the following sidebar for more information:

SERVICE LIFECYCLE AND SOFT-STATE APPROACH

The soft-state approach is related to Grid Computing service lifecycle management. Every Grid Computing service has a termination time set by the service creator or factory. A client with the appropriate authorization can use this termination time information to check the availability (lease period) of a service, and can request to extend the current lease time by sending a keep-alive message to the service with a new termination time. If the service accepts this request, the lease time can be extended to the new termination time requested by the client. This soft-state lifecycle is controlled by appropriate security and policy decisions of the service, and the service has the authority to control this behavior (for example, a service can arbitrarily terminate a service or can extend its termination time even while the client holds a service reference). Another decision to make is whether to really destroy a service or just make it unavailable. This is a service hosting environment decision.

Grid Computing Service Interfaces

The Grid Computing service interfaces are described by the OGSI specification. This section covers the basic and essential information: service interface behaviors, messages exchanges, and interface descriptions.

There are three sets of interfaces based on their functionality. Please note that in Tables 6.4 through 6.6, treatment is provided to describe the interface name, description, and service data defined for these interfaces, as well as the pre-defined static SDEs.

TABLE 6.4 portTypes that Support Grid Computing Service Behaviors

PortType Name	Interface Description and Operations	SDEs Defined by this portType	Default Service Data (Static) Values
GridService (required)	All Grid Computing services implement this interface and provide these operations and behaviors. Operations findServiceData setServiceData requestTerminationTimeAfter requestTerminationTimeBefore destroy	Interfaces serviceDataName factoryLocator GridServiceHandle GridServiceRefrence findServiceDataExtensibility setServiceDataExtensibility terminationTime	<ogsi: findServiceData Extensibility inputElement=" queryByService DataNames" /> <ogsi: setServiceDataE xtensibility inputElement=" deleteByService DataNames" />
Factory (optional)	To create a new Grid Computing service. Operation createService	createServiceExtensibility	Factory (optional)
HandleResolver (optional)	A service-provided mechanism to resolve a GSH to a GSR. *Operation* FindByHandle	handleResolverScheme	None

Of the interfaces listed in Table 6.4, all the Grid Computing services are required to implement the GridService portType and its behaviors. This interface and its associated service data set provide service-specific meta-data information for dynamic service introspection.

The second set of portTypes creates a notification framework for Grid Computing services.

TABLE 6.5 portTypes that Support the Grid Computing Service Notification Framework

PortType Name	Interface Description and Operations	SDEs Defined by this portType
NotificationSource (optional)	Enables a client to subscribe for notification based on a service data value change. Operation subscribe	notifiableServiceDataName subscribeExtensibility
NotificationSink (optional)	Implementing this interface enables a Grid Computing service instance to receive notification messages based on a subscription. Operation deliverNotification	None
NotificationSubscription (optional)	Calling a subscription of a NotificationSource results in the creation of a subscription Grid Computing service. Operations None defined.	subscriptionExpression sinkLocator

Following are concepts to remember about the notification process:

- The services or clients implementing the NotificationSink portType are not required to be Grid Computing services (i.e., no need to implement the GridService portType).
- The subscription Grid Computing services are created during runtime. Clients are used by subscription Grid Computing services to manage the lifecycle of a notification process.

The third set of portTypes provides grouping concepts for Grid Computing services. The Grid Computing services can be grouped based on certain classification schemes, or they can use simple aggregation mechanisms. Note that MembershipContentRule service data can be used for the classification mechanisms in Table 6.5. This "rule" service data is used to restrict the membership of a Grid Computing service in a group. This rule specifies the following:

- A list of interfaces that the member Grid Computing services must implement.
- Zero or more contents, identifiable through QName, that are needed to become part of this group. (i.e., the contents of QName are listed with the ServiceGroupEntry; please refer to Table 6.6).

This service data has a mutability value of "constant," hence these rules are created at runtime on ServiceGroup service startup (i.e., in most cases).

Another interesting feature introduced by the OGSI specification is the concept of extensible operations through the use of untyped parameters using the XML schema xsd:any construct. This feature allows Grid Computing service operations maximum flexibility and extensibility by allowing "any" parameters for an operation. However, this feature introduces confusion on the client side regarding the kinds of parameters it can pass to the service. To avoid confusion, the specification provides the clients a mechanism (query capabilities) to ask the service about the supported extensible parameter and its type (XML schema) information.

We have already seen that we can use findServiceData of GridService to query the service data of a service instance. A service can support different types of queries, including query by service data name and query by XPath. The specification defines the input parameter of findServiceData as extensible using ogsi:ExtensibilityType. You can retrieve all the possible queries supported by the service instance from the SDE values of the service's findServiceDataExtensibility. By default, a service provides a static query type called queryByServiceDataNames (defined using staticServiceDataValue).

TABLE 6.6 PortTypes that Support Grid Computing Service Grouping Behaviors

PortType Name	Interface Description and Operations	SDEs Defined by this portType
ServiceGroup (optional)	An abstract interface to represent a grouping of zero or more services. This interface extends the GridService portType. Operations None defined, but can use operations defined in the GridService portType.	MembershipContentRule entry
ServiceGroupRegistration (optional)	This interface extends the ServiceGroup interface and provides operations to manage a ServiceGroup, including adding/deleting a service to/from a group. Operations add remove	addExtensibility removeExtensibility
ServiceGroupEntry (optional)	This is a representation of an individual entry of a ServiceGroup and is created on a ServiceGroupRegistrati on "add." Each entry contains a service locator to a member Grid Computing service, plus information about the member service as defined by the service group membership rule (content). Operations None defined.	memberServiceLocator content

The OGSI specification handles required Grid Computing service behaviors, is extensible, is flexible, and achieves maximum interoperability across service implementations. Several software frameworks available today are based on the OGSI specification.

Grid Computing, Globus GT3, and OGSI

The discussion in the previous section gives Grid Computing service developers an introduction to a reference implementation of the emerging OGSI software model. In this section, we will cover the OGSI, the Globus GT3 software architecture, the programming model, the tools introduced by GT3, and the relationship between Web services and Grid Computing service software. Models, and help for dealing with new Grid Computing service behaviors, are also introduced into the world of Web services.

Introduction

GT3 is the major reference implementation of the OGSI specification (please refer to the previous section for details). This software is used by a number of technology initiatives, including Utility Computing, on demand business computing, virtualized resource sharing, distributed job schedulers, and similar applications, as a proof of concept for open standard interoperability using OGSI and the WSA.

One recent initiative surrounding this is the IBM alphaWorks Emerging Technology Toolkit (ETTK), which includes the Globus GT3 software and samples. This software and its surrounding architectures are still evolving, and this section concentrates on the recent and stable software preview (Globus GT3) from Globus.

This section will present two parts of the Globus software architecture and programming model.

1. The high-level architecture of this toolkit
2. The programming model and tools

GT3 Software Architecture Model

The discussion of the software framework used in the Globus Toolkit begins with a look at the server-side framework components. We will explore this aspect of the GT3 in this section.

Globus GT3 and the OGSI Specification Version

The Globus GT3 alpha 3 Toolkit is based on the OGSI draft published by the Grid Alliance, dated February 19, 2003.

Changes were introduced to the specification after that point in time. As a result, an earlier section in this book discussed the details of the new OGSI specification. As part of an ongoing effort to improve the functionality of the Globus Toolkit, GT3, the Globus Project launched a research and development program aimed at creating a next-generation toolkit based on OGSA mechanisms.

In this section, we will provide full treatment of GT3, and the current status and future plans for the toolkit. Because these plans for GT3 will continue to evolve, interested users should periodically check *www.globus.org/toolkit/* for updates. Please send feedback to *info@globus.org*.

The major architectural components of the server-side framework include the following:

- *Web services engine*—This engine is provided by Apache AXIS framework software and is used to deal with normal Web service behaviors, SOAP message processing, JAX-RPC[16],[17] handler processing, and Web service configuration.
- *Globus container framework*—GT3 provides a container to manage a stateful Web service through a unique instance handle, instance repository, and lifecycle management, including service activity/passivity and soft-state management.

Pivot Handlers

Pivot handlers are responsible for creating Web service instances and invoking operations on a service. There are different types of handlers available based on the style (e.g., document/RPC (Remote Procedure Call) of the SOAP message. GT3 uses "wrapped" style messaging, which is a special case of "document" style messaging, where each parameter is bundled and treated separately.

The RPCURIProvider supplied by GT3 is a special-case implementation of a JavaProvider of the AXIS framework to handle wrapped messages and contact the container repository based on the service instance handle, to find the correct service, and to invoke operations on it.

Currently, GT3 uses Apache AXIS as its Web services engine, which runs in a J2EE Web container and provides a SOAP message listener (AXIS servlet). It is responsible for SOAP request/response serialization and de-serialization, JAX-RPC handler invocation, and Grid Computing service configuration. As

16. JAX/RPC is a Java API for XML Remote Procedure Calls
17. Using a JAX/RPC one can use the Java API for XML-based RPC (JAX-RPC) to build Web applications and Web services, incorporating XML-based RPC functionality according to the SOAP 1.1 specification.

shown in Figure 6.2, the GT3 container provides a pivot handler to the AXIS framework to pass the request messages to the Globus container.

This container architecture is used to manage the stateful nature of Web services and their lifecycles (recall that Grid Computing service instances are stateful Web services). Once the service factory creates a Grid Computing service instance, the framework creates a unique GSH for that instance, and that instance is registered with the container repository. This repository holds all the stateful service instances and is contacted by the other framework components and handlers to perform the following services:

- Identify services and invoke methods
- Get/set service properties (such as instance GSHs and GSRs)
- Activate service or make service passive
- Resolve Grid Computing service handles to reference and persist a service

GT3 Architecture Overview

In previous sections, we were introduced to Grid Computing concepts, the new open standard software architecture model (i.e., OGSA), and the new programming model for Grid Computing (i.e., OGSI). GT3[18] is the major reference implementation of the OGSI standard.

The GT3 software is utilized by a number of worldwide technology initiatives, including Utility Computing, IBM's on demand business computing solutions, virtualized resource sharing, and distributed job schedulers. These activities serve as proof of open standard resource sharing and interoperability in Grid Computing. This software and its surrounding architectures are in the process of evolution; we will therefore concentrate our discussions in this chapter on the recent (very stable) release of GT3.

GT3 is the most widely utilized and explored infrastructure software for Grid Computing middleware development worldwide among Grid Computing practitioners. Hence, we will explore this toolkit in-depth. We will introduce the GT3 software architecture and programming model using sample code listings.

We will divide this chapter's discussion on GT3 into three subsequent sections. First, we will discuss the high-level architecture of this toolkit; in the next section, we will cover the programming model and tools introduced by GT3; finally, our discussion will end with an implementation sample of a

18. GT3 can be found at *www-unix.globus.org/toolkit/download.html*

Grid Computing service to explain the concepts we have learned. Figure 6.3 shows the GT3 core architecture.

GT3 Software Architecture Model

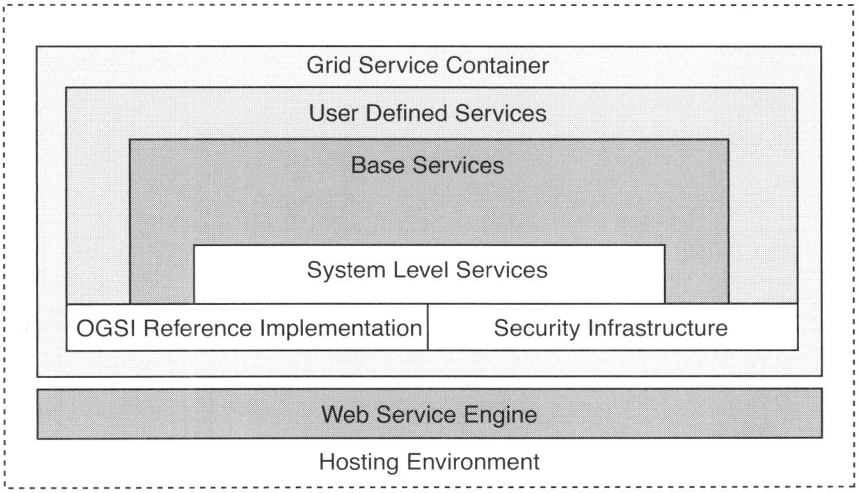

FIGURE 6.3 The Globus GT3 core architecture.

As shown in Figure 6.3, the GT3 architecture is a combination of:

- GT3 core system-level services
- Base services
- User-defined services

The GT3 Core System-Level Services

The GT3 core forms the basic building blocks for Grid Computing services. This core consists of:

- *OGSI reference implementation*—The OGSI reference implementation (as discussed in preceding chapters) provides OGSI-defined interfaces, messages, and Grid Computing behaviors. This implementation enables interoperability with the Web services engine and various hosting platforms. Tools provided with this implementation assist with Grid Computing services creation.
- *Security infrastructure*—The security infrastructure provides the basic Grid Computing security, including message- and transport-level protection, end-to-end mutual authentication, and authorization. This security framework works with the WS-Security specifications.

- *System-level services*—System-level services include logging services, administrative services, handle resolver services, routing services, and other important complementary services. These services are built on top of the OGSI reference implementation and security implementation. They provide system-level services to other OGSI services, for better manageability and customization.

Base Services The higher level base services are built on top of the GT3 core. Some of the services provided by the GT3 base are information services, data management services, and job execution services. This also includes services available within the GT3 software bundle, the information model exposed by these services, and its usage patterns.

User-Defined Services User-defined services are application-level services that are created to exploit the OGSI reference implementations and security infrastructure. These services may in turn work with other high-level services to provide an improved collective behavior related to resource management. Some of these services include meta-schedulers, resource allocation managers, and collaborative monitoring services.

The GT3 software introduces the notion of a Grid Computing service container, which forms an abstract notion of a runtime environment. This runtime environment provides capabilities for Grid Computing service persistence management, lifecycle management, and instance management. We call this container "abstract" because the real functionality implementation is likely to use some existing hosting environment capabilities. For example, a service implemented as an EJB may have a lifecycle managed by the J2EE EJB container.

The current GT3 container is implemented on a J2SE/J2EE Web container. At the same time, we can create Grid Computing services EJBs by using the delegation service programming model defined by GT3. Note that the service lifecycle and its management are still under the preview of the Web container.

Yet another important component requires our attention: the Web services engine, which is (again) responsible for managing the XML messages from a Grid Computing service client to the service. This functionality includes message decoding, un-marshalling, type mapping, and dispatching calls to the service methods. The layered architecture enables OGSI reference implementations to utilize any Web services engine of choice. Even though we can be selective in our choice of Web service engines, the current GT3 implementation is not entirely flexible. The current GT3 relies on the Apache AXIS Web services engine for some of its programming model (i.e., MessageElement and type mapping) and message flows (i.e., Grid Computing handlers).

We can further explore the details of the implementation architecture model with the default software framework implemented in the Globus GT3 Toolkit. We will begin our discussion with the server-side framework components.

Default Server-Side Framework

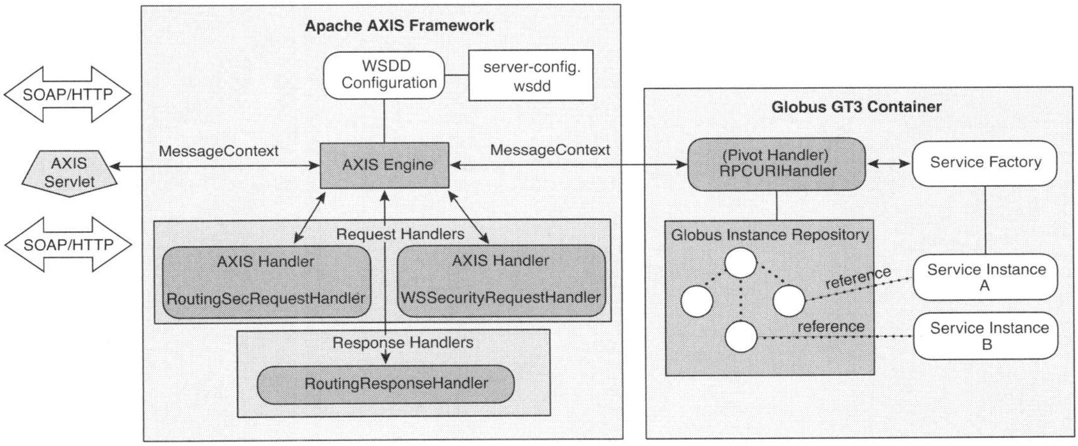

FIGURE 6.4 The current GT3 server-side reference implementation model.

Figure 6.4 shows the major architecture components of the server-side framework.

Web Services Engine The GT3 software Web services engine uses the Apache AXIS framework to deal with normal Web service behaviors, such as SOAP message processing, JAX-RPC handler processing, and Web services configuration.

Globus Container Framework The GT3 software provides a container to manage a stateful Web service through a unique instance handle, instance repository, and lifecycle management, which include the ability to make a service active/passive and soft-state management.

THE GRID COMPUTING HANDLER
The GT3 container provides a "pivot handler" to the Apache AXIS Web services engine framework to pass request messages to the Globus container. It is key to understand that Grid Computing service instances are stateful Web services, as is the respective service lifecycle. This pivot handler serves as a fundamental capability of GT3.

Message processing details are key. As already mentioned, GT3 utilizes the Apache AXIS Web services engine, which executes in a J2EE/J2SE Web container and provides a SOAP message listener (i.e., the AXIS servlet). Processing is the responsibility of SOAP request/response serialization and deserialization, JAX-RPC handler invocation, and Grid Computing service configuration. As shown in Figure 6.4, the GT3 container provides a pivot handler to the AXIS framework to pass request messages to the Globus container.

The Globus GT3 container architecture is utilized to manage the stateful nature of Web services. It is key to understand that Grid Computing service instances are stateful Web services, as is the respective service lifecycle.

Once a Grid Computing service instance is created by the service factory, a unique GSH is created for that instance by the framework. Note that the actual creation of a service is dependent on the service configuration properties. This stateful instance is registered with the container repository.

This container repository now acts as *the* repository for all stateful service instances. The other framework components and handlers contact the repository to invoke service methods, get/set service properties (e.g., instance GSH, GSR, etc.), activate a service or make it passive, and resolve GSHs to refer to and persist the service.

Based on the above discussion, we are now familiar with the high-level default framework implementation that comes with GT3. Let us now explore the GT3 architecture components in more detail, using the prior server-side implementation model as a reference point.

Globus GT3 Architecture Review

This discussion will review the Globus GT3 architecture concepts. The GT3 architecture is an open-ended reference framework, with specific implementations worthy of review.

Grid Computing Service Container The Globus Grid Computing service container model is derived from the J2EE managed container model, where the components are free from complex resource manageability and runtime infrastructure usage. These complex management processes include transactions, concurrency activity, networking services connectivity elements, data and state persistence, lifecycle state information, and security management.

In such a managed environment, the container is responsible for managing and controlling these attributes. This helps to achieve the all-important QoS requirements of the components. In addition, the programming model becomes less complex to implement.

The OGSI Specification The OGSI specification introduces and imposes a number of QoS requirements and behaviors on a Grid Computing service. The rendering of OGSI in a managed environment forces GT3 to reinvent a container model for Grid Computing services. We explained these default implementation behaviors in previous discussions.

The GT3 container is responsible for stateful instance management, persistence, lifecycle management, and activation/deactivation of Grid Computing services. The OGSI provides a repository for service instance identity and management.

In general, this container provides the following value-added features:

- Lightweight service introspection and discovery
- Dynamic deployment and soft-state management of stateful Grid Computing services
- Transport of independent, message-level security infrastructure supporting credential delegation, message signing, encryption, and authorization

OGSI Reference Implementation

The OGSI reference implementation is a set of primitives implementing the standard OGSI interfaces, such as: GridService, factory, notification (source/sink/subscription), HandleResolver, and ServiceGroup (i.e., entry/registration). Among these, the GT3 implementation provides the base implementation of the GridService and factory interfaces.

This forms the basis of the GT3 service implementation, container configuration, and service invocation design pattern. These implementations can be extended to provide more behaviors. We will see the implementation and configuration details of these interfaces in subsequent discussions.

GT3 provides implementations for all other interfaces defined by the OGSI. However, these implementations are dependent on the service requirements. In complex scenarios, these implementations can be replaced with more sophisticated services pertaining to the service requirements. For example, the simple point-to-point notification framework provided by GT3 may not always be sufficient, and should be replaced by asynchronous JMS[19] (Java Messaging Service) -based message queues.

Security Infrastructure

GT3 supports transport-level and message-level security. Another notable feature provided by GT3 is a declarative security mechanism for authentication and authorization using service deployment descriptors. This provides an extended plug-and-play security architecture using the Java Authentication and Authorization Service (JAAS) framework.

Transport-Level Security

GT3 is based on the Grid Security Infrastructure (GSI) security mechanism, as it exists today in the predecessor version, GT2. To communicate over this secure networking services transport layer, we need to utilize an invocation scheme other than HTTP; this is called *httpg*. The current transport-level security may be deprecated in favor of the message-level security introduced through the WS-Security architecture.

Message-Level Security

Web Services Security (WS-Security) is the focus of this discussion. WS-Security describes enhancements to SOAP messaging to provide *quality of protection* through message integrity, message confidentiality, and single message authentication. These mechanisms can be used to accommodate a wide variety of security models and encryption technologies.

WS-Security also provides a general-purpose mechanism for associating security *tokens* with messages. No specific type of security token is required by WS-Security. It is designed to be extensible (e.g. support multiple security token formats). For example, a client might provide proof of identity and proof that he or she has a particular business certification.

Additionally, WS-Security describes how to encode binary security tokens. Specifically, the specification describes how to encode X.509 certificates and Kerberos tickets as well as how to include opaque encrypted keys. It also includes extensibility mechanisms that can be used to further describe the characteristics of the credentials that are included with a message.

This specification proposes a standard set of SOAP extensions that can be used when building secure Web services to implement integrity and confi-

19. The Java Message Service (JMS) API is a messaging standard that allows application components based on the Java 2 Platform, Enterprise Edition (J2EE) to create, send, receive, and read messages. It enables distributed communication that is loosely coupled, reliable, and asynchronous.

dentiality. We refer to this set of extensions as the "Web Services Security Language" or "WS-Security."

WS-Security is flexible and designed to be used as the basis for the construction of a wide variety of security models including PKI, Kerberos, and SSL. Specifically, WS-Security provides support for multiple security tokens, multiple trust domains, multiple signature formats, and multiple encryption technologies.

This specification provides three main mechanisms: security token propagation, message integrity, and message confidentiality. These mechanisms by themselves do not provide a complete security solution. Instead, WS-Security is a building block that can be used in conjunction with other Web service extensions and higher-level application-specific protocols to accommodate a wide variety of security models and encryption technologies.

These mechanisms can be used independently (e.g., to pass a security token) or in a tightly integrated manner (e.g., signing and encrypting a message and providing a security token hierarchy associated with the keys used for signing and encryption).

- GSI secure conversation
- GSI XML signature

Secure Conversation

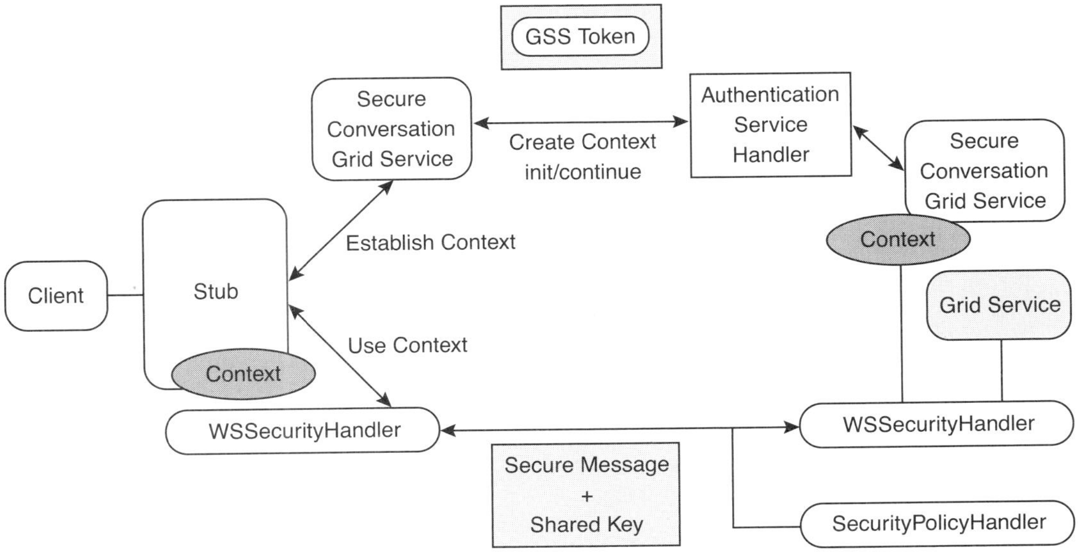

FIGURE 6.5 The establishment of a secure conversation.

Let us further explore a couple of points, as illustrated in Figure 6.5:

1. Initially, the client establishes a security context with the service, utilizing a system-level service known as the "secure conversation service." This security establishment scheme is accomplished utilizing the Generic Security Services (GSS) API.
2. Once the security context is established, the client will make use of this context to sign on, verify, encrypt, and decrypt request/response messages.
3. On subsequent calls, the client passes the shared secret key along with the message.

GSI XML Signature The GSI XML signature is a simple XML message encryption mechanism where the message is encrypted using the X.509 certificate capability. This provides additional flexibility, as any intermediary can now validate the certificate(s), and hence, the message.

We will now discuss the details of the message, the transport level security mechanisms, security programming, and the declarative security model for Grid Computing services.

- *Security direction*—The future of security in GT3 will be aligned with the Global XML architecture standards on security. WS-Security then becomes the default message-level security mechanism. The WS-Trust capability is used for any security context creation; WS-SecureConversation is utilized for secure exchange of the aforementioned tokens.

Figure 6.6 depicts this architectural direction:

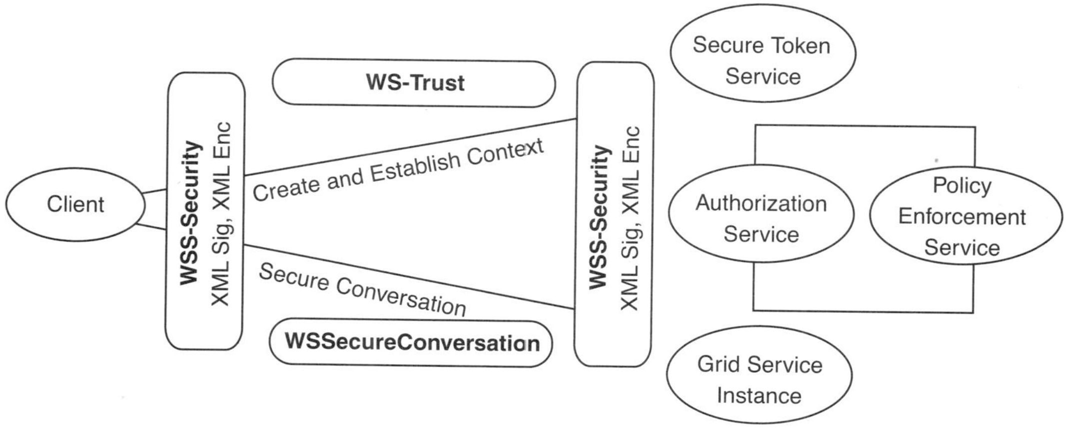

FIGURE 6.6 The emerging security standards and the r implementation with a Grid Computing service.

The plug-and-play nature of GT3 security enables us to create the above architectural model with our choice of service/security provider options. In addition, future security mechanisms must provide facilities for supporting WS-Federation, for federating across VOs and trust domains.

System-Level Services Some of the system-level services are introduced to provide a framework to achieve the required QoS in a production environment. These services must provide logging, management, and administration for a Grid Computing environment. The functions can be used as standalone facilities, or they can be used together. Some of the existing GT3 system-level services are:

- *Logging service*—This enables dynamic modifications of logging filters for the current runtime environment, or this can be persisted for subsequent executions. In addition, this service enables the grouping of log message producers, to adjust the size of backlogs, and to provide customizable message views.
- *Management service*—This service provides a facility to monitor the current status and load of a service container. It also provides functionalities to activate and deactivate service instances.
- *Administration service*—This service provides administrative activities, including pinging the hosting environment and shutting down the container.
- *Hosting environments*—The current GT3 code is developed in Java and supports the following types of hosting environments:
    1. Embedded utilities to be utilized with client and lightweight service deployments
    2. Standalone containers for service hosting
    3. Servlet-based environments
    4. EJB-based environments utilizing a delegation model to support existing EJB components implemented as services

Load Balancing Features in GT3 Generally speaking, a Grid Computing service is created in the same hosting environment where its factory is located. This is fine in most cases; however, there may be cases when the service needs to be created in a different hosting environment than the local hosting scenario.

The reasons may be many, including load balancing, user account restrictions, and backend resource requirements. In such situations, the factory needs to create a service in a hosting environment other than the local one, and the service calls must then be routed to the hosting environment where the service is operating. The local host now acts as a virtual hosting environment, but the client is unaware of this. The client performs operations as usual and the GT3 framework handles the routing. This routing information is embedded within the SOAP header.

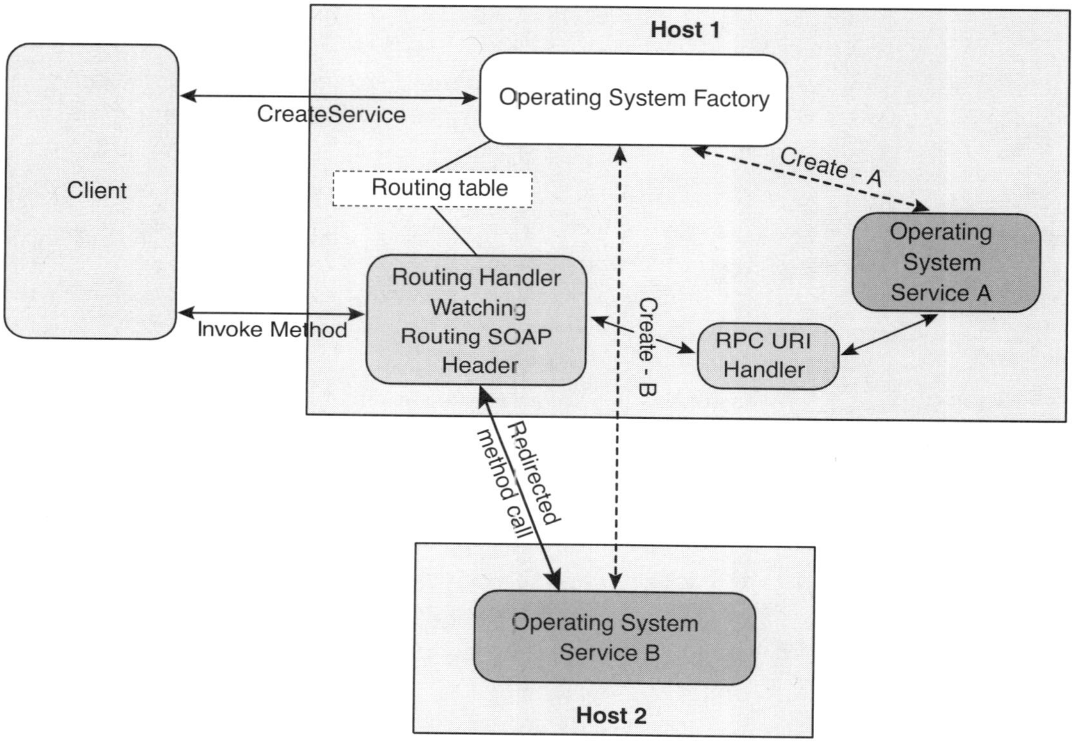

FIGURE 6.7 A virtual host and load balancing environment.

Figure 6.7 illustrates such a virtual hosting and load balancing process. The most prominent example is the GT3 Grid Resource Allocation Manager (GRAM) implementation. This provides a virtual hosting environment for user account restrictions and load balancing.

The Client-Side Framework

GT3 does not dictate an architectural framework for Grid Computing service clients. The default implementation comes with a number of tools supporting the Apache AXIS Web services engine and the corresponding Java code generation. The framework follows the JAX-RPC programming model.

The AXIS Framework

The AXIS framework provides the runtime JAX-RPC engine for client-side message processing and dispatching.

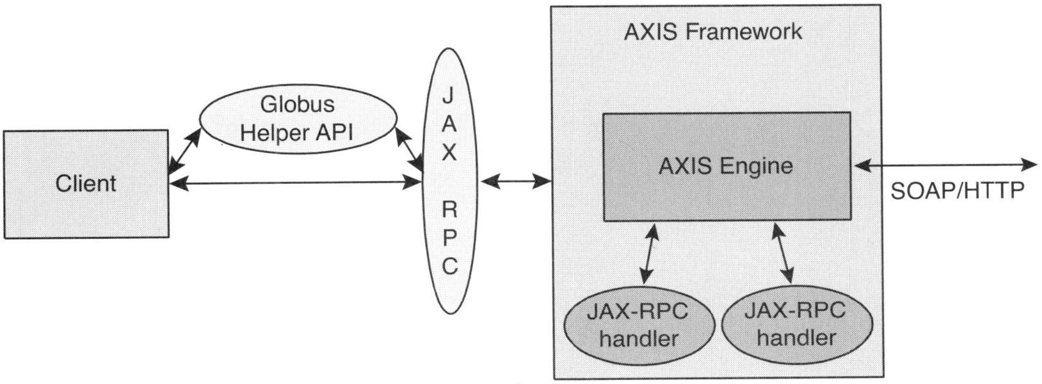

FIGURE 6.8 The GT3 software framework, focusing on aspects of the client-side architecture components.

As shown in Figure 6.8, GT3 uses the normal JAX-RPC client-side programming model and AXIS client-side framework.

In addition to the normal JAX-RPC programming model, Globus provides a number of helper classes at the client side to hide the details of the OGSI client-side programming model.

Message Preprocessing Handlers

Handlers provide a mechanism for processing raw SOAP messages. This is mainly utilized for handling SOAP headers that are implemented to communicate security, correlation, and transactional semantics about a message exchange.

The GT3 architecture utilizes JAX-RPC handlers and AXIS handlers to accomplish this SOAP message processing functionality. These handlers are applicable at both client- and server-side implementations. We will see further details of this in subsequent discussions.

GT3 Architecture Review

The GT3 architecture is a layered architecture with an emphasis on separating the functionality at each layer. This flexibility enables high-level services to utilize lower layer services in a plug-and-play manner to facilitate simpler system designs.

The provisioning for the hosting environment independence, and the provisioning required integrating into a Web services engine of choice makes the architecture more attractive. Even though this is highly desirable, this level of service extensibility is not yet available in GT3, where the current Web services engine of choice is Apache AXIS.

In the next chapter, we will explore some Grid Computing Solution Implementation Cases.

Grid Computing Solution Implementation Cases

This section contains several solution implementation references, which are currently realizing the values of Grid Computing solutions in a diverse set of application domains. According to the Gartner Group, many businesses will be completely transformed over the next decade by using Grid Computing-enabled Web services to integrate across the Internet to share not only applications, but also computer power. This section confirms this fact as seen by the nature of these innovative solutions being developed.

This section also intersects with previous sections describing how developers need to adhere to the basic idea of Grid Computing and the OGSA. Developers can also use the latest Globus Toolkit, GT3, to discover a Grid Computing service, create a Grid Computing service interface, and invoke a Grid Computing service instance. Some ideas to help application developers integrate Web services and Grid Computing solutions are highlighted by the success of other Grid Computing solutions described in this section.

The real and specific problem that underlies the Grid Computing concept is *coordinated resource sharing and problem-solving in dynamic, multi-institutional VOs*. The sharing that researchers are concerned with is not simply file exchange; instead, it concerns direct access to computers, software, data, and other resources by a range of collaborative problem-solving and resource brokering strategies. These are the foundations of the Grid Computing applications and implementations emerging today, in industry, science, and engineering disciplines. This sharing is highly controlled with resource providers and consumers defining exactly what is shared, who is allowed to share in the solution, and the conditions under which sharing is executed. A set of individuals and/or institutions defined by such sharing rules forms what is termed a VO.

Examples of VOs are noted by distinct on demand business needs, and can be characterized by the following simplistic examples of business endeavors:

1. Service Provider Delivery Environments
2. ASPs
3. Storage service providers

4. Machine resource cycle providers
5. Consultants engaged by a large manufacturer to perform scenario evaluation during planning for a new factory
6. A bio-terrorism crisis management team, plus the databases and simulation systems it requires for planning responses to emergency situations
7. Members of industrial consortiums bidding on new aircraft, ship, or space vehicle development efforts
8. Members of large, international, multi-year, high-energy nuclear physics collaboration efforts

These are examples of what a typical VO might appear to be from an organizational composure viewpoint. The notion here is that each of these VO examples represents the likely complex solution dependency addressed by Grid Computing, as well as complex problem-solving scenarios addressed by multi-machine resource collaboration in widespread computation- and data-intensive environments.

Most people will oftentimes think of a "grid" in a similar way. One idea that might come to mind is that of an interconnected system for the distribution of electricity (or electromagnetic signals) over a wide geographic area; for example, a network of high-tension cables, massive networks of poles and towers, and power stations.

This "electrical grid" model is a basic scenario we can easily understand due to the fact that industries, businesses, and home consumers have all understood for some time the infrastructure that utility companies maintain in delivering power grids and electricity into our homes and businesses. This same "grid" notion now stands true for Grid Computing, and when properly implemented, provides vastly powerful computer science technology integration solutions. Grid Computing solutions are not implemented by virtue of hardware or software alone. Grid Computing requires both hardware and specialized software programming techniques (as we discussed in the last section) to interconnect all the points within any given grid. It is by virtue of this unique combination of programming techniques, applications, toolkits, and hardware that successful Grid Computing environments are realized today across several global geographies. This chapter now continues to explore these powerful combinations of technologies and development strategies.

WHAT IS INVOLVED IN GRID COMPUTING AND PERSPECTIVES OF THE GRID COMPUTING VISION?

Around 1995, this same power grid "utility" concept was applied to computing. So what is a "grid," and what constitutes Grid Computing? According to IBM's definition, a *grid* is a collection of distributed computing resources available over a LAN or WAN that appears to an end-user or application as one large virtual computing system. The vision is to create virtual dynamic organizations through secure, coordinated resource sharing among individuals, institutions, and resources.

Grid Computing is an approach to distributed computing that spans locations, organizations, machine architectures, and software boundaries to provide unlimited power, collaboration, and information access to everyone connected to the grid.

In fact, Grid Computing technology is being widely used in areas such as finance, defense research, medicine discovery, decision-making, and collaborative design—on a worldwide basis. Currently, Grid Computing has started to implement advanced forms of combinatorial Web services, along with Utility Computing assets, to define standard interfaces for business services such as business process outsourcing. Grid Computing has allowed for innovative and advanced forms of networking services.

The typical Grid Computing implementation can (and often will) provide people from different organizations and locations a powerful ability to network with each other and to work together to solve specific problems, such as design collaboration or medical diagnostics (to name a few). This implies a very dynamic network resource sharing architecture, while implementing a highly orchestrated information exchange across the grid.

A Grid Computing environment requires resource discovery, resource sharing, and collaboration in a distributed, oftentimes continuously expanding, geographic environment. This tight integration of complex technologies and best practices enables this grid-like information exchange, which is notably purported to be one of the most powerful capabilities realized while implementing Grid Computing solutions.

Harnessing the power of a worldwide Grid Computing solution to fight against smallpox will now be explored. Smallpox and its cure (especially for post-infection and widespread infections) are key in today's volatile world of bio-terrorism.

The power of Grid Computing has shown promising benefits in this medical discovery example, where any individuals in the world may offer their computer's processing power to assist in the battle against smallpox. This indeed demonstrates how a Grid Computing environment allows for resource discovery, resource sharing, and massive collaboration in a distributed, constantly expanding geographic environment.

Smallpox is an acute, contagious, and sometimes fatal disease caused by the *variola* virus. It is oftentimes referred to as one of the viral terrors associated with current fears of bio-terrorism. And while there is a vaccine available to prevent infection by the smallpox virus, there is currently no *post-infection* cure available for smallpox to introduce a cure to those who have already (or very recently) been infected.

At the request of the U.S. Department of Defense, the IBM Corporation, United Devices, Accelrys, and others have joined together in a collaborative fight to alter and forever change this threat of smallpox bio-terrorism. This Grid Computing implementation team has initiated a global public-service project in the search for a cure to post-smallpox infections.

You too may join the fight by setting up a Grid Computing solution on your own computing device(s). To insert yourself into this Grid Computing implementation, download[20] and run the seemingly simple and very interesting screensaver developed by United Devices.[21] By performing this simple installation, you can donate spare CPU machine cycles to a worldwide "grid," a linked network of many computers collaborating to resolve a cure for post-smallpox infection.

In this simple-to-install example, Grid Computing applies the resources of many computers[22] to this single, and extremely important, medical problem—finding a cure for post-smallpox infection. Usually, a problem that requires an immense number of processing cycles, or access to large amounts of data, can be very difficult to solve and costly to operate, that is, unless the solution is defined exploiting Grid Computing concepts.

In this medical solution approach, Grid Computing enables resource devices, regardless of their computational operating characteristics, to be virtually shared, managed, and accessed across many enterprises, industries, and workgroups. This virtualization of resources places all the neces-

20. See *www6.software.ibm.com/reg/ls/ls-grid-i?S_TACT=103ALA1A* to download the smallpox-fighting Grid Computing screensaver.
21. Refer to the United Devices Web site at *www.ud.com*
22. All the machines utilized to write this book were operating this smallpox-fighting Grid Computing screensaver tool. This tool, once downloaded, installs automatically in less than three minutes. The graphical display utilized by the screensaver shows the molecular construction work being performed in which it has engaged your CPU for the smallpox problem resolution problem. This is an incredibly interesting screensaver, with a very useful purpose, where you can see the progress you are enabling and witness the power of this worldwide molecular computing grid.

sary access, data, and processing power at the fingertips of those (e.g., researchers and scientists) who need to rapidly solve complex problems related to smallpox, conduct computation-intensive research and data analysis, and engage all of this in real time.

Downloading this screensaver graphically shows your contribution and CPU power in helping the world to solve this difficult medical problem. Every time your computer is idle (or even if it is running, depending on how you choose to configure this), you contribute your computing resources (i.e., CPU cycles) to this medical grid. The massive computing grid—that you are part of—now speeds the analysis of approximately 35 million molecules against a series of protein targets related to smallpox, making the processing power of this virtual supercomputer greater than the most powerful supercomputers in use today. Try this, and you will see it; it is an amazing virtual collection of processing power coming together to cure a deadly disease.

Leading pox researchers at Oxford and Essex Universities in the UK, the University of Western Ontario, and other locations worldwide are now provided the opportunities that come with a more strategic use of time and technology. The result: Rather than spending years and tremendous amounts of money—neither of which is any more affordable—to screen hundreds of thousands of molecules, it is, in fact, now possible with this Grid Computing solution to screen hundreds of millions of molecules in just months.

JOIN THE PLIGHT FOR A SMALLPOX CURE

Smallpox was eliminated from the world in 1977 by a World Health Organization campaign. Despite this, stocks of the *variola* virus are known to exist today, and its use as a weapon of bioterrorism remains a frightening possibility. With vaccination having ended in 1972, the world population is highly susceptible. The availability of drugs to counter the virus would be a major defense.

Join the fight to cure smallpox! — Join this Grid Computing VO! — Become an endpoint!

This simple and free smallpox screensaver is based on the premise that there is a possible molecular target whose blockade would prevent the ravages of an infection. We (practitioners in technology) intend to use Grid Computing to screen millions of potential anti-smallpox drugs against this target. This will involve the use of the United Devices' global Grid Computing MetaProcessor Platform (MP), which we have successfully used in the past in cancer and anthrax research. The project is harnessing millions of computers (*including the one used to write this book*), belonging to people in over 200 countries, all of which will benefit from protection against smallpox.

While you are running the screensaver, you will be kept up-to-date on several aspects of the project—size of the grid, ligands[23] processed and their structure, and number of leads identified—so you can see the progress you (as an endpoint) are enabling. You will be able to see the power of this worldwide computing grid in a highly animated screensaver illustrating the construction of the experimental molecule. This also serves as a great learning instrument for the end-user, simply by the life sciences elements viewed.

How do IBM products enable this particular Grid Computing solution? United Devices' MP—the tool that aggregates the idle power of all these participating servers, PCs, and workstations into the existing worldwide grid—utilizes the IBM DB2 product exclusively as its host database system. The DB2 database manager efficiently controls an impressive 15 million SQL database queries per day as it seamlessly manages all aspects of data provided by the grid.

Overall, the smallpox research project grid is powered by an IBM on demand business infrastructure[24], which includes IBM p690 server machines running DB2 database software, using the AIX and Linux operating systems. Thus, the elements of Grid Computing have proven to be very powerful, and specific Grid Computing solution areas are limited only by one's own thinking.

Harnessing the power of a worldwide Grid Computing fight against cancer is another very important Grid Computing solution now underway. The United Devices' cancer research project is also asking anyone to volunteer his or her idle computing resources to help process molecular research being conducted by the Department of Chemistry at the University of Oxford[25] in the UK and the National Foundation for Cancer Research.[26]

To participate in this vital cancer research activity, one can simply download[27] a very small, free, non-invasive software program that (again) operates as a screensaver. It runs when your computer is idle and processes research as a background or foreground task, depending on how one chooses to configure the screensaver during the setup procedures. This

23. The term "*ligands*" deals with many types of DNA/RNA protein molecular structures. For more on examples of these ligands, refer to *www.biochem.ucl.ac.uk/bsm/pdbsum/ligands/ligindex.html*

24. For more information on the smallpox screensaver, refer to *www6.software.ibm.com/reg/ls/ls-grid-i?S_TACT=103ALA1A*

25. For more details, refer to *www.ox.ac.uk*

26. For more details, refer to *www.nfcr.org*

27. For more details, refer to *www.grid.org/download/gold/download.htm* to register to become a Grid Computing contributor for this innovative solution.

screensaver project never interrupts your usual PC activities, and can be installed by simply downloading the solution from a Web site.[28]

A Grid Computing solution for breast cancer diagnostics and screening is another very important example of a world-class Grid Computing solution. The University of Pennsylvania uses a powerful Grid Computing solution that works to bring advanced methods of breast cancer screening and diagnostic methods to patients across the U.S. All this is brilliantly accomplished in the Grid Computing solution, while at the same time helping to reduce computing operational processing costs at the university.

This innovative and effective solution is built using open standards. In this solution, the University of Pennsylvania Grid Computing team implements a massive distributed computer grid that delivers computing resources as a *utility* service over the Internet. Amazingly, this enables thousands of hospitals to store mammogram X-rays in digital form, while the Grid Computing solution also provides the analytical tools to help physicians diagnose individual cases and identify patterns of cancer "clusters" in the population. This innovative Grid Computing solution provides (for the first time) authorized medical personnel virtually instantaneous access to vast numbers of geographically distributed patient records, and therefore reduces the need for expensive X-ray film.

This University of Pennsylvania cancer-fighting Grid Computing solution combines IBM and Intel processor-based systems with IBM's DB2 Universal Database to define and create a unique and powerful solution.

The way this solution works is a patient's mammograms are loaded into the Grid Computing system, and once there, the X-rays are evaluated while powerful tools very quickly compare current X-rays with prior X-rays from previous examinations. This solution is part of the university's National Scalable Cluster Lab. Oftentimes, the traditional X-ray films of patients are distributed among many medical facilities, thus making them hard to locate when required. This Grid Computing solution helps ensure that all of a patient's vital data is provided to authorized physicians very quickly, efficiently, and securely.

Many hospitals are now connected to this cancer diagnostics grid via secure Internet portals that allow authorized physicians to upload/download and analyze digitized X-ray data. The advantages of this grid include:

28. For more details, refer to *www.grid.org/download/gold/download.htm*

- *Fast data retrieval*—Authorized physicians have immediate access to a patient's previous and current mammogram X-rays, no matter where or when the X-rays were produced.
- *Computer-assisted diagnosis*—X-rays can be scanned with powerful software that identifies potential cancer tumors and other respective problems, helping physicians and medical specialists to better diagnose patient illnesses.
- *Pattern identification*—Sophisticated algorithms can uncover problematic patterns that appear in the population, such as cancerous "clusters" or abnormal concentrations of diseased areas in a particular community.
- *Cost savings*—The average hospital spends $4 million annually to develop traditional X-ray films. Estimations indicate that participation in this Grid Computing solution could result in an average yearly cost savings in the millions of dollars.
- *Training*—A suite of educational tools is deployed on the grid to help doctors, medical students, and interns learn more about breast cancer and related diseases.

This University of Pennsylvania grid, in collaboration with a group from Oak Ridge National Laboratory, connects hospitals at the University of Pennsylvania, University of Chicago, University of North Carolina, and the Sunnybrook and Women's College Hospital in Toronto. The U.S. National Library of Medicine funds this impressive Grid Computing solution.

Earlier examples (2001) of Grid Computing solutions at IBM include the North Carolina Bioinformatics Grid, which was developed in collaboration with GlaxoSmithKline Inc., Biogen, the University of North Carolina, Duke University, and other organizations.

Also in 2001, IBM was selected by a consortium of four U.S. research centers to design, build, and power the world's most powerful computing grid. This solution, which continues today, is an interconnected series of Linux clusters capable of processing 13.6 trillion calculations per second. This Grid Computing system—known as the Distributed Terascale Facility (DTF)—enables thousands of scientists and researchers around the country to share massive computing resources over the world's fastest research networks. These scientists and researchers are in search of breakthroughs in the life sciences, climate modeling, and other critical disciplines.

Summary

Grid Computing, as mentioned in other chapters, is a key technology in some on demand business solutions. Grid Computing solutions sometimes can be combined with Autonomic Computing disciplines, and this combination will deliver powerful on demand business environments. However, on demand business environments do not depend on Grid Computing environments.

The specific examples discussed in this chapter illustrate exactly how some of these on demand business environments are created, and the variety of open-ended solutions that are noteworthy today. The power of applying Grid Computing to difficult problem-solving situations was also described in this chapter. On demand Operating Environment strategies and technologies can incorporate both computing disciplines described in this book: Autonomic Computing and Grid Computing.

Grid Computing Resources

This section refers to additional resources related to Grid Computing solutions and concepts.

1. For information on the *Global Grid Forum (GGF)*, see *www.ggf.org*
2. For information on the *Open Grid Services Architecture (OGSA)*, see *www.ggf.org/ogsa-wg/*
3. For information on the *Open Grid Services Infrastructure (OGSI) working group and specification*, see *www.gridforum.org/ogsi-wg*
4. For information on how to download the *Globus Toolkit*, see *www.globus.org/ogsa/*
5. For information on *applications development* for Grid Computing solutions, see *www-106.ibm.com/developerworks/library/gr-grid1/index.html*
6. For more information on Grid Computing technologies, see the white paper titled *"The Anatomy of the Grid"* at *www.globus.org/research/papers/anatomy.pdf*
7. For additional research information related to Grid Computing, see *www.globus.org/research/papers/ogsa.pdf*
8. For information about the *Web Service Description Language (WSDL)* versions 1.1 and 1.2, see *www.w3.org/2002/ws/desc*

9. For information related to *the Simple Object Access Protocol (SOAP)* versions 1.1 and 1.2, see *www.w3.org/2000/xp/Group*
10. For information related to the *eXtensible Markup Language* (*XML)* and respective language specifications, see *www.w3.org/XML*
11. For information related to *Web Services-Interoperability (WS-I)*, see *www.ws-i.org*
12. For information related to the *World Wide Web Consortium (WC3)*, a primary standards body, see *www.w3.org*

Further Reading on Grid Computing Topics

This section contains additional references for further reading on subjects related to Grid Computing.

13. Abramson, D., Sosic, R., Giddy, J., and Hall, B. "Nimrod: A Tool for Performing Parameterized Simulations Using Distributed Workstations," In Proc. 4th IEEE Symp. On High Performance Distributed Computing, 1995.
14. Aiken, R., Carey, M., Carpenter, B., Foster, I., Lynch, C., Mambretti, J., Moore, R., Strasnner, J., and Teitelbaum, B. "Network Policy and Services: A Report of a Workshop on Middleware," IETF, RFC 2768, 2000 (*www.ietf.org/rfc/rfc2768.txt*).
15. Allcock, B., Bester, J., Bresnahan, J., Chervenak, A.L., Foster, I., Kesselman, C., Meder, S., Nefedova, V., Quesnel, D., and Tuecke, S. "Secure, Efficient Data Transport and Replica Management for High-Performance Data-Intensive Computing," In Mass Storage Conference, 2001.
16. Armstrong, R., Gannon, D., Geist, A., Keahey, K., Kohn, S., McInnes, L., and Parker, S. "Toward a Common Component Architecture for High Performance Scientific Computing," In Proc. 8th IEEE Symp. on High Performance Distributed Computing, 1999.
17. Baker, F. "Requirements for IP Version 4 Routers," IETF, RFC 1812, 1995 (*www.ietf.org/rfc/rfc1812.txt*).
18. Barry, J., Aparicio, M., Durniak, T., Herman, P., Karuturi, J.,Woods, C., Gilman, C., Ramnath, R., and Lam, H. "NIIIP-SMART: An Investigation of Distributed Object Approaches to Support MES Development and Deployment in a Virtual Enterprise," In 2nd Intl Enterprise Distributed Computing Workshop, IEEE Press, 1998.
19. Baru, C., Moore, R., Rajasekar, A., and Wan, M. "The SDSC Storage Resource Broker," In Proc. CASCON '98 Conference, 1998.

20. Beiriger, J., Johnson, W., Bivens, H., Humphreys, S., and Rhea, R. "Constructing the ASCI Grid," In Proc. 9th IEEE Symposium on High Performance Distributed Computing, IEEE Press, 2000.
21. Benger, W., Foster, I., Novotny, J., Seidel, E., Shalf, J., Smith, W., and Walker, P. "Numerical Relativity in a Distributed Environment," In Proc. 9th SIAM Conference on Parallel Processing for Scientific Computing, 1999.
22. Berman, F., Wolski, R., Figueira, S., Schopf, J., and Shao, G. "Application-Level Scheduling on Distributed Heterogeneous Networks," In Proc. Supercomputing '96, 1996.
23. Beynon, M., Ferreira, R., Kurc, T., Sussman, A., and Saltz, J. "DataCutter: Middleware for Filtering Very Large Scientific Datasets on Archival Storage Systems," In Proc. 8th Goddard Conference on Mass Storage Systems and Technologies/17th IEEE Symposium on Mass Storage Systems, 2000.
24. Bolcer, G.A. and Kaiser, G. "SWAP: Leveraging the Web To Manage Workflow," *IEEE Internet Computing*, 1999.
25. Brunett, S., Czajkowski, K., Fitzgerald, S., Foster, I., Johnson, A., Kesselman, C., Leigh, J., and Tuecke, S. "Application Experiences with the Globus Toolkit," In Proc. 7th IEEE Symp. on High Performance Distributed Computing, IEEE Press, 1998.
26. Butler, R., Engert, D., Foster, I., Kesselman, C., Tuecke, S., Volmer, J., and Welch, V. "Design and Deployment of a National-Scale Authentication Infrastructure," *IEEE Computer*, 33(12), 2000.
27. Casanova, H. and Dongarra, J. "NetSolve: A Network Server for Solving Computational Science Problems," *International Journal of Supercomputer Applications and High Performance Computing*, 11(3), 1997.
28. Casanova, H., Obertelli, G., Berman, F. and Wolski, R. "The AppLeS Parameter Sweep Template: User-Level Middleware for the Grid," In Proc. SC 2000, 2000.
29. Chervenak, A., Foster, I., Kesselman, C., Salisbury, C., and Tuecke, S. "The Data Grid: Towards an Architecture for the Distributed Management and Analysis of Large Scientific Data Sets," *J. Network and Computer Applications*, 2001.
30. Childers, L., Disz, T., Olson, R., Papka, M.E., Stevens, R., and Udeshi, T. "Access Grid: Immersive Group-to-Group Collaborative Visualization," In Proc. 4th International Immersive Projection Technology Workshop, 2000.

31. Clarke, I., Sandberg, O., Wiley, B., and Hong, T.W. "Freenet: A Distributed Anonymous Information Storage and Retrieval System," In ICSI Workshop on Design Issues in Anonymity and Unobservability, 1999.

32. Czajkowski, K., Foster, I., Karonis, N., Kesselman, C., Martin, S., Smith, W., and Tuecke, S. "A Resource Management Architecture for Metacomputing Systems," In The 4th Workshop on Job Scheduling Strategies for Parallel Processing, 1998.

33. Czajkowski, K., Foster, I., and Kesselman, C. "Co-allocation Services for Computational Grids," In Proc. 8th IEEE Symposium on High Performance Distributed Computing, IEEE Press, 1999.

34. Dierks, T. and Allen, C. "The TLS Protocol Version 1.0," IETF, RFC 2246, 1999 (*www.ietf.org/rfc/rfc2246.txt*).

35. Dinda, P. and O'Hallaron, D. "An Evaluation of Linear Models for Host Load Prediction," In Proc. 8th IEEE Symposium on High-Performance Distributed Computing, IEEE Press, 1999.

36. Foster, I. "Internet Computing and the Emerging Grid," *Nature Web Matters*, 2000 (*www.nature.com/nature/webmatters/grid/grid.html*).

37. Foster, I. and Karonis, N. "A Grid-Enabled MPI: Message Passing in Heterogeneous Distributed Computing Systems," In Proc. SC '98, 1998.

38. Foster, I. and Kesselman, C. "The Globus Project: A Status Report," In Proc. Heterogeneous Computing Workshop, IEEE Press, 1998.

39. Foster, I. and Kesselman, C. (eds.). *The Grid: Blueprint for a New Computing Infrastructure*, Morgan Kaufmann, 1999.

40. Foster, I., Kesselman, C., Tsudik, G., and Tuecke, S. "A Security Architecture for Computational Grids," In ACM Conference on Computers and Security, 1998.

41. Foster, I., Roy, A., and Sander, V. "A Quality of Service Architecture that Combines Resource Reservation and Application Adaptation," In Proc. 8th International Workshop on Quality of Service, 2000.

42. Frey, J., Foster, I., Livny, M., Tannenbaum, T., and Tuecke, S. "Condor-G: A Computation Management Agent for Multi-Institutional Grids," University of Wisconsin-Madison, 2001.

43. Gabriel, E., Resch, M., Beisel, T., and Keller, R. "Distributed Computing in a Heterogeneous Computing Environment," In Proc. EuroPVM/MPI '98, 1998.

44. Gasser, M. and McDermott, E. "An Architecture for Practical Delegation in a Distributed System," In Proc. 1990 IEEE Symposium on Research in Security and Privacy, IEEE Press, 1990.

45. Goux, J.-P., Kulkarni, S., Linderoth, J., and Yoder, M. "An Enabling Framework for Master-Worker Applications on the Computational Grid," In Proc. 9th IEEE Symp. On High Performance Distributed Computing, IEEE Press, 2000.

46. Grimshaw, A. and Wulf, W. "Legion—A View from 50,000 Feet," In Proc. 5th IEEE Symposium on High Performance Distributed Computing," IEEE Press, 1996.

47. Gropp, W., Lusk, E., and Skjellum, A. *Using MPI: Portable Parallel Programming with the Message Passing Interface*, Cambridge, MA: MIT Press, 1994.

48. Hoschek, W., Jaen-Martinez, J., Samar, A., Stockinger, H., and Stockinger, K. "Data Management in an International Data Grid Project," In Proc. 1st IEEE/ACM International Workshop on Grid Computing, Springer Verlag Press, 2000.

49. Johnston, W.E., Gannon, D., and Nitzberg, B. "Grids as Production Computing Environments: The Engineering Aspects of NASA's Information Power Grid," In Proc. 8th IEEE Symposium on High Performance Distributed Computing, IEEE Press, 1999.

50. Leigh, J., Johnson, A., and DeFanti, T.A. "CAVERN: A Distributed Architecture for Supporting Scalable Persistence and Interoperability in Collaborative Virtual Environments," *Virtual Reality: Research, Development and Applications*, 2(2), 1997.

51. Linn, J. "Generic Security Service Application Program Interface Version 2, Update 1," IETF, RFC 2743, 2000 (*www.ietf.org/rfc/rfc2743*).

52. Litzkow, M., Livny, M., and Mutka, M. "Condor—A Hunter of Idle Workstations," In Proc. 8th Intl Conf. on Distributed Computing Systems, 1988.

53. Lopez, I., Follen, G., Gutierrez, R., Foster, I., Ginsburg, B., Larsson, O., Martin, S., and Tuecke, S. "NPSS on NASA's IPG: Using CORBA and Globus to Coordinate Multidisciplinary Aeroscience Applications," In Proc. NASA HPCC/CAS Workshop, NASA Ames Research Center, 2000.

54. Lowekamp, B., Miller, N., Sutherland, D., Gross, T., Steenkiste, P., and Subhlok, J. "A Resource Query Interface for Network-Aware Applications," In Proc. 7th IEEE Symposium on High-Performance Distributed Computing, IEEE Press, 1998.

55. Papakhian, M. "Comparing Job-Management Systems: The User's Perspective," *IEEE Computationial Science & Engineering*, April–June 1998 (see also *http://pbs.mrj.com*).
56. Steiner, J., Neuman, B.C., and Schiller, J. "Kerberos: An Authentication System for Open Network Systems," In Proc. Usenix Conference, 1988.
57. Stevens, R., Woodward, P., DeFanti, T., and Catlett, C. "From the I-WAY to the National Technology Grid," *Communications of the ACM*, 40(11), 1997.
58. Thompson, M., Johnston, W., Mudumbai, S., Hoo, G., Jackson, K., and Essiari, A. "Certificate-based Access Control for Widely Distributed Resources," In Proc. 8th Usenix Security Symposium, 1999.
59. Tierney, B., Johnston, W., Lee, J., and Hoo, G. "Performance Analysis in High-Speed Wide Area IP over ATM Networks: Top-to-Bottom End-to-End Monitoring," IEEE Networking, 1996.
60. Vahdat, A., Belani, E., Eastham, P., Yoshikawa, C., Anderson, T., Culler, D., and Dahlin, M. "WebOS: Operating System Services For Wide Area Applications," In 7th Symposium on High Performance Distributed Computing, July 1998.
61. Wolski, R. "Forecasting Network Performance to Support Dynamic Scheduling Using the Network Weather Service," In Proc. 6th IEEE Symp. on High Performance Distributed Computing, Portland, OR, 1997.

Part 3
Service Providers and Customer Profiles

This part of the book will continue to explore various aspects of the operational environments for on demand business computing by describing industry examples. The on demand Operating Environment will be described by a variety of concepts dealing with service provider environments.

This chapter contains robust discussion on an ecosystem that is capable of providing advanced on demand business services for consumers and creators of advanced Web services and content—provided via a carrier environment (such as a telecommunications firm) or business enterprise.

We will explore many industry issues driving on demand business transformations. The issues are not specific to any particular industry; rather, these studies indicate on demand business transformation patterns across many industries. Customer profiles will be presented, with telling details of why these businesses decided to begin the on demand business journey, along with aspects of how they started the journey, and the results of the implementations they have embarked on.

We will conclude the book with proof that it is simply a matter of time before every business embarks on the on demand business journey in one form or another. So, sit back and enjoy the final chapters.

It is a fact that—worldwide—IBM, from a number of different perspectives, utilizing a number of different technologies and strategic approaches, has been implementing and continues to implement a wide variety of very efficient on demand business environments. It is paramount that every business enterprise considers the factors presented throughout this book: There is a reward.

7 The On Demand Business Service Provider Ecosystem

This chapter introduces the IBM on demand business ecosystem from a service provider viewpoint. In the second chapter of this book, we explored the on demand Operating Environment, which is reflected in this chapter as it relates to service providers. Service providers can be thought of as business enterprises; and in many cases, telecommunications companies (Telcos) are service providers due to the advanced services the telecommunications (telecom) firms provide. As stated by Alex Cabanes, IBM Software Group's Industry Marketing Manager, *"In today's rapidly changing marketplace, the ability of a service provider to sense and respond to ever-changing customer demands will determine who survives, and who does not. This requires a technology fabric that can adapt on demand business to these ever-changing requirements."* This chapter addresses service providers and this fabric.

As we have discussed throughout this book, the IBM Corporation defines on demand business as: *An enterprise whose business processes—integrated end-to-end across the company and with key partners, suppliers and customers—can respond with agility and speed to changing customer demands, market opportunity, or external threat.* We have also explored several strategic perspectives of on demand Operating Environment, namely Autonomic Computing and Grid Computing.

While originally the term "e-business on demand" was used in IBM to define activities in "sourcing" initiatives, today we are using a much broader definition of the term to encompass what it takes to be an on demand business. For a business to be capable of providing on demand business services, it must respond in real time to its markets and customers. An on demand business must be variable in its abilities to provide services, focused on its core differentiating capabilities, and it must be resilient to both internal and external interruption. It is also important that the enterprise have an on demand Operating Environment. An on demand Operating Environment, as discussed in previous chapters, is *an open standards-based, heterogeneous world, integrated and freely enabled with self-managing capabilities*. As stated by Gordon Kerr, IBM Corporation Distinguished Engineer and IT Architect in the Global Telecommunications Sector, it is important to note that "when the on demand business ecosystem is carefully scrutinized, the companies, organizations, and end-users can successfully interoperate, worldwide, when open standards are fully sanctioned and embraced." The IBM Corporation is a leading global organization helping to define these open standards for on demand business.

This term "*ecosystem*" will be used throughout this chapter to describe a three-part service-enabling environment. This three-part ecosystem, as shown in the following illustration (Figure 7.1), includes content providers, Telco carriers and enterprises, and consumers. This ecosystem delivers services to global consumers of all industries, with a myriad of pervasive devices.

The context of the term "ecosystem" includes many types of end-users, plus a variety of consumer telephonic and computing devices. It includes several complex strategies involving flexible hosting and delivery of on demand business autonomic processes, information anytime and anywhere, and advanced Web services. These on demand business ecosystem services deliver information to an environment where end-user devices are not dependent on proprietary technologies; rather, they are receptive to many standards of technological integration techniques, devices, platforms, and business operations. These services will be demanded by the individual; the persistent services will be those that meet new human factors that are resilient and focused.

Figure 7.1 shows this ecosystem, which many global service providers are now addressing using innovative and industrial strategies. Note that the ecosystem has essentially three parts: On the left, there are the end-users (or consumers) and the many devices they utilize; on the right is the origination point of a wide variety of content, applications, and services targeting specific markets and being provided to the end-users; and in the middle is the enterprise (or, in some cases, the telecom carrier) serving the end-users as the information services "carrier" of sorts. The importance of this ecosystem model is that it is the single monolithic model that enables on demand business services, with a plurality of autonomic operations, to be delivered to global markets. This ecosystem is valid in virtually every country throughout the world, regardless of whether one is involved as a telecom service provider or an enterprise service provider.

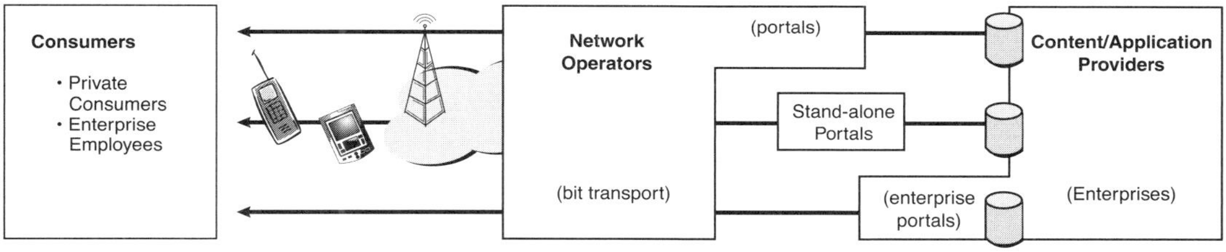

FIGURE 7.1 The topology of this Telco on demand business ecosystem is one that many global Telcos are actively pursuing.

The service provider on demand business ecosystem delivers a diverse set of business capabilities to virtually any market. This is achieved by an open standards (or open source) approach to fortifying and sustaining the on demand Operating Environment. In Figure 7.1, the content providers on the far right deliver a wide variety of content and advanced services to the end-users on the far left, while the telecom carrier in the center (in this example) can provide a multitude of on demand business capabilities and services to enable this vast ecosystem.

The general plight of those delivering advanced Web services into this global ecosystem is consistent around the world. The overall objective is a two-fold endeavor, enabling content creators to reach out to more consumers utilizing a myriad of methods to enable this on demand business ecosystem. Part one of the objective is to enable service providers to supply autonomic solution capabilities to their consumers, both private and public. These consumers are business enterprises (including government consumers) and private individuals. Part two of the objective is a by-product of sorts, yielded from part one. This manifests in the sense that once the service provider or enterprise is capable of providing on demand business solutions, it too needs to enable its partners to become on demand businesses, thus fortifying the overall ecosystem. This strategy is being realized by a pattern of trends that is prevalent across almost all industry sectors. One of the most prominently noted industries in this space is the telecom service provider as an industry-leading carrier of information and services delivery capabilities.

A Telco is often referred to as a "service provider." This is in part due to the fact that the telecom industry does serve as the primary carrier(s) of network data transport (enhanced by content, Managed Services, Business Communications, and wholesaling models), thereby placing the Telcos in the center of this vast ecosystem, touching virtually each and every one of us in our daily lives. This, however, is changing as key industry "services" leaders are now in the midst of providing a multitude of competing enterprise services, while at the same time utilizing the carrier transport mediums for the transmission of networked data. This is sometimes accomplished through the formation of key strategic "service" alliances between the Telcos and some of the industry-valued partners they support. They may oftentimes work together to provide advanced services. As Component Based Models more fully represent industries, the role of these alliance partners will become more critical.

In the next discussion, we will explore interesting strategy perspectives related to providing New-Generation Operations Software and Systems (NGOSS). These strategic perspectives are key to understanding the critical challenges that service providers must overcome, while at the same time maintaining optimal costs and efficiencies.

New-Generation Operations Software and Systems (NGOSS)

NGOSS are very important in the service provider on demand business ecosystem, the worldwide telecom industry, and across many other vertical industries. The objectives of NGOSS include the simplification and transformation of many types of service providers (not simply Telcos) to establish a more automated, more cost-effective approach to integrating operational support systems.

The TeleManagement Forum[1] (TMF) is the founding organization of the innovative NGOSS approach. The TMF is providing hard-hitting leadership

1. For more information on the TMF, refer to its Web site at *www.tmforum.org*

initiatives worldwide, while at the same time providing world-class expertise in this complicated area of NGOSS. This section discusses the many successful steps, concepts, and objectives the TMF has taken toward NGOSS initiatives, and some of the many worldwide partnerships it has established with respect to its many accomplishments.

THE TMF IS THE AUTHOR OF THE NGOSS STRATEGY

The objectives of NGOSS include the simplification and transformation of many types of service provider operational support systems/business support systems (OSS/BSS) environments, to establish a more automated, more cost-effective approach to integrating operational support systems. The IBM Corporation is in partnership with the TMF to support the NGOSS mission, while leveraging specific deliverables from the TMF.

This important work is important for the simplification of on demand business solution integration elements related to OSS/BSS initiatives.

The TMF is involved in many strategic and tactical corporate endeavors addressing the complexities of operational support systems (OSS) integration. This complicated systems environment is absolutely key to the service provider ecosystem.

For purposes of this discussion, it is important to understand that the TMF works diligently with many telecom firms and other service providers worldwide (including IBM) to create standardized processes, architectures, and a common language. These cross-industry integration and simplification initiatives then become key enablers in reducing the costs incurred by the integration of OSS/BSS.

Members of the TMF include participants from many vertical industries, but a strong presence is made by many of the largest Telcos in the world. It is in the interests of the Telco and cross-industry enterprise, as well as in the consumers' interests, that the industrial sectors work together to reduce today's excessive costs of integrating and managing complicated OSS solutions. This simplified approach is at the core of the NGOSS strategy.

The NGOSS initiative (among many other benefits) includes a comprehensive, integrated reference framework for developing, procuring, and deploying OSS/BSSS and software. OSS/BSS are key dependencies in the on demand business ecosystem for the service providers. NGOSS today are available[2] as a toolkit of industry-standard specifications and guidelines that cover key business and technical areas. These areas include:

- The business process automation elements delivered in the enhanced Telecom Operations Map (eTOM)

2. For more information on the TMF's NGOSS, refer to *www.tmforum.org*

- The common entities and their standard definitions and relationships delivered in the Shared Information and Data (SID) model
- The systems analysis and design delivered in the contract interface and Technology-Neutral Architecture (TNA)
- The solution design and integration delivered in the components and Technology-Specific Architecture (TSA)
- The conformance testing delivered in the NGOSS compliance tests
- The procurement and implementation guidelines delivered in the ROI model, the RFI (Request for Information) template, and the implementation documents

The NGOSS approach enables all players in the OSS/BSS supply chain to use the elements appropriate for their business, while maintaining the confidence that they will all fit together with a reduced level of skills required and funding.

NGOSS-based solutions are strategic and in several ways complement the on demand business strategies for service providers. This is based on the fact that NGOSS utilizes mainstream IT concepts and technologies to deliver a more productive development environment, and thus, a more efficient management infrastructure. NGOSS are prescriptive for only those few "cardinal points" where interoperability is key, while enabling easy customization across a wide range of functionality. This allows NGOSS-based systems to be tailored to provide a competitive advantage, while also working with traditional legacy systems.

The value proposition toward on demand business here is that NGOSS must deliver tangible values to the service provider supply chain through enabling business process automation and nimble operations, reducing costs, and improving customer service. In other words, streamlining processes and making business functions autonomic are the key underpinnings.

Major Telcos generally have thousands of discrete business processes they use to run and sustain their operations. To automate a subset of these processes, they have at a minimum many hundreds and sometimes several thousand discrete OSS/BSS software applications. Adding further complexity, most processes require integration of multiple applications to achieve end-to-end automation. As a result, the scope of the process automation problem gets large and expensive at a rapidly compounding rate.

While service providers can identify the processes they have today, they would like to automate the limiting factor in making these process changes. This limiting factor is the complexity of changing the software systems fast enough to meet evergreen production demands. And because of this multi-faceted complexity, changes in the ecosystem are often not cost-effective enough to deliver an attractive ROI. As a result, the effort to continuously automate existing processes and further simplify processes with additional automation is challenging and only occurs on a limited basis. Solving this problem is the mandatory foundation for NGOSS.

Clearly, service providers understand that if they automate processes, they can reduce operational costs and improve service to end-customers, becoming "lean" operators. They know they can become nimble in meeting the needs of customers and deliver easily and affordably the plethora of new technology and service combinations that are becoming affordable to the mass markets. However, exactly how to get from today's status quo to tomorrow's well-integrated, easily changeable, market-driven business and operations systems is certainly the challenge most of the world's service providers face. The cycle time to transition is compressing due to the increased density of ecosystem changes.

The Problem NGOSS Addresses Is Complex

NGOSS forms an industrial effort to identify and automate existing core processes, predicated on the fact that to further simplify these core processes with additional automation is too costly and only occurs on a limited basis. Thus, solving this problem of vast complexity is the foundation for NGOSS, which is a primary mission of the TMF.

The innovative solutions evolving from NGOSS involve an industry-wide participation from service providers and business enterprises worldwide.

To offer the industry a blueprint for a less expensive, more responsive, and significantly more flexible way of performing business, a group of operators, software suppliers, consultants, and systems integrators has been working on a common framework for delivering notably more efficient operations for service providers worldwide. The work has been coordinated by the TMF and developed by TMF member companies working on multiple, collaborative teams. The result of several hundred person-years of development work is impressive; this is the NGOSS framework.

Automating processes requires a multi-step approach, from understanding existing processes to designing how systems will integrate. Typical activities include:

- Defining and engineering/re-engineering business processes
- Defining systems to implement processes
- Defining data in a common information model
- Defining integration interfaces
- Defining integration architecture

The elements of NGOSS align directly with the steps in this process automation approach. As a result, NGOSS gives service providers the tools they need to undertake automation projects with confidence.

Business Benefits As with Autonomic Computing, process transformation and automation is the cornerstone of NGOSS. Many of the business benefits revolve around and intersect with the direct and indirect stages in automating telecom and other service provider operations. However, global service providers are not the only beneficiaries of the standard language and specifications that NGOSS will define. Large enterprises and other businesses will realize this benefit delivered through the adoption of NGOSS approaches.

NGOSS offer service providers tangible business benefits that positively impact the bottom line. These benefits include:

- Having a well-defined, long-term, strategic direction for the integration of business processes and OSS/BSS implementation reduces investment risk. When new systems and services are purchased, if they fit in with a well-defined strategy and detailed set of requirements, their longevity is more assured than in an environment with ill-defined definitions.
- Being first to market is important in all competitive environments. Being first to market and not spending a lot of money to get there is a well-known recipe for success. Being nimble and exact is key to preparing for the combined broadband and wireless services onslaught that is approaching.
- Moving to an environment where process definitions, interfaces, and architecture are all standard allows for true competitive bidding environments.

Sense-and-respond organizations are, today, learning how to leverage these benefits.

The NGOSS initiative delivers measurable improvements in development and software integration environments. These improvements include:

- The fact that with NGOSS, large portions of process language, requirements, data models, interfaces, and tests are already defined, significantly reducing development costs.
- Standard building blocks, software modules, and even standalone products can be built once and sold many times, increasing ROI with every sale.

- The integration cycles for software with standard interfaces are significantly shorter, reducing the costs of bringing a new software system into an existing environment. In addition, integration using NGOSS interfaces becomes a repeatable process, therefore saving time and money on each project while improving success rates.
- The clear definition of "use cases" and requirements becomes easier across service provider/supplier and supplier/supplier partnership relationships. This is by virtue of the fact that when a common language emerges, as provided by the TMF eTOM and the SID, a new and simpler language can be implemented.

By utilizing NGOSS, ongoing savings are realized across operational environments, specific to the daily churn of tasks to keep networks running and customers satisfied. These savings include:

- The realization of automation, which in turn enables lower operational expenditures. The NGOSS approach, tackling the tasks of introducing additional automation to operational environments, brings with it a clear blueprint to follow and guidelines to step through the changes. The task may still be large, but much of the work has already been accomplished within the NGOSS elements.
- The realization that once NGOSS automated systems are in place, making changes in a well-designed, well-understood environment is straightforward. Therefore, reacting to a need to change a service offering, a billing option, or a QoS requirement (etc.) becomes an easy-to-follow process rather than struggling with significant changes that require many resources and weeks of testing.

NGOSS as a Framework The NGOSS framework is a sound technical solution developed by industry leaders with hundreds of combined years of Telco, OSS/BSS, and enterprise experience. Recognizing the need to create a common integration environment for software systems, the TMF member companies have contributed the time of their senior architectural and engineering resources to make the NGOSS framework a success. Additional benefits noted by adherence to NGOSS principles are:

- The NGOSS framework is real, documented, and ready for consideration by any interested party. The NGOSS components all consist of detailed definitions, and are ready for implementation.
- The NGOSS principles and many detailed NGOSS documents draw from existing industry standards and recommendations wherever pos-

sible. The NGOSS leadership team has engaged the best available resources to create the best possible solutions.

- The NGOSS framework is defined in such a way that it provides a coherent, long-term direction and allows for specifications to be made available to developers and implementers of complex (or straightforward) OSS/BSS systems. Whether a service provider is targeting strategies for the long term or a software supplier is establishing a product roadmap, the NGOSS framework provides structure and details to work toward a common goal.
- Utilizing the pre-defined elements of NGOSS allows development efforts throughout the telecom supply chain and other enterprises to focus on solving value-added problems, not defining processes, data models, and architectures.

NGOSS Components

The elements of NGOSS intersect to provide an end-to-end system for OSS/BSS development, integration, and operations. The elements of NGOSS may also be used as an end-to-end system to undertake large-scale development and integration projects, or NGOSS may be utilized separately to solve very specific problems. The NGOSS components are as follows:

1. *Enhanced Telecommunications Operations Map (eTOM)*—The eTOM provides the map and common language of business processes that are used in telecom service provider operations and other similar enterprises. In addition, process flows are provided for an ever-expanding list of key processes. The eTOM can be used to inventory existing processes at a service provider, act as a framework for defining the scope of a software-based solution, or simply to enable better lines of communication between a service provider and its systems integrators.
2. *Shared Information and Data Model (SID)*—The SID provides a "common language" for software providers and integrators to use in describing management information, which will in turn allow for simpler and more effective integration across OSS/BSS software applications provided by multiple vendors. The SID provides the concepts and principles needed to define a shared information model, the elements or entities of the model, the business-oriented Unified Modeling Language (UML) class models, as well as design-oriented UML class models and sequence diagrams to provide a system view of the information and data models.

3. *Technology-Neutral Architecture (TNA) and contract interface*—These two components comprise the core of the NGOSS integration framework. To successfully integrate applications provided by multiple software vendors, the connectivity paths and intersections of the system must be common. The TNA defines architectural principles to guide OSS developers to create OSS components that operate successfully in a distributed environment; the contract interface defines the APIs for interfacing those elements to each other across the architecture. This architecture is specifically called "Technology-Neutral" as it does not define how to implement the architecture, rather it outlines the principles that must be applied for a particular technology-specific architecture to become NGOSS-compliant.
4. *NGOSS compliance*—To improve the probability that OSS components will truly integrate with each other, NGOSS provides a suite of tests for compliance to the eTOM, SID, TNA, and contract interface components. The NGOSS compliance approach can be achieved by any or all of these components either singly or in combination with other components.

NGOSS Audience

Across the telecom supply chain, OSS/BSS has solved many problems, but at the same time, it presents many challenges looking ahead into the future. Each of the major players in the supply chain is seeking a new approach to OSS, and in each case, there are compelling reasons to achieve this state of operations.

Service Providers

In the financially sensitive markets of today, service providers need cost-effective OSS/BSS implementations. These OSS/BSS systems must automate complex business processes to solve difficult operational issues in the short term, while at the same time demonstrating rapid returns for the investment. In addition, service providers require a long-term strategy for their IT systems. The OSS systems in many carriers today were constructed without a long-term view, and are now proving difficult to expand to handle the challenge of managing more complex networks, complicated services integration needs, and automated business processes.

OSS Software Vendors

The OSS vendor marketplace has blossomed to many hundreds of companies. This expansion means that in each market sub-segment, numerous companies are competing for the same business within the service provider communities. This competition, along with price pressure toward the service providers to reduce capital expenditures and operational costs, is driving software vendors to likewise reduce their development costs to be profitable. In addition, the OSS marketplace has become a conglomeration of companies that solve niche problems for service providers. As a result, software vendors must be prepared to fit into the OSS puzzle presented to them by each service provider customer they engage.

Systems Integrators While custom integration projects are typically the boon of telecom systems integrators, mounting pressures from service providers to cut costs put the systems integrators into a position where they must make their projects more predictable and repeatable—and less customized—to retain their margins. Systems integrators are looking to reuse elements across projects while utilizing less staff to accomplish their end results. In addition, with the large number of software suppliers across the industry, and service providers using an ever-increasing variation of software components, systems integrators are now forced to continually learn how to integrate new elements into service providers' IT environments in a simple fashion.

Network Equipment Providers Equipment vendors are extremely influential in the world of OSS. Often, service providers rely on the management systems from their preferred vendors to do much more than manage their own network elements. And oftentimes, equipment vendors see the value in becoming a one-stop-shopping experience for their customers, offering diversity and a broad range of OSS solutions. In addition to this, network equipment vendors are often faced with bidding situations, where they are placing their equipment into pre-existing multivendor environments with established OSS solutions. In these cases, providing NGOSS-based solutions enables these vendors to integrate their hardware and systems with third-party NGOSS-enabled systems faster and more easily, deriving substantial business benefits.

NGOSS Applications

NGOSS may be used as an integrated system end-to-end or as components to solve particular problems. NGOSS can be applied throughout telecom organizations by operations staff, software developers, and systems integrators. Example applications of NGOSS include:

- *Business process redesign*—Telecom service providers utilize the eTOM to analyze their existing business processes, identify redundancy or gaps in their current strategies, and re-engineer business processes to correct deficiencies while introducing automation into the environment.

- *Development of an OSS migration strategy*—The NGOSS provide clear direction for migration of legacy OSS systems and solutions to a future-proof, maintainable, flexible OSS environment. Service providers may utilize the elements of NGOSS end-to-end to define common infrastructure for the future.
- *OSS solution design and specification*—The NGOSS approach defines detailed data models, interfaces, and architectural specifications that service providers can utilize to specify and procure future OSS solutions.
- *Software application development*—At the core of NGOSS, the business process maps, information models, and integration framework are all designed to guide software engineering organizations through the process of creating NGOSS-compliant OSS components.
- *Systems integration*—When faced with integration challenges, the NGOSS' well-defined language, its interfaces, and its architectures provide the systems integrator with a clear direction for repeatable and cost-effective integration of multi-vendor, disparate systems.

In summary, the NGOSS strategic approach is being widely adopted throughout the global telecom industry by service providers, software vendors, and systems integrators. The TMF is, therefore, a world-class organization with major contributions toward improving cost efficiencies for the service provider's on demand business strategies.

The Need for Persistence and Advanced Forms of Communications by Service Providers

The persistent need for an advanced means of end-user content communications and advanced forms of content delivery is both a benefit and a detriment to the telecom industry. It is a benefit in the sense that on demand business transformations are opening up new and innovative services markets for the Telcos and other business enterprises around the world. It is a detriment in the sense that the demands placed on the networking services sustaining these environments are introducing challenging technological situations, which oftentimes require very forward-thinking technology and networking strategies to maintain customer satisfaction and optimal market presence. Traditional, non-Telco business enterprises acting as service providers are also noting similar benefits and technological challenges introduced by the on demand business ecosystem. Figure 7.2

shows some of the patterns and trends that are driving this transformation of businesses and services.

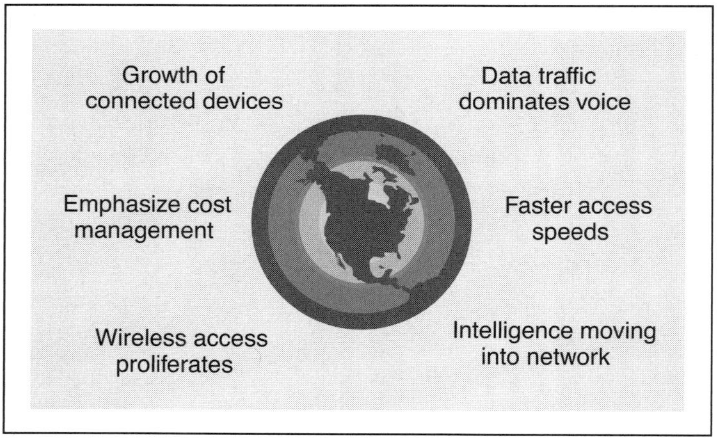

FIGURE 7.2 The global trends underlying on demand business have evolved from end-user personal and professional network services demands.

The telecom trends driving the on demand business ecosystem clearly underpins the environments for global next-generation networks, advanced next-generation Web services, plus new generations of OSS/BSS. These focus areas accommodate and continue to rapidly stimulate lateral growth within the global on demand business ecosystem, allowing it to proliferate and enable a wide variety of innovative on demand business services.

The on demand business ecosystem has been realized through a variety of end-user needs and network services behaviors. As mentioned earlier, the illustration depicted above implies that in the "center of the universe" is a telecom firm or business enterprise of some sort: This is the carrier or service provider delivering information across the networks in the ecosystem. Another rendition of this same illustration could yield a service provider that is a non-Telco business enterprise as the center of the universe—both image scenarios would be valid. This center of the universe serves as the ecosystem's lifeline and is the transport medium—often referred to as a pipe—delivering next-generation on demand business services. These service delivery enterprises and/or Telcos enable the transport of key end-user information across the vast landscape of on demand business networks in the ecosystem. This is a daunting and massively intensive challenge, unless approached from a strategic and highly automated fashion.

Ecosystem Dynamics and Business Drivers

The on demand business ecosystem is the result of technology evolutions that have delivered us into a world of advanced computing strategies and sophisticated service provider environments. Many of today's advanced computing strategies are related to a variety of topics surrounding pervasive computing solutions, telecom service providers with robust capabilities, Autonomic Computing business transformations, and complex forms of Grid Computing. Many technologies have also evolved to a state of "open standards" and technology foundations that support cross-industry interactions between global enterprises, enabling many of these advanced strategies while working as partners. It is this fascinating combination of strategy, technology, and business partnerships that has enabled the on demand business ecosystem.

The dynamics driving these evolutionary services solutions and combinations of next-generation strategies are the result of a more demanding global culture of end-users, finding more and more information needs being met by the Internet, while using an ever-widening selection of pervasive devices. The Internet has indeed served as an incredible change agent, forcing many global business enterprises to feverishly work to manage this evolutionary change toward a new ecosystem of on demand business services.

The business drivers are evolving toward complex business alliance models, allowing many enterprises and service providers to team together to bring to market value-added services in a very diverse global field of needs. Large enterprises team with other large enterprises to show value-adds to customer demands for "on demand" information. Small businesses team with large enterprises to leverage the advantages of the large enterprise customer base, while enhancing the small business' customer base by virtue of the alliance. These opportunities are wildly successful in many cases, and the innovative strategies in combination with a never-ending choice of robust technologies are driving the dynamics of transformation, business wealth, and a state of autonomic business operations to manage these demands.

Ecosystem Dynamics

The on demand business ecosystem has dynamics that have demonstrated some interesting characteristics as the ecosystem has evolved over the last few years. Keeping in mind that in 1995, no one could even perform a secured financial transaction across the Internet, now we are able to perform highly secured government business in the Internet, involving large sums of money, over wireless technologies. This is quite an achievement in technology and business operations in such a short period of time. Today, by virtue of this on demand business ecosystem, companies are able to mitigate risk while providing highly advanced applications to virtually every type of business consumer in the world. Today, multi-tenant business processes are provided by emerging service providers, delivering, for example, HR processes, sales process management, contract management, and other on demand business services.

In the past, large global service providers attempted to define what they called "killer applications." These killer applications were competitive Web services predicted to be highly utilized by a wide variety of consumers. Oftentimes, these applications were time-intensive to create and somewhat costly to maintain. The on demand business ecosystem affords creators the ability to quickly create and push out to their consumers advanced Web services, virtually eliminating risk from the large enterprise in determining what best killer application was the target, to provide to their consumer base the most robust service.

The content applications/services creators on the far right of this ecosystem (illustrated earlier in this chapter) are very competitive and they seem to identify with consumer groups indicating demands for certain kinds of services. These are oftentimes services that meet on demand business types of needs. The content creators then create and provide these services, by pushing these services through the transport mediums (i.e., the Telco or business enterprise) to the consumers that are using a multitude of devices, as shown in Figure 7.3.

This creator-to-consumer protocol virtually eliminates investment risks in the large enterprise or Telco creating what they believe to be the consumer "killer" applications, and potentially missing the actual consumer on demand needs. These smaller content creator companies are important because they target specific consumer markets. In this ecosystem, content creators are able to quickly create and deliver content for these advanced business services, while utilizing the tools and technologies of the larger on demand Operating Environment. The creators are able to quickly create and deliver these applications/services, due to the fact this ecosystem environment provides many innovative tools and delivery mechanisms for the delivery of this content. This effectively relieves the Telco or enterprise from the financial burden of becoming the creator of the advanced business services, yet leveraging the innovations provided in the market by the small-to-medium businesses. The Telco enterprise is now the service provider delivery environment.

This new on demand model strategy depends on the information and data transport carriers, or enterprise carriers, providing advanced technological environments, thus enabling content creator tools and environments to quickly develop and push to their own consumers those Web services that deliver "on demand" types of values. Thus, they mitigate the risks of the carriers or enterprises in spending vast amounts on content delivery strategies, instead providing the underpinnings of the advanced on demand Operating Environment for others to strategically address a wider variety of consumer informational needs. The Telcos can maintain on demand Operating Environment for customers, and deliver them.

Content providers are those enterprises, businesses, or individuals that develop Web services/applications and/or provide data or advanced forms of content for open markets to leverage by virtue of the Internet. One example of a content provider might be a music/audio/video company, able to deliver advanced media content through Internet-based business services. Commercial book resellers and retailers such as Amazon.com, Borders, and Barnes & Noble all have tremendous amounts of content with very innovative means of service delivery to their customers. Other forms of content providers are evident in advanced billing system solutions, such as those provided by companies like Convergys, CSG, Portal, and others.

Many types of business service content providers deliver very strategic elements of the on demand business ecosystem. These providers deliver services through very streamlined solution approaches, such that any solution interfacing to their services is transparent to any end-user demanding this type of service. However, there are a tremendous number of very efficient business enterprise operations an on demand business solution can leverage. Configuration management by companies like Cramer, or event correlation and root cause analysis by companies like System Management Arts (Smarts), plays a huge role in an on demand business ecosystem. This discussion is not limited, however, to these few business providers; this list is long and the partnerships of any on demand business will vary depending on the mission.

This discussion quickly enters an exploration of what is referred to as OSS/BSS. As an example of OSS in an on demand business, solution elements are required for network monitoring, trouble ticketing, event correlation, configuration management, and more. Likewise, BSSs refer to order fulfillment, supply chain, and so on. In many ways, the TMF's NGOSS strategy endorses the absolute importance of this type of strategic positioning, solutions design, and what some of the major challenges are today surrounding OSS/BSS.

To articulate the nature of the OSS/BSS strategy challenge for the on demand business ecosystem, consider the fact that today, to integrate a content creator or service provider element into a on demand business solution, tremendous resource efforts and large sums of money are often required in the area of system integration to enable the connection of the content or service provider element. In today's world, we can plug a credit card-sized network adapter card into a computer and instantly be connected to a network without any need for system integration, or travel in a wireless world from Internet carrier to Internet carrier. So why can we not simply plug a system service module (e.g., billing, configuration management, event correlation, network monitoring, etc.) into an on demand business solution while incurring very minimal resource and system integrations costs? This deals with NGOSS, the notion of industry standards[3] and exactly how the content and service provider companies are creating the OSS/BSS systems.

Service Provider Delivery Environment (SPDE)

This section describes the IBM Service Provider Delivery Environment (SPDE). SPDE is a key strategy and solution delivery ecosystem provided by the IBM telecom industry. SPDE has proven itself very successful as a leading-edge solution framework. Several key industry organizations in the IBM Corporation work with many strategic partners in global telecom solutions, providing elements of the SPDE framework and respective on demand business solutions. Autonomic Computing approaches, combined with other industry-specific IBM strategic services approaches, enable advanced on demand business solutions.

SPDE is an advanced multimedia, data, and content service delivery environment that embodies a multitude of products and services that are intended to best match the telecom customer's strategic service delivery strategies. SPDE is an open standards framework. It allows for global telecom corporations to provide for their customers' customers—creating an on demand Operating Environment for performing application delivery and mission-critical, fully integrated, day-to-day business operations.

3. The worldwide TMF addresses these challenges with industrial precision, and is focused on sustaining and expanding its global board of service providers, all with a common need. More on this strategic challenge and the TMF is found at the following Web site: *www.tmforum.org/browse.asp?catID=0*

While this section will include details about the ability of SPDE to enable the introduction of Internet and advanced multimedia content capabilities to mobile phones and other pervasive computing devices (PvCs), SPDE itself is not limited in this way. SPDE, coupled with the increasingly attractive costs of mobile phones and PvC, can make these handheld devices the preferred choice for accessing computer servers and global next-generation Web services.

There is a vast amount of business at stake in the service provider domain environments. For example, a leading telecom corporation in Japan expects to spend $10 billion (USD) to build the networks needed to support its new wireless initiatives. The reason is simple: increasing demand. Twenty million subscribers signed up in just the first year of this company's mobile Internet access service. The provisioning of content and advanced multimedia services will likely surpass the opportunity for basic services.

SPDE Highlights IBM is in a unique position to leverage its industry leadership, expertise, global reach, extensive offerings, and relationships with industry leaders to help service providers in this embryonic market find ways to bring new revenue-generating services to market faster and more cost-effectively. To answer these needs, IBM developed SPDE, an open standards-based framework designed to give mobile and fixed network operators and service providers the flexibility to introduce new voice, text, and fully integrated Internet services to their customers faster, easier, and at a lower cost.

With SPDE, new strategic services can be created and managed independently of the underlying wireless, wireline, or Internet-based communications network, and can be delivered to virtually any computing device that allows for the delivery of data. Based on extensive work with other technology and service providers, and customers at the IBM Network Innovation Labs, SPDE's integration capabilities have enabled easier introduction and modification of content, applications, and location-based services and their supporting processes, such as billing, customer relationship management (CRM), provisioning, assurance, and resource management. With SPDE, service providers can potentially reduce development time for new services from months or years to days or weeks, and quickly react to changing customer demand. SPDE also helps to create new revenue opportunities by opening up service creation and management to third-party application vendors.

The SPDE ecosystem encourages the fluid integration of new capabilities and processes with the rest of the existing IT infrastructure through an open hub and services distributorship (or brokerage) integration approach. SPDE provides this comprehensive environment for next-generation service delivery through multiple transport environments that do not require significant modification to existing infrastructure. SPDE is based on open telecom and IT standards, including telecom standards such as Parlay and Open Service Architecture (OSA), as well as key Internet standards important for Web services such as Universal Description, Discovery and Integration (UDDI), WSDL, and SOAP. SPDE is also designed to take advantage of new standards as they emerge. SPDE is ideally suited as a base for building service delivery platforms and applications proposed by the recently announced Open Mobile Alliance (OMA) initiative.

IBM SPDE Reference Architecture The SPDE reference architecture described in this section is based on a global solutions framework that can be instantiated with a combination of hardware and software products and robust middleware technologies (from IBM Corporation and/or third parties). The SPDE achievements as described in [Boisseau01] in this section provide full treatment of the IBM Telco industry strategy. This open framework can be integrated into any service provider's infrastructure without tremendous change to the installed base.

To enable an infrastructure solution for the next generation of service delivery, and for purposes of this discussion, a SPDE functional taxonomy was developed. This taxonomy is based on domains, services, components, and products:

- A *domain* is a grouping of related services.
- A *service* is a grouping of related components.
- A *product* is an instantiation of a service or component. In some cases, a product is a software package from an ISV or partner company.

Now, in general *domain* terms, let's further explore the SPDE ecosystem:

- The *User domain* includes anyone who accesses the User Services domain. These access points may include visitors (anonymous users), subscribers (consumers and/or employees), customer service representatives (CSRs), and external third parties (providers).
- The *Device domain* includes the means by which users access the User Services domain. These may include servers, laptops, wireline/wireless phones, PDAs, and PvCs.
- The *Network Delivery domain* includes the transportation mechanisms between the devices and the User Services domain. These may include mobile, broadband, or any other kind of delivery.

- The *Services Brokerage domain* contains the components that select, render, and aggregate requested content and applications. This domain resides between the Network Delivery and User Services domains, and to some degree, hides the complexity of different mobile and fixed devices and networks from other applications.
- The *Services Management domain* contains the components that fulfill, assure, and bill for user services. Such a capability corresponds to OSS/BSS.
- The *User Services domain* contains matters related to content and applications consumed by users.
- The *Financial Services domain* includes mechanisms for performing financial transactions. These transactions will normally be performed by banks, but may also be performed by service providers. In this case, the financial transactions will go back to the billing services element in the Services Management domain.

SPDE is a Reference Architecture.

The SPDE ecosystem's architecture is robust. As much as SPDE deploys architecture point solutions in a variety of products, it is also a reference architecture, allowing diversity in solution component selections, thus permitting the closest and most cost-effective architectures for any solution.

The IBM SPDE reference architecture incorporates multiple domains and key telecom and Internet standards such as Parlay, OSA, UDDI, WSDL, and SOAP. These open standards, combined with aggressive efforts by many global service providers to standardize an OSS/BSS integration standards base, place SPDE as a leading solution contender in the global markets of telecom service provider strategies.

IBM clearly understands the importance of quickly and efficiently locating specific content and/or applications as close as possible to end-user request(s). The IBM SPDE framework allows for bringing content and applications to the edge of the network domain. The capability to bring content and application to the edge of the network also requires the capability to manage the content and applications, that is, the capability to manage distribution of these elements. Figure 7.3 provides a graphic representation of the domains included in the IBM SPDE framework that help make this possible.

As shown in Figure 7.3, the SPDE framework clearly separates the content and application enablers from the content and applications. The Services Brokerage and Services Management domains are the two main enablers, typically owned and operated by any SPDE service provider, while the actual content and applications are owned and often operated by third parties. However, a single organization can play in any (or all) of these domains. SPDE is an open and non-limiting framework.

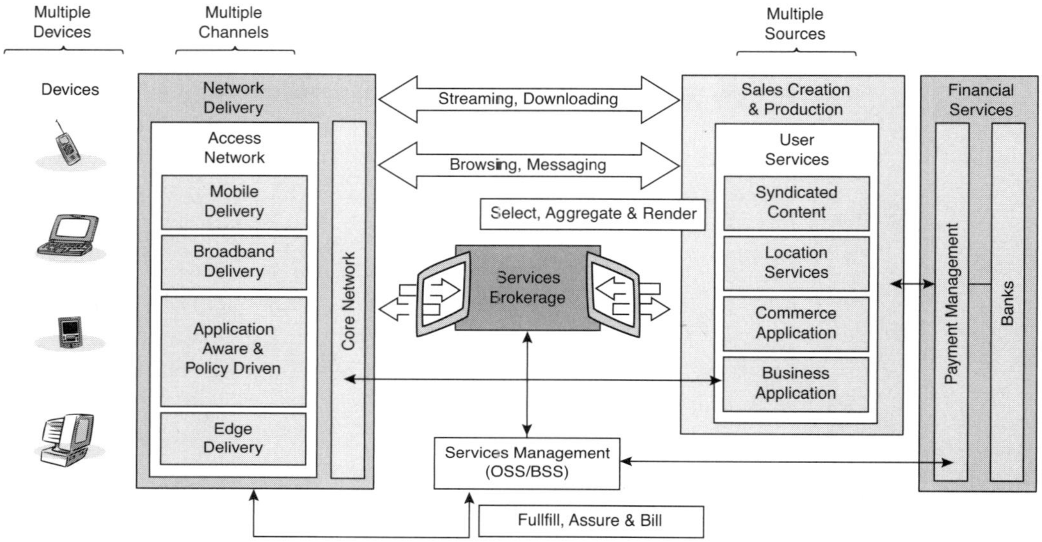

FIGURE 7.3 The "Services Brokerage" domain in the center of the ecosystem communicates with the consumers on the far left, the content creators on the right, and the carrier and/or enterprise provider in the center.

The SPDE framework identifies at least four pivotal areas: User Services, Services Brokerage, Services Management, and Network Delivery. A basic scenario might include the following:

1. An ASP or content provider (in the User Services domain) has an application or some form of content to offer, and publishes it on a User Services portal part of the Services Brokerage domain.
2. Any interested User Services portal in other Services Brokerage domains receives notification of the new user services offering(s).
3. The User Services portal adds the new user services offering to its retail portfolio.
4. When a customer surfs to the portal and subscribes to the user services offering, customer subscription(s) are created for this new user services offering.
5. This customer request for user services triggers the provisioning of the necessary resources with the appropriate service levels to provide for the newly subscribed user services.

The provider and consumer of such user services interact (in simple business terms) at the top within a business layer that includes the portal where service providers publish user services. In the middle, interactions occur within a service layer that translates the business agreement into managed policies, which are then deployed onto the physical elements at the bottom, which act as a resource layer of sorts to activate and monitor the service(s) for the consumer, according to agreed-upon service levels.

Many different vendors, with a vast array of products, could be involved in the creation of this particular SPDE framework. Application integration is a well-known way to combine a multitude of products without the need to build separate interfaces between each (the "*n*-squared" approach). Integration at the application levels (for example, by using middleware) among the User Services, Services Brokerage, Services Management, and Network Delivery domains comprise the foundation of the SPDE framework. This is where the OSS/BSS and NGOSS integration standards become increasingly important.

As described earlier in this section, each domain within the service provider ecosystem is divided into services and components that can oftentimes be instantiated by commercially available products from a multitude of partners or vendors. This functional breakdown has been created to allow for the evaluation of best-of-breed products that might be integrated into a particular customer solution to deliver an operational solution based on the proposed architecture being explored for that portion of the environment. As a reference point, IBM works with a number of content and application services companies that have already implemented products compliant within SPDE, which in turn are targeting the larger OSS/BSS global solution environments. These same companies can and often will be the most desirable for becoming providers of many advanced SPDE solutions, while at the same time embellishing Autonomic Computing disciplines for establishing on demand business ecosystem environmental characteristics.

A fundamental challenge in distributing service and component functions is in assessing the gaps and overlaps between products. Methods utilized in instantiating Autonomic Computing environments using SPDE ecosystem approaches help ensure integrity. That is, where degrees of overlap between products are identified, there are architectural principles in place that help to determine the placement of the boundaries. Subsequently, in-depth engagement(s) with vendors can be planned before entering the final physical design.

Impact of IBM SPDE on Market Ecosystem The SPDE solution is highly dependent on the customer set targeted by the potential buyer. Three different areas are addressed by SPDE, which will, in turn, lead to specific product selections. Let us review a previous illustration of the SPDE ecosystem now that additional context has been placed on various SPDE elements in the ecosystem strategy. As previously described, Figure 7.4 shows the consumers on the far left, the content providers on the far right, and a telecom carrier in the center. This constitutes the entire ecosystem; however, there are many complexities surrounding this broader view.

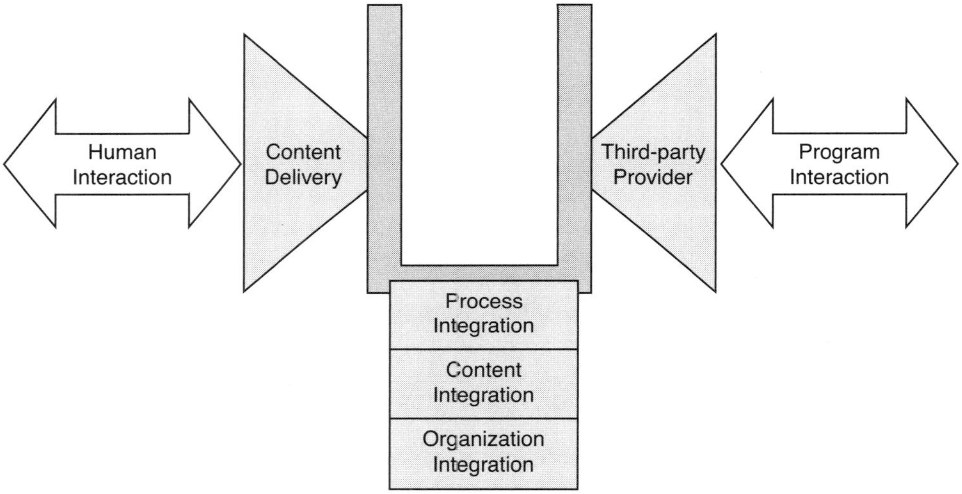

FIGURE 7.4 The "ecosystem" is simple, as it focuses on three areas: creators on the right, carrier or enterprise in the center, and the consumers on the left with a multitude of devices.

The Services Brokerage domain includes components that provide the necessary authentication, authorization, charging, and privacy schemes to support seamless interaction and trust between related business entities. Such entities include users, mobile and fixed network operators, service providers, content providers, and enterprise environments. The SPDE framework provides the necessary functionality to assemble the user services and publish them as Web services in a manner that is technically familiar to the application developers to help speed development time.

Services and applications, whether HTTP-based or not, will be owned by the enterprise for services targeting medium and large businesses. The network operators and service providers will mostly provide an enabling service infrastructure.

The SPDE framework helps support the following business drivers:

- Provide service independence from network technology (such as radio access and core networks). Provide services for customers seamlessly, regardless of whether the session is voice or data: 2G, 2.5G, or 3G, digital subscriber line (DSL) or cable, or Internet-centric.
- Enhance services from third-party service providers and/or content providers with information from mobile and fixed network operator networks (such as presence, location, and availability), as well as provide customers with a greater range of innovative services. This will, in turn, help open streams of new revenue to mobile and fixed network operators.
- Facilitate rapid service deployment by mobile and fixed network operators, partner content/application providers, third-party content/application providers, and/or end-customers.
- Support the deployment of global services so that customers receive the same services, wherever they may happen to be located in the world.

SPDE Components The SPDE ecosystem is composed of two fundamental concepts. These concepts in implementation are essentially a hub and spoke type of utility and an application delivery type of utility. The hub and spoke functionality is referred to as the IBM SPDE Integration Hub, and the application delivery utility is referred to as the IBM SPDE Application Delivery Environment.

The SPDE Integration Hub integrates functions and processes that manage relationships with customers. This enables capabilities such as taking orders for products and services, provisioning and activating network elements, and helping assure and bill for those products and services.

The SPDE Application Delivery Environment is implemented for a variety of key technological functions, including collecting and aggregating content from a number of sources, providing a plug-in environment for network-enabled applications, selecting, aggregating, and delivering personalized content and services without dependence on network and devices, and finally, ensuring a secure, scalable, and flexible ecosystem environment.

SPDE ecosystems depend on these two key strategic underpinnings for the ecosystem to operate in a seamless and holistic manner.

The SPDE Integration Hub The IBM SPDE Integration Hub integrates new-generation OSS/BSS applications with existing systems.

The IBM SPDE Integration Hub is a major component of the SPDE reference architecture. The Integration Hub is composed of new-generation technologies: open standard OSS/BSS reference architectures, autonomic business processes, a common business data model, automated best practices, and hardware and software. The Integration Hub essentially integrates new-generation OSS/BSS applications with existing systems to:

- Better manage relationships with customers
- Provide assisted or self-service selection and ordering either through the Web or a call center
- Take orders for existing and next-generation products, services, and solutions
- Provision and activate network transport, application, and content elements
- Help assure and bill for these services using a variety of business models

The SPDE ecosystem's OSS/BSS integration layer of the reference architecture is primarily achieved through IBM WebSphere software. IBM DB2 Universal Database, Tivoli, WebSphere, IBM hardware and software, and third-party or partnership software components further optimize it. The Integration Hub extends the reach of service providers to include third-party partners' and suppliers' processes and information. Again, the successful integration of these OSS/BSS partnerships largely depends on their abilities to be able to adhere to an open standard for integration. This OSS/BSS standards approach is a key strategic initiative in line with the global TMF advancements being established in the areas of open system integration standards.

The SPDE Application Delivery Environment

The IBM SPDE Application Delivery Environment allows service providers and mobile network operators (MNOs) to rapidly develop, intersect with, and deploy new on demand business services.

The IBM SPDE Application Delivery Environment is an IT infrastructure that helps mobile and fixed network operators and service providers quickly and easily deliver new applications, content, and services to the end-users of any on demand business. The SPDE Application Delivery Environment is based on open industry standards such as Parlay, OSA, and Web services, as well as other key open standards. It is composed of reference architecture, processes, technological interfaces, middleware, hardware, and defined service delivery capabilities.

The SPDE Application Delivery Environment supports end-user applications; it also supports content, content delivery, and many Web services provided either by in-house development or by ISVs that meet industry open standards. The capabilities of the SPDE Application Delivery Environment are delivered through the use of IBM WebSphere software and IBM Global Services Business Innovation Services. Key capabilities include:

- The ability to collect, manage, and aggregate content from a number of sources
- Industrial open, flexible adapters for network-enabled applications and on demand business services
- Selection, aggregation, and delivery of personalized content and services with a complete transport network and device independence
- Design points to help ensure a secure, reliable, scalable, and flexible environment
- Connections to the SPDE Integration Hub for customer support and service management

The Pivotal Role of the Services Brokerage Domain The Services Brokerage domain helps a service provider or MNO leverage its assets in both the Network Delivery and Services Management. This is accomplished by allowing each respective area to be accessed from applications in the User Services domain. Such assets include user location information (or contact information) and billing relationships with the user.

The Services Brokerage domain is designed to provide access to content and applications, and interact with components in the Services Management domain for care, fulfillment, assurance, and billing. Access to content and applications supports what the service provider sells to its many subscribers. Interactions with the Services Management domain support the way in which subscribers enroll and modify their subscriptions, whether by themselves or through a CSR. The Services Brokerage domain specifies how services are fulfilled, assured, and billed.

The presentation for most of the care and billing functions is provided by components within the Services Brokerage domain. The care and billing functionalities are provided by components within the Services Management domain. For example, the way in which subscribers view their accounts is supported by components in the Services Brokerage domain (for the presentation), while the billing component in the Services Management domain provides the functionality.

The Services Brokerage domain is at the hub of the Network Delivery, Services Management, and User Services. As such, it includes three key sets of interfaces:

- Interfaces to applications and content residing in the User Services
- Interfaces to applications residing in the Services Management
- Interfaces to the underlying network technologies residing in the Network Delivery domain

The Services Brokerage includes interfaces to content, applications, and underlying network technologies. The fundamental concept in SPDE for this next-generation service delivery is the underlying application integration "delivery" services that surround the Services Brokerage. This brokerage concept is the only provider needed for establishing the SPDE interfacing capabilities.

Application Integration Services Let's now examine the intra- or inter-enterprise integration challenge with a magnifying glass. There are typically three fundamental issues that need to be addressed:

1. Process integration (control flow)
2. Content integration (data flow)
3. Organization integration (common directory and security infrastructure)

In addition, there are two communication dimensions:

1. Human interaction (via portals supporting a variety of client devices over multiple channels)
2. Program interaction (via gateways supporting various business-to-business [B2B] protocols).

Application integration deals with process, content, and organization integration issues, as well as with human and program interactions. Figure 7.5 illustrates these fundamental aspects of application integration, which are at the core of the delivery mechanisms in SPDE. Figure 7.6 illustrates how the fundamental aspects of application integration map onto the IBM SPDE reference architecture. This allows for the final enablement of SPDE as a next-generation services delivery framework.

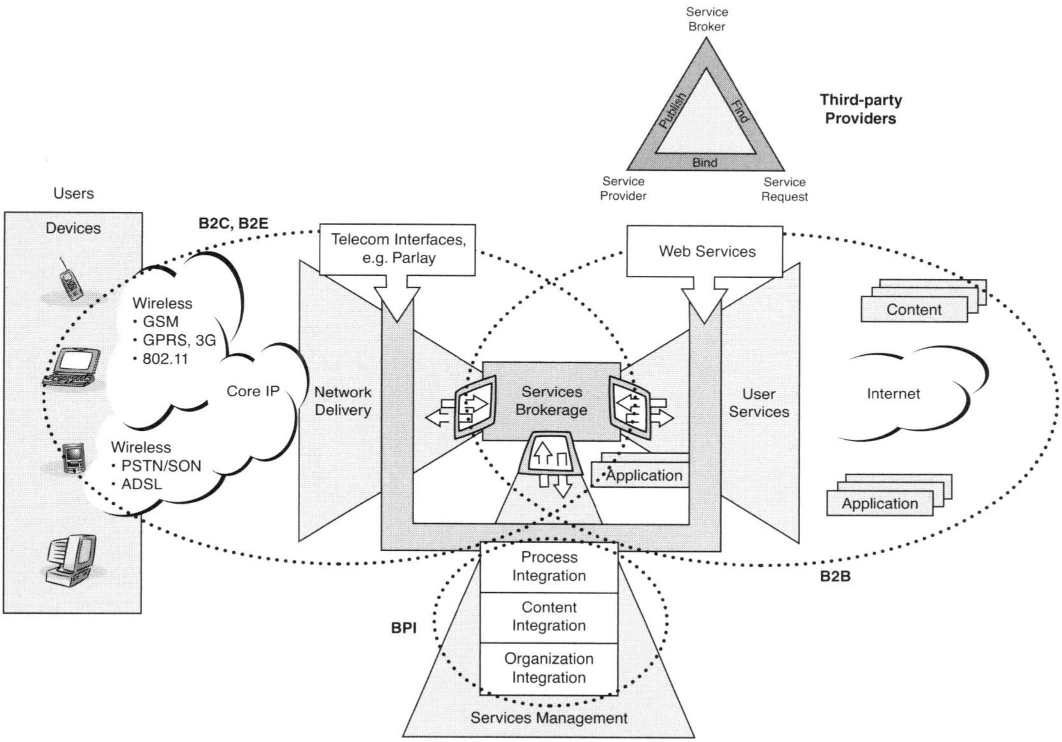

FIGURE 7.5 The fundamental aspects of application integration are at the core of the delivery mechanisms in SPDE.

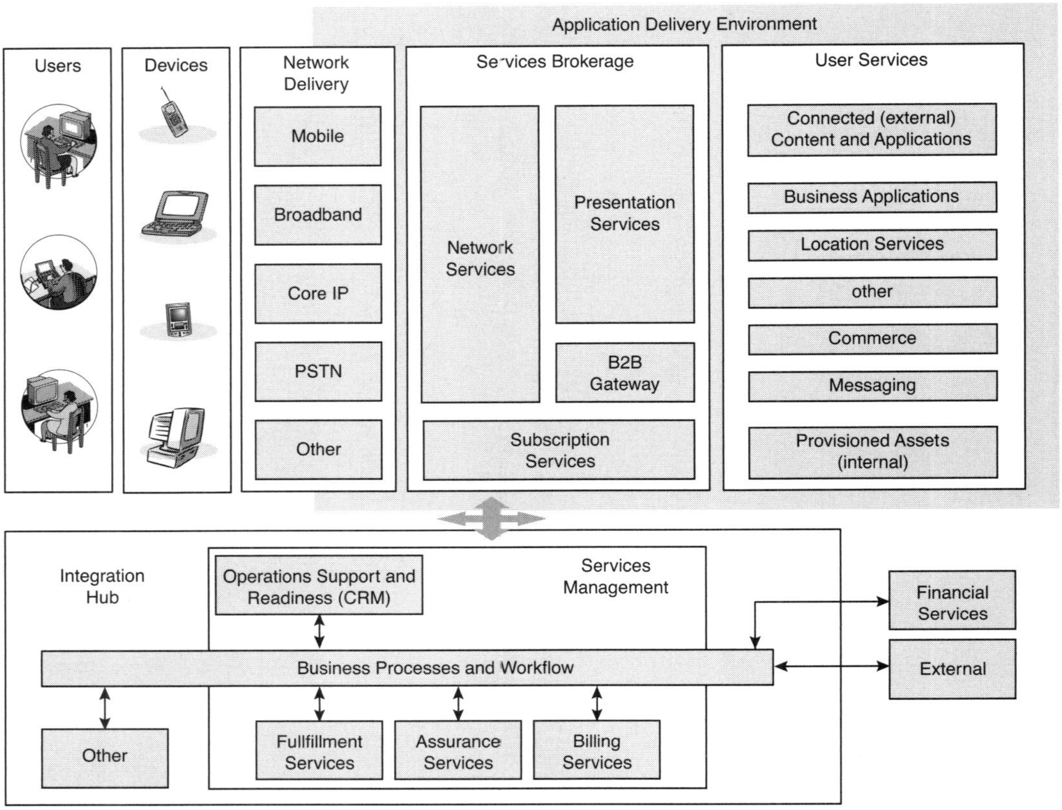

FIGURE 7.6 The fundamental aspects of application integration map onto the IBM SPDE reference architecture.

Open APIs (i.e., adapters or connectors) in IBM SPDE help speed deployment of new content and applications. Interfacing and integrating components residing in the Services Management and Services Brokerage domains involve process, content, and organization integration. This functionality is achieved through business process integration (BPI) technologies. This BPI capability is automated in the sense of workflows, data translations, and business process redirections, and it remains independent of other on demand business functions throughout the state of any given consumer transaction.

Open APIs are the key enabler in SPDE that facilitate the deployment of new content and applications. These open APIs include connections to network services capabilities. Applications may combine these network APIs with other Internet APIs to create innovative multimedia services.

Note that open APIs do not necessarily mean standard, although these are included. In fact, the provision of non-standard, open APIs is one of the ways in which a service provider could differentiate itself from the competition.

Open technologies such as XML, UDDI, WSDL, and SOAP (Web services) are increasingly common in the IT industry and provide the base for this complex program interaction. Web services provide the interfacing technology of choice between the Services Brokerage and User Services domains. Web services can be used as the external B2B interface between a service provider and other parties, while also providing content and applications. These technologies can also be used as internal interfaces for company-owned content and applications to be provided to consumers.

Interfacing with components in the Network Delivery domain requires consideration of both standards and de facto telecom interfaces. Despite the shift toward open and network-decoupled application architectures, network operators still require control and careful engineering of their core networks. To ensure that applications (potentially developed and managed by third parties) do not adversely affect the reliability, robustness, and capacity requirements of their core networks, operators must continue to exercise careful engineering practices and put in place secure network access control measures.

Network APIs (such as connectors or adapters) provide the bridge between these complementary domains (that is, Signaling System 7 [SS7] centric services currently embedded in the network and IP-based services). These APIs provide a controlled interface point through which applications gain access to the core capabilities provided within the Network Delivery domain.

At the same time, these APIs provide sufficient control and security to ensure that the applications do not adversely degrade the operation and performance of any operator's network(s). For example, in a mobile environment, these APIs make network components such as home location registers (HLRs) or mobile location centers (MLCs) visible from an application viewpoint (and enable applications to benefit from information available at the HLR or MLC level). An API on each relevant network component (such as an HLR or MLC) aligns with the evolving Third-Generation Partnership Project (3GPP) OSA, Java in Advanced Intelligent Networks (JAIN) Service Provider API (SPA), and/or Parlay standards.

The IBM SPDE framework enables SS7-centric services to cooperate with IP-based services. Access to APIs is managed and monitored, especially when third parties are invoking the APIs. The provision of service management components within the Services Brokerage domain allows the use of these APIs to be securely granted. These components are in addition to, and distinct from, the normal OSS part of the Services Management domain.

Figure 7.7 illustrates how the SPDE framework can enable SS7-centric services to cooperate with IP-based services:

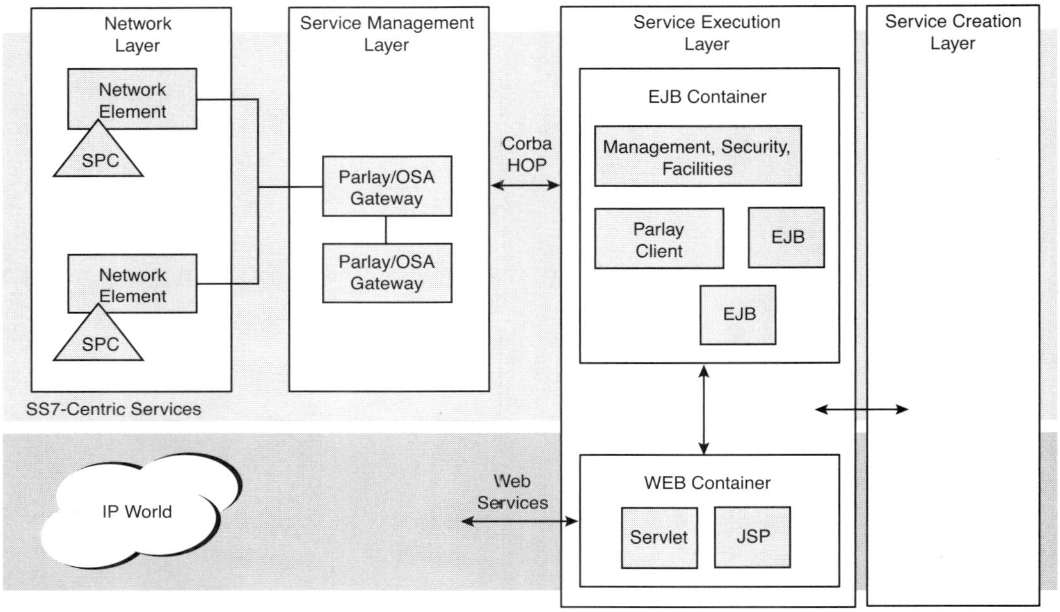

FIGURE 7.7 The SPDE framework enables SS7 in the larger ecosystem.

Standard execution environments help ensure the seamless integration of components from multiple vendors. The use of Java 2 Micro Edition together with Mobile Information Device Profile (J2ME/MIDP) helps providers deliver high-demand, feature-rich applications to devices with limited memory. Proprietary and/or technical barriers should be carefully assessed and monitored (i.e., prohibited) so that both the application development and content development communities can maximize involvement and achieve full and open participation. This approach will facilitate the maintenance of an open, flexible framework. The use of standard execution environments, such as J2EE, helps components from multiple vendors to be seamlessly integrated, which is conducive to the creation of a complete, robust service delivery platform. This, in turn, allows the application development communities to benefit from a uniform programmatic environment that helps ensure their ability to easily and rapidly create and deploy revenue-generating user services.

Technologies such as J2ME/MIDP are important technology enablers to consider, in particular for devices that have limited memory available for running Java environments and applications. J2ME/MIDP provides application developers with the ability to develop feature-rich applications that enable diverse sets of users to download terminal-resident content and applications to personalize their terminals. Figure 7.8 illustrates the systems software architecture associated with SPDE:

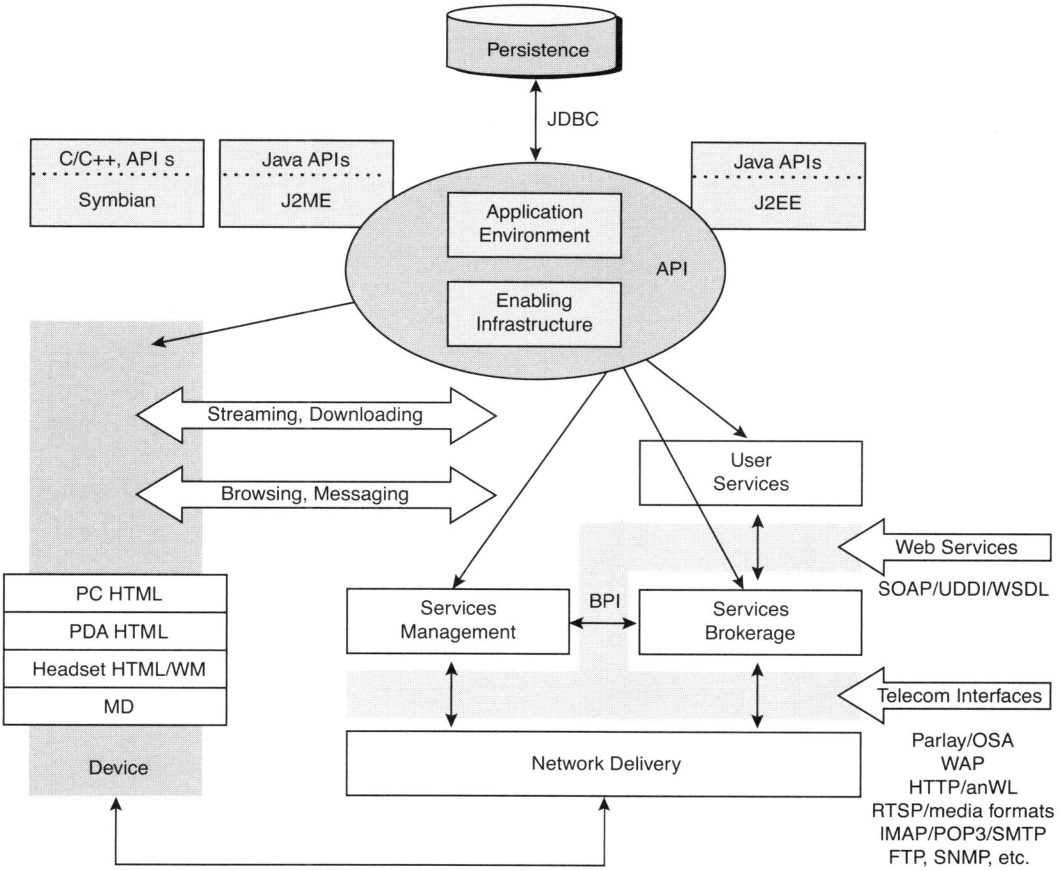

FIGURE 7.8 The systems software architecture associated with IBM SPDE reference architecture is robust.

Components in the Services Brokerage Domain The Services Brokerage domain is designed to provide secure selection of user services based on authentication, authorization, and accounting (AAA) processes. IBM SPDE provides authentication and authorization functionality, which in turn can provide robust functional security features.

The components within the Services Brokerage domain:

- Allow the selection of user services based on user AAA, extending from the lower layer of connectivity services (such as VPN) to the upper layer of content and application services.
- Enable the support of applications that use browsing, messaging, streaming, and downloading technologies.
- Enable the platform owner to manage a wide community of business partners.
- Provide associated management capabilities, such as device management, subscription management, Digital Rights Management (DRM), content management, and more.

Note that SPDE is a functional reference architecture and it does not contain components that support operational capabilities such as load balancers (scalability and/or availability), firewalls (security), and more. However, security belongs to both operational and functional architectures. Authentication and authorization are functional capabilities, while firewalls and intrusion detection are part of the on demand business operational models.

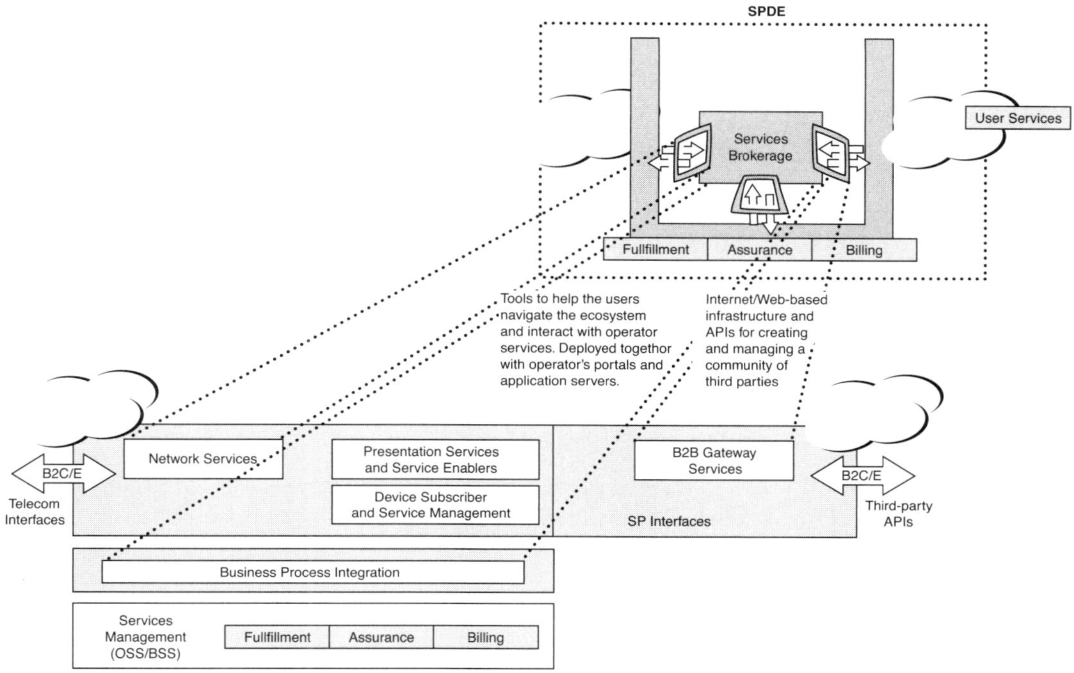

FIGURE 7.9 The services brokerage has several components delivering the core functions.

As shown in Figure 7.9, there are four main types of components in the Services Brokerage domain:

- Network services
- Presentation services and service enablers
- Subscription management services (device, subscriber, and service management)
- B2B gateway services

Business integration services provide consistent process control between the Services Brokerage and Services Management areas. Components within the Network Services domain are intended to convert the various communications protocols of the delivery channels into HTTP (or secure HTTP [HTTPS]), which is the base protocol element within the Services Brokerage domain. This includes products such as Wireless Access Protocol (WAP) gateways that convert WTP/WSP (Wireless Transaction Protocol/Wireless Session Protocol) into TCP/HTTP, voice gateways that convert GSM (Global System for Mobile communications) voice into voice XML, and more.

The Services Brokerage domain is designed for protection against unauthorized use through firewalls, intrusion detection systems, and AAA servers. Because the Services Brokerage domain is connected to external systems such as the Internet, its services should be adequately protected against unauthorized use.

AAA servers are designed to identify users, control access to content and applications, and achieve an SSO environment. AAA services disclose some of the user's private data to various user services in a controlled and confidential manner so that personalization services can be provided. When a session is established, accounting information from the content and applications is routed through accounting services to billing services, which then manage user accounts in the Services Management domain. AAA is based on information held as tuples[4] on relational databases, and makes use of the LDAP (Lightweight Directory Access Protocol) and/or RADIUS[5] protocol.

Content adaptation servers convert markup languages as needed and provide for data synchronization for users working in a disconnected mode. Presentation services include markup, content, formatting, and scripting languages. Content adaptation servers within the Network Services domain are responsible for the conversion of one markup language to another, to match the specific requirements of a channel, device, or business protocol. This SPDE design point also supports data synchronization, allowing users and/or processes (e.g., device management) to work in a disconnected mode. The SyncML standard allows the User Services domain and mobile devices to store the data they are synchronizing in different formats, or make the data available when different software systems are being utilized as the data target.

Presentation services are designed to aggregate and render information for the user/customer in a compact and easily consumable form. These services provide users with personalized service interfaces to both applications and content. Presentation services are typically accessed from graphical interfaces, but voice interfaces can also be used. The essential elements of presentation services are page description (markup) language, content formatting language (style sheets), and scripting language. The use of the W3C's xHTML[6] (eXtensible HyperText Markup Language) helps ensure the way in which content is rendered. Presentation services can also include

4. A *tuple* is a virtual record of data elements, assembled by relational database systems, wherein the data originates from one or more distributed databases as opposed to a single database system residing on a single machine.
5. RADIUS servers are responsible for receiving user connection requests, authenticating the user, and then returning all configuration information necessary for the client to deliver service to the user.

management capabilities that support the deployment of service providers' content and applications. In fact, presentation services and their related management capabilities are generally the primary ingredients of any portal.

Service enablers are modules associated with presentation services that can be invoked by content and applications in the User Services domain to execute a particular function. For example, a notification service enabler can provide the proactive distribution of personalized content. It then uses the appropriate gateway in the Network Services domain to push information to users over various networking channels, such as the Short Messaging Service (SMS).

IBM SPDE device management helps enable MNOs to remotely provision and configure mobile devices. These elements, combined, strengthen and help to position the SPDE framework as a powerful and unique ecosystem across all global industries.

Subscription management services offer the ability to capture, retain, and handle information about users and their delivered services. Subscription services are designed to store and access subscriber and services profiles in support of a variety of non-voice services. Subscription services capabilities include enrollment, care, device management, and provisioning of external systems.

As part of the subscription management services, device management can allow fixed and mobile network operators to remotely provision and configure devices on behalf of subscribers. It can also provide for the initial provisioning of parameters to the devices, as well as continuous provisioning and management capabilities.

Subscriber management services can be enabled with the emerging SyncML device management (SyncML DM) standard, which will be executed by a SyncML server part of the Network Services domain.

Web services APIs and "bolt-on" OSS types of systems (as suggested by the TMF) provide extreme power and flexibility for creating and managing a wide community of business partners.

B2B gateway services are designed to provide a unified mechanism for the discovery (by third parties) of service enablers available on the chosen platform, and a Web-based infrastructure with APIs (i.e., adapters or connectors) for creating and managing a community of business partners. These services include policy-based access control and management of API invoca-

6. xHTML is a family of current and future document types and modules that reproduce, subset, and extend HTML 4 [HTML4]. xHTML family document types are XML-based, and ultimately are designed to work in conjunction with XML-based user agents.

tion by third parties. The B2B gateway also provides for any necessary format conversions of data or other information. Power and flexibility are provided through the use of APIs based on Web services (UDDI/WSDL/SOAP).

Figure 7.10 illustrates the relationships that link the components within the Services Brokerage domain. Note that not all the possible components are represented here. This illustration also shows the primary lower level components within the Services Brokerage domain.

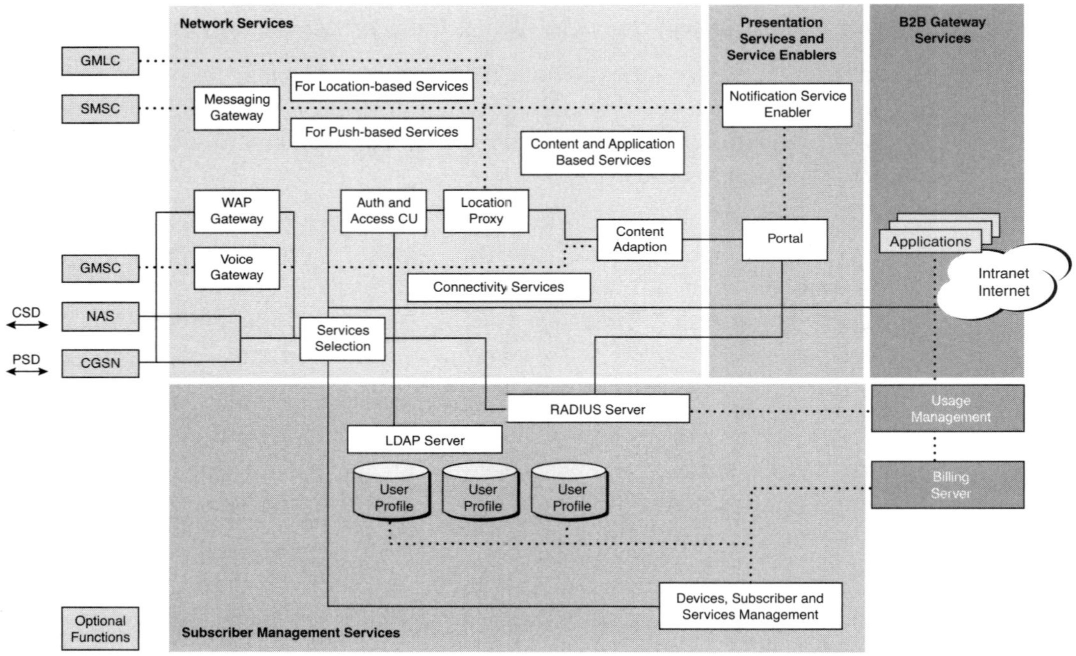

FIGURE 7.10 The services brokerage contains multiple components within the functional core.

Media Streaming and Downloading Services The previous discussion and illustration describes capabilities that define a view often referred to as *services enablement*, which is the foundation for all types of user services in the SPDE framework. Media streaming and downloading services use technologies that deliver media content such as entertainment and business applications to a wide variety of end-user devices. In fact, by virtue of these diverse and open technologies employed in the SPDE reference architecture, the choice of end-user devices is virtually limitless due to the compliance approach of adhering to open standards.

For example, let us explore aspects of the content delivery and media ingestion capabilities of SPDE. To support media streaming and downloading services, SPDE requires a *content distribution* component that is embedded into the service provider's IP network to bring content to the edge of the network for a better user experience.

This includes:

- A content distribution manager
- A content distribution network
- Surrogate media servers

This scheme also embodies a media ingestion technology component with the following capabilities:

- Content transcoding, relaying, storing, and archiving capabilities
- Meta-data enrichment, creation, and processing
- Streaming of live and/or on demand business media content to the user over the network
- Downloading of Business on demand business media content to the user over the network
- DRM[7], for protecting the legal rights associated with any media content to be streamed or downloaded

Figure 7.11 illustrates the SPDE reference architecture for media streaming and downloading services.

With this part of the SPDE architecture, fixed network operators and MNOs can deploy download servers that help other participating content providers leverage the operator's billing capabilities to cost-effectively sell content.

7. The subject of *DRM* is one area in which IBM provides tremendous technology direction and leadership; however, the subject itself extends beyond the scope of this book.

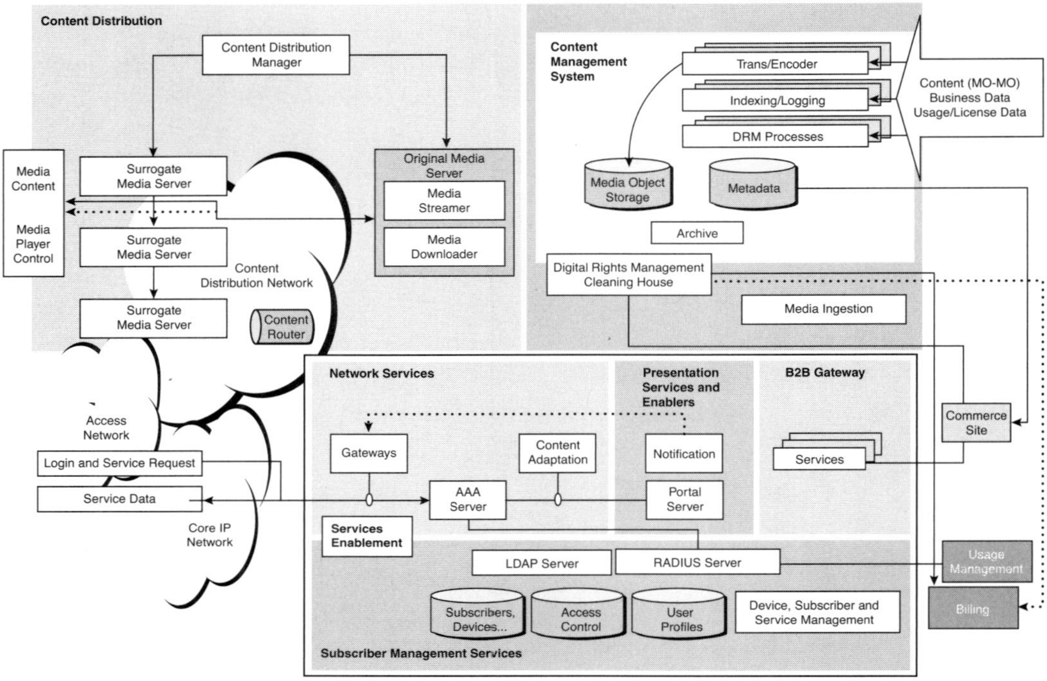

FIGURE 7.11 The media streaming architectural reference is rich in function, as shown by the components in this illustration.

Multimedia Messaging Services This section provides an example of the communications dynamics provided by the SPDE framework. The fundamental point of the discussion in this section is to illustrate the pervasiveness of potential computing devices and the dynamics of communications capabilities to these devices. These dynamics are achieved by virtue of several open standards in the SPDE framework, advanced networking services, and innovative software/hardware approaches.

A part of the IBM SPDE framework includes a Multimedia Messaging Service (MMS)-supported, rich media content messaging capability across a variety of network types, which integrates with existing messaging systems. While currently dominated by pure text using SMS, mobile messaging is now evolving to rich media content using the advanced MMS. IBM SPDE can enable the MMS service; MMS is a store-and-forward messaging service for personal communication that uses any available data bearer channel. MMS allows for a combination of network types, as well as integration with existing messaging systems (e.g., fax, e-mail, voice mail, and instant messaging). MMS requirements are realized with existing transfer protocols such as:

- WAPP[8] (Wireless Application Protocol)
- SMTP[9] (Simple Mail Transfer Protocol)
- ESMTP[10] (Extended Simple Mail Transfer Protocol)

These protocols depend on lower layers to provide push, pull, and notification. MMS does this by using existing message formats such as Audio/Modem Rister (AMR), MPEG-1 Audio Layer-3 (MP3), Joint Photographic Experts Group (JPEG), Graphics Interchange Format (GIF), MPEG-4, and/or synchronized multimedia integration language (SMIL).

Most importantly, MMS assumes the basis for connectivity between different networks to be IP and associated messaging protocols. MMS also supports e-mail addresses or mobile subscriber ISDN numbers (MSISDNs) to address the recipient of a multimedia message. Address resolution ensures that messages can be successfully routed regardless of the address format used.

A carrier's MMS approach should be based on Internet technologies with proven stability, functional completeness, and comprehensive systems integration. The standard MMS architecture provides the necessary elements for message delivery, storage, and notification, including:

- The MMS server for the storage and handling of incoming and outgoing messages.
- Several messaging servers (e.g., an MMS server, e-mail server, SMS center, and fax servers).
- An MMS relay for transcoding multimedia message formats, interacting and operating with other platforms, and enabling access to various servers residing in different networks. The MMS relay also recognizes the capabilities of the device and reformats or transcodes the message according to the device's capabilities. MMS-based services depend on user preferences, device capabilities, and available network resources.

8. The Wireless Application Protocol (WAP) is an open, global specification that empowers mobile users with wireless devices to easily access and interact with information and services instantly.
9. SMTP provides mechanisms for the transmission of mail, directly from the sending user's host to the receiving user's host when the two hosts are connected to the same transport service, or via one or more relay SMTP-servers when the source and destination hosts are not connected to the same transport service.
10. ESMTP is a user configurable relay-only Mail Transfer Agent (MTA) extension with a *sendmail* compatible syntax. ESMTP provides the capability for a client e-mail program to ask a server e-mail program which capabilities it supports and then communicate accordingly. Currently, most commercial e-mail servers and clients support ESMTP.

- MMS user databases for storing user-related information such as subscription and configuration data (e.g., user profiles, HLR information for mobility management, and more).
- An MMS user agent for a client module on the device handset to view, compose, send, receive, or delete MMS messages.

MMS types of on demand business components help enable delivery, storage, and notification for rich media messages on mobile devices. Message routing is a core component for determining the rules for when, where, and how users receive messages, particularly when combined with notification features. The message routing component manages the "presence," which encompasses user activity and current device capabilities.

For example:

- A user may be currently active, that is, logged into the network for data services.
- A device may be connected to the network—and have an IP address assigned to it.
- A given capability (such as the availability of an instant messaging client) may be present in (or absent from) the device. The presence of capabilities in the device may be dynamic; for example, the instant messaging client may be turned on or off at any time.

MMS message routing checks the user activity and device capabilities to determine when, where, and how a user receives an MMS message. Message routing in this scenario depends on the device management functionalities present on the user's device, and the device capability to interoperate with the various open standards technologies. Figure 7.12 illustrates how MMS components relate to the SPDE framework:

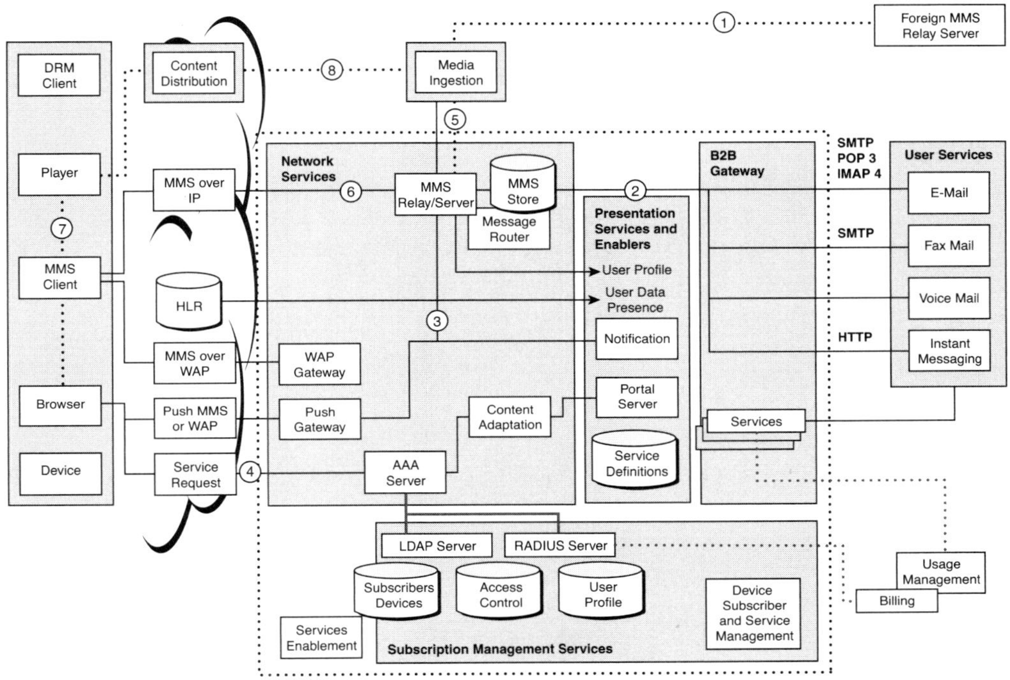

FIGURE 7.12 The MMS components are noted as an integral part of the SPDE framework.

MMS provides the user with an On Demand choice of either downloading or streaming multimedia content.

A typical MMS scenario involves the following:

1. An incoming message with media content is sent to the MMS relay.
2. The MMS relay/server queries the MMS server for terminal capabilities and/or user preferences.
3. The MMS relay/server notifies the user of the arrival of the message/content.
4. The user requests the content.
5. The MMS relay copies a media content request to the media ingest manager and gets a unique reference (such as an MSISDN or uniform resource locator [URL]).
6. The reference is sent to the terminal.
7. The MMS client invokes the player and opens a media session (for streaming or downloading).

8. The media content is streamed or downloaded from the media ingest server to the player.

The media component of the message can be kept on the storage associated with the media ingest manager (note that only its reference [MSISDN or URL] will be embedded in the MMS message). The user does not have to download the entire multimedia message. The actual choice of downloading or streaming the multimedia content is on demand. This also allows a user to download or stream content protected by DRM, such as MIDI[11] (Musical Instrument Digital Interface) ringing tones, animated GIF screensavers, and downloadable games (to name only a few).

Delivery Business Values with SPDE and IBM Software

As described in [Cabannes01], IBM software plays a significant role in the enabling of IBM SPDE and on demand business services for many worldwide customers. This section describes this role and the key technology underpinnings related to the software enablement environment.

Today's telecom service provider's business strategy is increasingly linked to its ability to rapidly leverage software to effectively and efficiently execute business operations. IBM software technologies, when effectively applied through an industry-specific reference point, help service providers and enterprises to deploy and integrate complex solutions that address a broad range of business and operational challenges. An on demand business solution involves the synthesis of technology, autonomic business processes, and a multitude of industry insights.

The IBM telecom software portfolio is more than technology; it embraces telecom industry and IT standards, as well as the industry-specific blueprint (SPDE) to create a flexible on demand business infrastructure designed to support service providers' emerging business needs. In addition to this combination, the implementation of the IBM Autonomic Computing Blueprint (described earlier in this book) is also extremely critical to some successful on demand business transformation efforts. In addition to overcapacity and declining revenues for increasingly commodities-centric voice-based services, service providers face many challenges, such as:

11. MIDI enables people to use multimedia computers and electronic musical instruments to create, enjoy and learn about music. There are actually three components to MIDI, which are the communications Protocol (language), the Connector (hardware interface) and a distribution format called Standard MIDI Files.

- OSS/BSS integration costs, capital expenditures, and operational expenditures
- Convergence and management of legacy and new-generation OSS/BSS
- Speed to market with new revenue-generating services
- Risks associated with OSS/BSS integration
- Implementation of commercial, off-the-shelf (COTS) software that involves multiple partners
- Simplifying the process and systems for managing security
- Churn and low customer satisfaction
- Decreasing costs and improving efficiency across the board

This section outlines how the IBM software portfolio leverages the IBM SPDE reference architecture to help service providers and other enterprises implement solutions that address the myriad of challenges they face today. The intent of this section is to provide the basis for ongoing explorations with service providers, including positioning IBM software offerings in the context of SPDE. In this section, we will outline the role of standards organizations, on demand business capabilities, and demonstrate how IBM software products effectively serve the unique requirements of service providers and other enterprises.

SPDE Reference Architecture The IBM SPDE, as previously discussed, is a scalable and flexible reference architecture designed specifically for telecom service providers. The foundations for the SPDE framework include:

- Adoption of industry and IT standards, for instance, the TMF, OMA, Parlay, Linux, W3C, etc.
- Numerous customer engagements with some of the world's largest and most innovative wireless, wireline, and broadband providers
- Extensive research conducted at IBM Network Innovation Labs on industry-leading tools, methods, and business processes

The IBM SPDE framework defines a consistent delivery and management environment for services, applications, and content across multiple networks and devices. Service providers can implement and integrate all or parts of the IBM SPDE framework within their existing environment. SPDE enables a service provider to:

- Introduce and manage new services faster, easier, and more cost-effectively.

- Facilitate access for business partners, to deliver "value-added" services in a consistent and secure manner. This enables partners to develop new and innovative revenue-generating offerings in cooperation with the service provider.
- Integrate new capabilities and content consistent with the service provider's legacy environment.
- Take advantage of new commercial applications, programs, and technologies as they become available in the market.
- Reduce the duplication of systems to lower operating expenses.
- Rapidly respond to changes in a business environment through a tight coupling of proven business processes with a flexible technology infrastructure.
- Provide a metering, billing, and settlement infrastructure to maximize revenue.

IBM software solutions created with SPDE increase business responsiveness, improve operational efficiency, and provide rapid ROI in two key areas:

- IBM software helps service providers reduce operating and capital expenses through the enablement of a flexible services delivery infrastructure for both consumer- and corporate-oriented content and applications (i.e., Applications Delivery Environment).
- IBM software reduces and manages complexity by providing a platform for technology and business integration of disparate software-based systems (i.e., the Integration Hub).

IBM continues to develop industry-specific software products and solutions for the telecom industry. IBM actively collaborates in the development of numerous industry standards. IBM also invests in research and development to meet the emerging requirements of service providers.

Summary

This chapter has provided full treatment of many service provider on demand business perspectives.

IBM works with many of the world's largest telecom firms (and other broadband providers), and a tremendous number of the world's largest business enterprises in implementing on demand business solutions. It is through the implementation of many of the technologies and strategies presented in this chapter, as well as in the rest of this book, which assist our customers in their own on demand business transformations.

The next chapter will review some case studies with regard to many of the on demand business accomplishments of our customers. Also, we will explore the industry issues that are forcing these on demand business transformations to occur.

8 Industry Matters and Customer Profiles

Why are so many companies turning into on demand business enterprises? In this chapter, we will explore this question and the specific results that some companies have achieved in the on demand business transformation landscape. We will explore the pressing industry climate driving these on demand business transformations across many vertical industries. It is, in fact, these issues and success stories that describe several transformations and explain why many are moving into the on demand business world.

We will discuss how our customers started the on demand business transformation, why they felt this transformation was necessary, and the benefits they achieved in the transformation. We will also explore interesting economic and industrial issues that are driving many of these transformation efforts.

CURRENT PRODUCT AND SERVICES OFFERINGS MAY NOT BE REPRESENTED IN THESE CUSTOMER SOLUTION REFERENCES, DEPENDING ON THE SOLUTION TIMEFRAME

It is important to note that only certain products and services have been identified in this section. These products and services were specific product offerings from the IBM Corporation during the timeframes in which each of the solutions was delivered. New IBM products and services related to on demand business solutions are introduced into the market at a frequent rate, according to ongoing internal development plans and strategies.

This fact means that reliance on these offerings may or may not represent the most current premier offerings/services available from IBM.

There are literally hundreds of IBM on demand business customers that could have been reviewed in this section of the book; however, we will only focus on a few examples that relate to some of the discussions presented in the book. All the IBM on demand business solutions are very important in each of their vertical industries; however, this discussion had to be limited for reasons of content allocation.

Industry Sector Issues Driving On Demand Business Transformations across Vertical Industries

This section will openly discuss industry sector issues that are driving on demand business transformations. The order of the following issues is random, and although some specifics are related to certain vertical industries, several on demand business problem patterns apply (in some form) across all industries.

The Retail Industry

The retail industry is facing increasing pressures to enhance the in-store experience through customization, personalization, and differentiation by leveraging next-generation technologies. At the same time, it must operate in an environment of increasing competitive intensity and financial pressures. In a classic "the rich get richer" storyline, those companies that can afford to take advantage of technology to enhance the customer experience will be able to reap disproportionate gains relative to their smaller, less-endowed competitors.

This industry has also noted that empowered consumers—who are driven by time, price, and information—are demanding high-value, multi-channel shopping, and personalized care. Global expansion and rapid industry consolidation have also affected the competitive arena. Retailers must now find ways to differentiate themselves with highly focused value propositions or immense scale.

Advancements in technology are also enabling retailers to adapt quickly to new business models, taking advantage of operational and cost efficiencies. However, despite technological benefits, the recent global economic unpredictability has forced companies to exhibit true and rapid economic returns, to maintain their competitive spaces in the marketplace.

With increasing competitive pressures, more demanding consumers, and shrinking profit margins, retailers are tiring—some even retiring—of the chase. Knowing that the future promises an even more frenzied pace, it is time for retailers to consider trading in common practices for on demand business processes that provide the speed and flexibility needed to stand out in a crowded retail marketplace.

Customer relationship management (CRM) is becoming a mainstream practice for retailers looking to increase sales and profitability. But many current initiatives do not have the customer-centric focus that today's market requires. In their rush to make the next sale, many retailers have lost sight of the first step to successful CRM: understanding what all customers really want from retailers—raising the bar to create the first "intelligent" value chain. Imagine a store where your product is on the shelf each and every time a consumer reaches for it.

Imagine a warehouse where perishables do not perish, where shipping, receiving, and picking are performed with enhanced accuracy and reduced labor expense, where theft is truly detectable, where systems exactly match the physical inventory counts, and where transfers with trading partners are automated and precise. Think about the next-generation on demand business retail industry.

The Airline Industry

The airline industry is facing several key strategic imperatives. The challenges include the impact of economic and social events, the industry's own realignment, the emergence of low-cost providers, regulatory issues, and the heightened need to focus on security. These challenges are forcing the major, established airlines to reduce overhead/operations and enhance their customers' satisfaction at the same time.

Airlines must focus on the complete travel experience to develop a compelling value proposition for why customers should choose them over their competitors. While this approach may not be new, put into context with the challenges stated above, the airline industry is facing a tremendous business trial.

In the past, travel was about a destination. In 2010, travel will be all about engaging in powerful personal experiences that are carefully tailored to certain tastes, and delivered on demand business solutions to the individual guests. Successful hospitality and leisure companies will understand target consumer segments, and therefore entice them with differentiated travel experiences and services customized to meet everyone's individual needs.

The Media and Entertainment Industry

The media and entertainment industry has always been high-risk, unpredictable, often a bit mad, and always looking for the next "hit." But, it has never been in the turmoil it is today. As consumers spread attention and spending across many options, "hits" are more difficult to predict and revenues from offerings are declining. The media and entertainment firms, attempting to support all the routes of getting to market—mostly through mergers and acquisitions—have contracted an acute case of cost complexity syndrome. The good news, however, is that a Hollywood "ending" is possible.

IBM's on demand business prescription for the media and entertainment industry is two-fold: maximize revenue to improve asset management and optimize cost to reduce complexity and inefficiency. Fueled by the rapid spread of high-speed connectivity and standards, information today is increasingly delivered in digital formats.

Digital solutions combine traditional content (e.g., entertainment, product information, marketing collateral, process components, and application software) with the personalized services that deliver what customers need, when they need it, on demand business solutions, to the devices and channels they specify. The possibilities for creating and capturing significant values for media and entertainment companies are enormous—as are the challenges and new competencies that must be mastered. Such a transformational shift demands a much deeper insight into customer preferences and much greater exchanges of previously unstructured data on customer needs, available supply, and business terms.

The Pharmaceutical Industry

The pharmaceutical industry is finding it difficult to sustain expected revenue growth rates. Growth through new products has come under pressure as blockbuster drugs have become more difficult to develop. In addition, many companies are finding that the benefits from new technologies to develop chemical entities may still be years away.

Although companies have sought in-licensing deals to heal ailing pipelines, large pharmaceutical companies have bid up the price of late-stage products to potentially unprofitable levels. If the pharmaceutical industry adopts a new business model based on "targeted treatment solutions," it will be able to achieve breakthrough growth. These integrated solutions will be used to diagnose, treat, monitor, and provide healthcare support to the benefit of patients, payers, physicians, and the industry itself.

The Transportation Industry

The transportation industry sector involving trucking and shipping—with regard to logistics, globalization, and the evolution of on demand business and e-commerce—has challenged transportation logistics with demands for smaller loads, more shipments, and faster delivery. At the same time, technologies such as business analytics, global positioning systems (GPSs), decision support, biometrics, and online collaboration have opened up new opportunities to increase shipping efficiency and reliability while maintaining requirements for controlling and safeguarding shipments.

The Construction Industry

The construction industry employs more people, contributes more to the Gross National Product, and consumes more raw materials than any other single industry. There is an unusually high rate of failure among construction contractors. One of the most pervasive causes of failure among construction contractors, and one that can be most easily corrected, is the inability to properly manage the financial and administrative aspects of the business. The construction industry is very competitive with margins averaging as low as 2 percent for some segments of the industry.

The Consumer Packaged Goods (CPG) Industry

The consumer packaged goods (CPG) industry is showing a gentle decline into a sedentary old age threat, requiring a powerful counterpunch. The prescription is to realign the core goals of CPG companies to establish a much-improved importance and presence with consumers, and to increase linkage with retailers. These are the cornerstones of the "Play Big" approach, and the impetus for the strategies by which influence can be recaptured by the industry.

The Insurance Industry

The insurance industry, including *automobile* insurance lines of business, has realized that the economic slowdown has adversely affected many automobile insurers. Most face increased claims even as premiums drop. Insurers need a way to better calculate and assign appropriate premiums to motorist-based usage. In recent years, industry and market dynamics have attached greater strategic significance on a service-based differentiation approach.

Slowing market growth has led to aggressive price-cutting as insurers nationwide have worked hard to increase their market share. With the consumer market increasingly willing to shift providers for the sake of lower rates, the need to increase customer loyalty has never been more acute than it is now (e.g., healthcare, automobile, life, etc.).

In addition to this, competitive conditions are forcing insurers to focus on customer retention, new market distribution strategies, and further cost reductions in the insurance industry. Issues like privacy, consolidation, and channel marketing are changing the global insurance landscape as well.

The Freight and Logistics Industry

The freight and logistics industry segment reflects the health of the economy in general. It is also highly customer-driven. Customers are demanding partner relationships rather than traditional shipper/customer relationships. These relationships include easier access to shipment information in real time, increased levels of service, reduced shipment handling times, and on-time delivery.

This is an industry segment where technology is important because the services these companies provide to their customers are based largely on information about shipments. Having instantaneous, up-to-date, and accurate information—about where goods are, what condition the goods are in, who signed for the receiving shipment, when it was picked up, and when it was delivered—is just as important as transporting the goods. In the shifting sands of the global market, we see the bulk of high technology devoted to moving information around the world—but that is not to say that moving solid goods is any less complex.

Logistics—the discipline of procurement, maintenance, and transportation of material, facilities, and personnel—is a large business. Everything from your morning mail to multi-ton cargo loads needs to get from point A to point B safely, efficiently, economically, and reliably. In this era of narrow tolerances and razor-thin profit margins, many businesses rely on logistics companies to get their products to market on time.

The Banking Industry

The banking industry has noted that with recent value creation strategies stalled, banks are undertaking fundamental changes to their business structures; meanwhile, the industry itself is taking on a new shape. As banking institutions reconstruct, and the industry itself deconstructs, these two seemingly divergent paths will naturally converge—and yes, they are on the road to on demand business, enriched banking environments.

The Financial Services Industry

The financial services industry has noted that with deregulation in the 1990s, it has broken down the 60-year-old ownership walls among banks, insurers, and financial services firms. This has allowed cross-industry mergers and consolidation.

Investment banks, retail brokers, and asset managers did well. The rise of 401K retirement funds increased the profiles of financial services among middle-class Americans. This was enlightening news for many consumers.

One dark side of this enlightenment period was, however, the collapse of the energy trader Enron, and of course, its far less-than-remarkable auditor, Arthur Andersen, which was found guilty of obstructing justice. This illegal act cast a harsh light on the accuracy of all the big accounting firms. Difficult questions about accounting and corporate accountability in turn opened up a floodgate of concerns, skepticism, and investigations into the standards and practices of the American financial services industry.

This skeptical climate lingers on even today. Customer trust and loyalty, however, remain at the forefront of consumer concerns. The financial services sector contains a $20 billion worldwide market opportunity, with a renewed focus on the customer and distribution, automation of the trading process in financial markets, customer retention, and overall cost reductions. There is also a very strong push to automate the trading process in the financial markets industry. Likewise, mega-mergers and acquisitions have changed the shape of today's financial markets, resulting in a focus on technology investments, outsourcing, straight-through processing, risk management, globalization, and compliance.

The Government

The government has noted that across its segments, including levels and jurisdictions, leaders now face an imposing set of challenges. Citizens want convenient and real-time access to data. Businesses want integrated services and reporting mechanisms. Governments have to transform in many cases.

Economic downturns lower the amount of collected revenues, and dramatically influence budget allocations. Information and physical security concerns abound, and employees need to improve skills to tackle changing needs and trends. This environment drives the government to simultaneously promote customer-centric strategies while improving internal efficiencies and information sharing. A new economy is at hand; for example, the U.S. Internal Revenue Service is re-modernizing itself with a new agenda and innovative means of tax processing.

The Public Sector

The public sector recognizes that its constituents have experienced ease of information access within the private sector, and now they are demanding the same type of service from governments. At the same time, tight IT budgets have forced governments to develop more efficient processes to lower costs and more effectively provide advanced forms of services, or face stiff competition from the private sector.

The Communications Industry

The communications industry, most specifically telecom markets around the world (as service providers), is looking for options to manage its way through the current economic downturn. Many carriers amassed a significant amount of debt to build their network capacity in the late 1990s, and many of these networks remain underutilized because anticipated new service markets failed to materialize.

Now, Telcos are generating insufficient revenues to service their debt and are consequently facing debt defaults and bankruptcy. These providers are aggressively seeking ways in which to reduce capital and operational expenditures, while at the same time increasing revenues and stock ROI to their investors.

Fortunately, the current trends in the telecom industry are creating opportunities for new players to compete directly with incumbents. These trends drive chief executive officers in the industry to address several fundamental issues as follows: reducing costs while improving Telco services; identifying, acquiring, and retaining the most profitable customers; reducing and/or increasing customer churn (as desired); and offering unique value-added services. The foremost and most notable key trends include:

- Significant price reductions in wireline and wireless voice services
- Advances in wireless and Internet technologies
- Growing need for data network services because of increasing Internet access and corporate intranet usage
- Addition of high-speed access to the last mile
- Bundled value-added services to attract and retain customers
- A changing regulatory environment

The Automotive Industry

The automotive industry is particularly challenged by ever-increasing customer demands and expectations. It must think collaboration.

This emerging business model is largely unknown in the traditional automotive original equipment manufacturer (OEM), supplier, and distribution network relationships. On demand business principles, in fact, began the process of linking internal data systems to reduce manufacturing costs and improve and streamline procurement between parties—but this did not mean collaboration was yet fully adopted.

In the new competitive era, market leadership will not be contested by a handful of global automotive manufacturers acting on their own. Instead, competition will be characterized by a race between closely integrated collaborative clusters, or communities of manufacturers and key suppliers.

This collaborative approach in both technology and the transformation of business processes will help meet the customers' continued demands for simplicity and inclusiveness in their relationship between "car" and "driver." Increasingly, the automotive industry is expected to respond quickly to the unpredictable demands of its customers, suppliers, partners, and employees.

Whether it is enhancing order and supply chain management, transforming design and manufacturing processes, managing products through their lifecycles, or providing more personalized service to customers, auto manufacturers now require solutions that enable them to respond with agility and speed.

The Petroleum Industry

The petroleum industry as a whole has held up well from the market's perspective through the implosion of technology/Internet companies. Ongoing political turmoil in the Middle East does impact market price in the near term, but the longer term issues facing this industry continue to be the ability to manage its upstream and downstream costs associated with finding, drilling, and supplying oil and natural gas.

The Healthcare Industry

The healthcare industry, with its rising clinical and administrative costs, is a major concern for many hospitals, especially smaller hospitals that are often less efficient due to a lower adoption of technology. These cost constraints are compounded by increasingly tighter compliance and security mandates that require additional resources from both staff and cost perspectives.

While maintaining their focus on maximizing patient care, smaller hospitals are struggling to update their IT infrastructure and perform basic functions such as digitizing records, all helping to reduce the risk of errors. With the consumer market increasingly willing to shift providers for lower rates, the need to increase customer loyalty has never been more acute. In addition, competitive conditions are forcing insurers to focus on customer retention, new distribution strategies, and further cost reductions in the insurance industry. Issues like privacy, consolidation, and channel marketing are changing the global insurance landscape as well.

The Utility and Energy Services Industry

The utility and energy services industry is not foreign to transformation. As a result of recent deregulation legislation, the utility and energy services industry is in a state of transition from a monopolistic environment to a competitive environment.

The industry as a whole is changing structurally as existing utilities are forced to merge, acquire other companies, and develop partnerships to remain competitive. With deregulation, new companies are entering the industry and adding further competitive pressure to existing companies. Because of the pressure to keep prices low, utility companies might have to diversify their product lines to keep their ROI at an acceptable level.

Customer Profiles

This section discusses a number of on demand business customer case studies. In a few cases, the customer name has been withheld; however, in each case, we will explore the type of industry, a brief description of the customer's line of business, why the customer felt the need for on demand business transformation, how the customer started the transformation process, and the benefits the customer achieved from the on demand business transformation.

The following introduces us to a wide variety of on demand business customers, across a wide spectrum of industries.

Lands' End: Integrating Business and Technology for a Single Customer View

APPAREL BUSINESS SEEKS TO IMPROVE CUSTOMER MARKETING CAMPAIGNS AND ELIMINATE REDUNDANCY

Lands' End, one of America's most well-known direct merchant retailers and the world's largest apparel Web site based on business volume, needed help with its customers. Fragmented customer information was leading to inefficiency in marketing campaigns, high production costs for diminishing return, and an exposed weakness that competitors could quickly target and exploit.

Why become an on demand business? Faced with stiff competition in the apparel business, catalog retailer Lands' End needed to do a better job of identifying customers likely to buy particular products. It needed more accurate, timely information about customers' buying behavior. The company also wanted to centralize the marketing departments of its five divisions to minimize duplicate offers and conflicting marketing campaigns.

How and where did Lands' End start? Lands' End needed an integrated technology infrastructure to promote a single, unified view of customer behavior for its five divisions. The company adopted a unified CRM application, Affinium Campaign, from IBM Business Partner Unica Corporation. It created a customer data mart for campaign management based on IBM DB2 Universal Database Enterprise Edition for AIX. The solution runs on IBM eServer pSeries and IBM Global Services handled the implementation.

What benefits did Lands' End achieve? Lands' End can identify customers' interests more accurately, resulting in lower customer acquisition and retention costs. Improved customer information/visibility also resulted in higher per-transaction revenue and decreased time-to-market for new products. Manual, labor-intensive processes were automated, freeing valuable marketing staff for additional revenue-generating customer initiatives.

Michelin: Increasing Tire Company Efficiency via On Demand Business

EXTRANET INCREASES TIRE COMPANY'S CUSTOMER SATISFACTION AND COMPANY EFFICIENCY

Michelin, based in Clermont-Ferrand, France, is the world's largest tire manufacturer. Michelin needed a way to allow tire dealers to order tires 24 hours a day, without forcing this burden on the customer call centers in Europe.

Why become an on demand business? The company knew that continued profitability depended on responding to tire dealers' needs by making it easier for them to order Michelin tires 24 hours a day. At the same time, the company needed to relieve its seven customer service call centers in Europe of the order-taking burden. To limit expenses, Michelin wanted to leverage its existing hardware and database infrastructure.

How and where did Michelin start? Michelin wanted to set up a self-service, online order system for its European dealerships. IBM Global Services and Michelin deployed an extranet, delivering secure, real-time access to inventory and order information. Michelin opted for a centralized ordering application rather than distributed applications for each major European market, to save deployments costs and distribute the order processing load more smoothly. The extranet is built with IBM WebSphere Application Server Enterprise Edition V3.5, IBM Studio Site Developer, IBM WebSphere MQ, IBM WebSphere Studio Application Developer, IBM HTTP Server, and IBM CICS Transaction Server. It runs on IBM DB2 Universal Database for zSeries and an IBM eServer zSeries 900.

What benefits did Michelin achieve? Overall, order volume and dealer satisfaction have increased due to the quicker, easier order process. The value of the average order has grown considerably. Michelin is realizing substantial operational cost savings from lower call center volume. The extranet takes orders 24 hours a day, even when the database goes offline for nightly updates.

Mostransagentstvo: Staving off Customer Flight and Restoring Confidence with On Demand Business

TRANSFORMING BUSINESS PROCESSES TO PROVIDE REAL-TIME INFORMATION ON REQUEST TO CUSTOMERS

Mostransagentstvo started out as a moving company, but quickly branched into a number of other transportation and travel areas such as cargo hauling, railway transport, vehicle repair, and travel reservations. In the rush to expand, however, the company's business processes didn't keep up and customers found better service from Mostransagentstvo's competitors. Mostransagentstvo turned to IBM and its business partners to design a new online system that gives customers real-time access to information with an integrated Web site and call center.

Why become an on demand business? It's not much of an exaggeration to say it was either respond more efficiently to customer needs or go out of business for Mostransagentstvo, a Moscow-based logistics and travel company. Mostransagentstvo lost 40 percent of its customer base in a single season because of inefficient business processes and a weak customer service communication infrastructure. To dramatically improve service, Mostransagentstvo decided to transform its business processes and give customers anytime access to information they wanted—either via a new Web site or an improved call center.

How and where did Mostransagentstvo start? Mostransagentstvo worked closely with Computer Age (an IBM business partner) to build an e-commerce Web site that, among many features, allows customers to book airline reservations, schedule a moving truck, or reserve a railway transport car. At the same time, the system lets the company track orders and analyzes customer buying patterns.

What benefits did Mostransagentstvo achieve? The company has unified its numerous services into a single, cohesive business. Just one month after the system went online, the company added 1,000 new clients. The other side of that statistic is that the lost client rate dropped to zero. The new technology slashed sales order processing time and greatly reduced the time customers have to wait on the telephone for answers. Another plus: Mostransagentstvo has raised its airplane and train ticket sales to 1.5 million per year. The company expects full payback from its IT investment to take 18 to 24 months.

Senshukai: Introducing Shopping On Demand Business

Customers Can Buy Anytime, Anyplace With This Mobile Solution

One of Japan's most successful mail order companies, Osaka-based **Senshukai**, is a leading name in home furnishings, clothing, household merchandise, and children's goods. In the late 1990s, the company complemented its traditional catalog sales with an e-commerce Web site. The next step in Senshukai's evolution centers on mobile-enabling this site, allowing customers to shop anytime and anyplace using their mobile phone handsets.

Why become an on demand business? Spurred by fierce competition in the mail order sector, and given Japan's expensive dial-up Internet costs, Senshukai needed to find other channels to reach its customers, secure their loyalty, and enable them to shop on demand. Given that 90 percent of its customers are women in their twenties and thirties—who are also prolific mobile phone users—Senshukai exploited wireless technologies to bring e-commerce quite literally to its customers' fingertips. By expanding the business in this way, the company realized it could gain new customers and significantly reduce indirect costs compared with traditional voice phone-based channels.

How and where did Senshukai start? Working with IBM Global Services, Senshukai implemented bellne.com, an integrated e-commerce site supporting both desktop and mobile phone users. Via bellne.com, i-mode users can order products from Senshukai's catalog via their mobile handsets. The service also enables Senshukai to promote special offers and items not listed in printed catalogs.

What benefits did Senshukai achieve? Senshukai has gained members at a rate of 500,000 per year. It has reduced indirect costs, while increasing revenues by 65 percent in 2002, with continued growth anticipated. The comprehensive multi-channel support of WebSphere Commerce enables Senshukai to comply completely with the currency, data formats, and taxation and shipping rules used in Japan. It also makes it very simple to extend the company's mobile-enabled e-commerce site to other countries where it has business interests, for instance, Hong Kong and Thailand.

Whirlpool: Looking for an Innovation Strategy

Company Uses On Demand to Enhance Responsiveness and Resiliency to Support Customer Demands

In a small community in southwestern Michigan, homeowners are testing a Web portal as part of an integrated home solution that allows them to control their home appliances remotely over the Internet. It isn't a networking or computer company that's pioneering this innovative service; it is **Whirlpool**, the world's largest home appliance manufacturer. The company is making a

name for itself in on demand business through its astute application of innovative technology.
Why become an on demand business? One of its challenges as a growing company was the difficulty of growing its IT systems while maintaining availability and reliability to meet customer demands. Whirlpool did not have a business process to handle the problems of a heterogeneous environment. As the system grew beyond what it could monitor, the IT staff discovered that customers were alerting the company to problems before it knew about them. Whirlpool knew it needed to increase resiliency, or risk losing competitiveness in the marketplace.
How and where did Whirlpool start? Whirlpool installed a solution to cost-effectively achieve high availability and performance in its IT environment. The new solution helped all systems and platforms work together and centralized the components at reduced cost and with reliable support for critical on demand business applications. This integration allowed for a single view to provide centralized event management, allowing the IT staff to use problem cause analysis to identify issues before any system disturbance affected customers.
What benefits did Whirlpool achieve? Whirlpool found enhanced productivity with automated workflow and remote control of systems and applications. The company is saving money through improved asset management and reducing the costs of purchasing redundant equipment. Whirlpool's Global Midrange and Distributed Services Department achieved a 46 percent increase in productivity in 18 months and application availability stands at 99.998 percent. Tivoli Remote Control has already saved $750,000 (USD) in application licensing costs over a one-year period. Whirlpool estimates a 60 percent payback on its $6.5 million (USD) investment in two years.

AirToolz Software: Renovating Communications in the Construction Industry

STREAMLINING AND AUTOMATING COMMUNICATIONS PROVIDE MAXIMUM MARKET FLEXIBILITY AND RESPONSIVENESS IN CONSTRUCTION OPERATIONS

AirToolz Software, a joint venture between LMC Design and Construction, Inc. and IBM Business Partner Unity Software, saw a need to streamline and automate the communication processes among personnel. Since scheduling changes in the construction industry are frequent and affect all parties involved, improving the communication process provides maximum flexibility and responsiveness to changing needs and requirements.
Why become an on demand business? For years, building superintendents have complained that they spend at least 60 percent of their day calling subcontractors to make scheduling changes. Since each change can throw off the schedule by a few days, builders often miss deadlines, resulting in greater costs. AirToolz decided to create a new wireless scheduling solution for construction superintendents that enables remote access to scheduling information for site management and real-time updates on changes.
How and where did AirToolz start? AirToolz provides a field force automation solution in which construction workers located around different sites can communicate with each other and access information via wireless Palm handhelds. IBM's approach to embedded Java is unique

in that it extends existing business processes, using open standards, out to PvCs by integrating enterprise components such as databases and message queuing in its runtime. This allows for seamless integration with enterprise applications and provides AirToolz with the advantage it needs to deliver wireless productivity solutions to a very demanding crowd.

What benefits did AirToolz achieve?

1. AirToolz customers can focus their efforts on construction rather than rearranging schedules; as a result, 80-day homebuilding cycles have been reduced by two weeks.
2. Customers save about $3,000–$4,000 (USD) per home in labor, materials, and completion penalties.
3. AirToolz speeds payments to subcontractors, increasing user satisfaction.
4. Easy-to-use, resilient communication tool works in wireless connected or disconnected modes, eliminating the need for paper schedules and multiple phone calls.

Arena Holdings: Preventing Perishables from Spoiling Profits

Food Distributor Turns to Wireless for Real-Time Product Information and Market Conditions

Arena Holdings is a leader in the Italian market of food production and distribution, distributing more than 500 types of products from four production sites. Arena needed a wireless solution to support 100 mobile sales representatives during their daily activities away from the office. The solution needed to give the sales representatives the ability to search a product catalog, see customer information and profiles, consider the price list associated with a particular customer profile, and compose product orders by using tools hosted on PocketPC devices.

Why become an on demand business? The company knew that perishable products lose value the longer they sit on the shelf. However, its field sales representatives were having trouble closing sales quickly and profitably because of unstructured processes to convey continually changing customer and product information. Arena needed an integrated channel for rapid distribution of important product and promotional data to its sales force, so it could become more responsive to changing market conditions, and therefore more competitive.

How and where did Arena Holdings start? Arena decided to create a system to automate the real-time delivery of information, including product catalogs, customer account records, and details of current promotions. IBM designed a sales force automation solution based on IBM DB2 Everyplace, IBM DB2 Everyplace Sync Server, IBM WebSphere Studio Site Developer, and a third-party Java application on the PocketPC platform. It allows sales reps to instantly retrieve information and transmit electronic order forms back to Arena headquarters. In addition, the solution is continually updated and accessible by wireless links on handheld computers.

What benefits did Arena Holdings achieve? IBM's solution provided five primary benefits to Arena Holdings:

1. On-the-fly product and customer information, enabling Arena's sales reps to be more productive and close sales faster
2. Improved demand projections, decreasing the amount of unsold food produced and increasing Arena's profits
3. Wireless gives the sales representatives greater mobility, making them more flexible and productive
4. Open standards platform promotes quick, smooth integration with Arena's three separate manufacturing software systems, reducing the company's IT administration burden and increasing efficiency
5. The company has saved millions of dollars in the solution's first year, equivalent to 400–600 percent ROI

Aviva: Receiving a Premium Rating

UK INSURER USES INFORMATION ON DEMAND TO CALCULATE MOTORIST PREMIUMS

Aviva is the United Kingdom's largest insurance group, and the world's seventh largest insurance group. Aviva selected IBM to provide the technology for its "Pay As You Drive" pilot insurance program. The program will collect data from customers' vehicles and calculate insurance premiums based on when, where, and how often the vehicles are used.

Why become an on demand business? To establish this two-year research and development pilot program, Aviva needed an IT solution that would provide an infrastructure for information collection and analysis, as well as data security to protect customers' confidential information. Aviva chose IBM for its expertise in Telematics technology, as well as its knowledge of the insurance and automotive industries.

How and where did Aviva start? IBM will provide the hardware and software, as well as the actual in-car "black box" Telematics devices. The in-car devices will use GPS information to record the speed and location at which a vehicle is driven. The black box will measure vehicle usage and transmit the real-time data for analysis using a mobile phone network from Orange. The program will also allow examination or replay of any trips taken by the vehicle, enabling Aviva to conduct usage and risk analysis to refine its billing process.

The information the system collects is personal and private to the driver in question. IBM will provide data security expertise so Aviva can ensure privacy protection and assure drivers that their data is protected from internal or external attack.

What benefits did Aviva achieve? The majority of Aviva's customers have indicated that they would prefer their auto insurance to reflect the use of their car. A successful "Pay As You Drive" program will provide motorists with fairer insurance rates based on actual use of their vehicles and the ability to locate and track vehicles reported as stolen. The program will also provide the ability to request emergency and other services using a specially designed, multi-button console.

Beijing Golden Trucking Technology Company: Starting the On Demand Business Journey

MOBILE SOLUTION PUTS A MAJOR TRANSPORTATION BUSINESS IN THE FAST LANE

China's importance as one of the world's major production centers is growing fast, and opportunities for freight and logistics services in the region are on the rise as a result. With increased deregulation as well, competition is revving up to capture opportunities. Gaining a jump on its rivals, the **Beijing Golden Trucking Technology Company** is accelerating into the e-logistics fast lane by deploying an integrated mobile technology solution that will give it added flexibility, speed, and customer responsiveness.

Why become an on demand business? With an ambitious goal to be the number one logistics service provider in China, the company needed to find new ways to orchestrate its cargo and warehouse logistics services—to speed up operations while reducing costs and improving customer satisfaction. It also needed to be more flexible and responsive to the changing demands of customers, distributors, trucking fleets, warehouses, and freight owners.

How and where did Beijing Golden Trucking Technology Company start? A mobile communications system—integrating wireless, GPS, and SMS technologies to form a supply chain system—gives the Beijing Golden Trucking Technology Company the opportunity to overtake its rivals. With a new, real-time logistics dispatching and management system transforming paper-based processes, the company is able to improve route tracking, planning, and delivery, and to respond to customers through Web-based solutions, which provide enhanced information services.

What benefits did Beijing Golden Trucking Technology Company achieve? Faster distribution, instant dispatch capabilities, and the introduction of new services have already helped the company to expand its business to cover a network of more than 200 distributors and partners. More efficient route planning—supported by IBM xSeries servers—has reduced downtime of trucks and made pickups and deliveries easier to schedule. The company's improved flexibility has enabled higher levels of productivity and lower operating costs, thanks to a reduction in paperwork and time savings. It now offers a far more responsive, efficient, and reliable service to its customers, which is fuelling business growth.

Cahoot: Getting in "Cahoot" with IBM

CAHOOT BANKING PUTS CUSTOMERS IN CONTROL WITH IBM ON DEMAND BUSINESS BANKING SOLUTIONS

Cahoot, a division of Abbey National, one of the UK's leading retail banks, built an initial online customer base of more than 160,000 accounts in less than one year. It is now over 300,000 and continuing to rapidly expand. To continue this expansion and retain existing customers, Cahoot needed to satisfy demands for high security, availability, and transaction processing speed, plus enable easy integration with a wide range of consumer devices.

Why become an on demand business? Abbey National started Cahoot in 2000 as a way to compete more effectively in the deregulated banking market by offering consumers convenient financial services through the Internet, cell phone, and other electronic channels. As a startup, Cahoot also had to provide attractive rates, which meant it had to minimize its operating costs, including capital investment and staff requirements, while still ensuring high stability and redundancy. By transforming its business to be more responsive, variable, and resilient, Cahoot pushed itself further in the on demand journey.

How and where did Cahoot start? Cahoot decided to outsource technology management to minimize up-front and ongoing hardware and IT staff costs without limiting its ability to serve growing numbers of customers. Now, Cahoot can focus on developing appealing new financial services and bringing them to market sooner. IBM Managed Hosting provides a hosted solution based on IBM eServer pSeries and iSeries machines, IBM WebSphere Business Components, IBM Visual Banker-Internet, and software from IBM Business Partner Fiserv. Cahoot conducted collaborative research at the Centres for IBM e-business on demand innovation to develop new products and channels.

What benefits did Cahoot achieve? The IBM Global Services group handles all IT issues so Cahoot can focus on growing its business and providing richer services. At the same time, Cahoot has competitive IT expenses, enabling it to offer better rates and be more competitive. Cahoot's hosted solution is highly resilient, assuring customers at least 99.6 percent uptime, supporting the increasing number of customers and transactions. The infrastructure supports electronic transaction methods, including straight-through processing, which is faster and more cost-efficient than manual check processing, the standard method employed in traditional banks. Finally, the hosted solution's open standards support new access channels, including Web-enabled cell phones and interactive TV, and new products such as Web cards, expanding Cahoot's business opportunities.

Charles Schwab: Enabling Information On Demand

A WORLD-CLASS INVESTMENTS COMPANY TWEAKS ITS SYSTEMS TO ALLOW ON DEMAND RESPONSES TO CUSTOMER INQUIRIES

Charles Schwab, a financial services company and long-time IBM customer, wanted to investigate Grid Computing technology and what it might mean for its business. It defined an On Demand challenge for IBM to solve: How can Schwab employees provide immediate, real-time help to customers within an IT infrastructure that currently necessitates customer callbacks?

Why become an on demand business? When customers phone Schwab with questions, they frequently can't get immediate answers because the application that employees use is too slow in responding. Instead, customers have to wait for a return call. Speeding up the application would increase customer satisfaction.

How and where did Charles Schwab start? IBM and Charles Schwab's Advanced Technology Group took an existing application that ran on non-IBM systems and enabled it for Grid Computing with the Globus Toolkit running RedHat Linux on IBM eServer xSeries 330 machines.

What benefits did Charles Schwab achieve? This solution reduced the processing time on the application from more than 4 minutes to 15 seconds: a 94 percent improvement. Since this particular application scaled extremely well in a Grid Computing environment, proving the potential for Grid Computing-based services in a financial services environment, Schwab hopes to implement the solution sometime this year. The two companies are now looking at ways to expand Grid Computing activities to other areas of Schwab's business.

eArmy University: Providing Technology for Educating Soldiers

ENABLING WEB-BASED LEARNING FOR THE ARMY, TO ALLOW SOLDIERS TO "BE WHAT THEY CAN BE" IN A GLOBAL ON DEMAND BUSINESS INFORMATION THEATER

eArmy University provides online education for enlisted soldiers around the globe, helping them further their professional and personal goals, and also providing the Army with top preparation for its forces. eArmyU offers Web-based courses that can be completed anytime, anywhere, allowing soldiers to study at times that are most convenient for them. It brings together a collaboration of colleges and universities by offering approximately 137 programs from 32 different educational institutions.

Why become an on demand business? The military wanted to develop a more skilled and dynamic force to respond to the complexities of the 21st Century battlefield. To accomplish this, it needed to develop a reliable and effective approach for educating troops without being restricted by time and location.

How and where did eArmyU start? IBM designed an e-Learning virtual university with 32 participating colleges and universities. It also integrated applications from 12 different technology and software partners, built the IT platform, and launched the Web portal on which eArmyU runs. IBM hosts the eArmyU portal and offers centralized help desk support 24 hours a day, 7 days a week.

What benefits did eArmyU achieve? eArmyU now has the ability to deliver education from 32 different colleges and universities to soldiers in 36 countries, 4 U.S. territories, and 49 states through a single Web portal. The new portal has become the preferred method of education in the Army and will open up educational opportunities to 60 percent of soldiers who currently do not participate in Army education programs.

Finnair: Turning to IBM for On Demand Business Transformation

AIRLINE SEEKS TO SURVIVE INDUSTRY CONSOLIDATION AND DOMINANCE BY BIG AIR CARRIERS

Finnair, one of the world's oldest operating airlines, wants to be the travel industry's digital champion, estimating that more than half its passengers will soon be using the Internet for airline services, from making ticket reservations to clearing check-in. To meet this anticipated customer demand, Finnair came to IBM, not only to transform its service chain and IT systems, but to create an innovation center for incubating new solutions for the airline industry, such as wireless check-in, e-ticketing via the Internet, and wireless ticket sales.

Why become an on demand business? Finnair had been using IBM products for years. So when it chose to explore a real business transformation, Finnair chose IBM for its broader business value, its depth in research and technology, and long airline industry IT experience.
How and where did Finnair start? IBM is working with Finnair to move to a digital service chain so that every customer contact becomes more personalized, available as a record, and integrated with related records. This requires integrating its internal business processes so that the effect of variables in one process can be seen across the enterprise as a whole and integrating its middleware so that Finnair's different computer systems could work together. By using open standards, Finnair can interact with any company, supplier, and partner.
What benefits did Finnair achieve? Moving to an on demand business utility model will allow Finnair to scale up or down according to demand and pay only for the computing capacity it uses. Additional benefits Finnair will enjoy from this approach include the following: the flexibility for responding dynamically to any kind of variable across its systems or in the marketplace; improving its competitiveness; moving its cost structure from fixed to variable; reducing IT-related expenses and release capital for its core airline business; and reinvesting a significant portion of resulting cost savings in business transformation to help realize its vision of being the digital travel industry champion.

Harry and David: Meeting the Holiday Season Rush with On Demand Business Operations

USAGE-BASED PRICING PROVIDES FLEXIBILITY AND RESPONSIVENESS, WHILE ALSO REDUCING OPERATIONAL RISKS

Harry and David is an American mail order legend, marketing and distributing America's best-loved gifts: fruit baskets, rich chocolates, gourmet meats, confections, and Royal Riviera Pears. About 65 percent of its annual sales occur during the holiday season, beginning in mid-November and climaxing during a peak week prior to Christmas, where transactions can spike to 90,000 orders per day. With many customers submitting multiple orders, shipment transactions can run as high as 300,000 per day.
Why become an on demand business? High volumes of activity could overload servers, resulting in slow deliveries, unhappy customers, and lost revenue on future sales. Harry and David wanted to have access to capacity as needed on a usage-based pricing model, making it more resilient and responsive during intense periods of sales activity.
How and where did Harry and David start? The company began a steady migration to all IBM hardware—a mix of eServer pSeries, xSeries, and zSeries servers—and IBM software, including the IBM DB2 Universal Database. The WebSphere platform and pSeries Web servers were added to power Harry and David's popular e-commerce site launched in 1998. Harry and David also implemented a usage-based pricing approach to help the company improve system reliability, availability, and efficiency in a cost-effective manner.
What benefits did Harry and David achieve? Usage-based pricing allowed Harry and David to handle traffic while significantly lowering the risk of downtime from overloaded servers, thus

protecting sales revenues. The company realized a savings of approximately $550,000 (USD) in a single holiday period, largely through cost avoidance by importing capacity on demand. The organization also became more responsive to higher and unpredictable levels of customer activity, enabling the company to grow with less concern for increasing IT requirements. Finally, Harry and David eliminated the cost of carrying underutilized infrastructure during the non-holiday months.

Honda: Steering You in the Right Direction On Demand

ON DEMAND NEEDS DRIVE INVESTMENTS IN VOICE-BASED NAVIGATION SYSTEMS TO STRENGTHEN CUSTOMER LOYALTY AND SURPASS THE COMPETITION

To maintain its competitive position as the best-selling car in America, **Honda** set out to develop a first-of-its-kind voice-based navigation system based on IBM's voice recognition technology. Providing customers with voice directions on demand, Honda seeks to attract new customers, improve customer loyalty among repeat buyers, and create a competitive advantage.

Why become an on demand business? With a reputation built on offering customers quality products and premium options often seen only in higher end markets, Honda sought to mass-market the rapidly growing field of Telematics. Positioning the company as an early adopter, Honda hopes to sustain its competitive advantage and continue to expand its market share.

How and where did Honda start? IBM's deep industry expertise helped Honda to identify the market for the solution. Working closely with Honda's team, the revolutionary new speech recognition solution complements the existing navigation system, utilizing 180 preset commands and 7 million points of interest.

What benefits did Honda achieve? Honda's responsiveness to and adoption of wireless technology establishes the automotive giant as a technology innovator, thus strengthening its competitive advantage. Partnering with IBM, Honda's "Touch by Voice" navigation system integrates previous research and development investment in satellite-linked navigation systems. Ultimately, Honda's push to bring emerging technology to the mass market will improve customer satisfaction and loyalty.

Kohl's: Erasing the Worry of Seasonal Demands

RETAILER TAPS INTO DYNAMIC CAPACITY TO IMPROVE BUSINESS OPERATIONS AND SUPPORT GROWTH STRATEGY

Kohl's department stores encounter more business on the frantic Saturday before Christmas than on a lazy Tuesday in mid-July. Moving Kohl's to on demand business, IBM upgraded the company's overall IT infrastructure and eliminated costly downtime while pegging expenses to the amount of capacity used.

Why become an on demand business? In the past, the company had to maintain costly IT capacity for seasonal surges. To support its growth strategy both on the ground and on the Web, Kohl's wanted to upgrade its IT mainframe infrastructure with hardware that offers dynamic upgrade capabilities and reduces software-related outages.

How and where did Kohl's start? Kohl's replaced its online store hardware using IBM Capacity Upgrade on Demand. As a result, the retailer taps into additional processors and storage when traffic suddenly increases and scales back when demand subsides. Kohl's upgraded its existing hardware to an IBM eServer zSeries 900, pre-configured with a single-footprint IBM Parallel Sysplex for uninterrupted availability during software updates.

What benefits did Kohl's achieve? Customer satisfaction and online revenue opportunities have increased with Kohl's Web site up 100 percent of the time, and average system response time has dropped from 9.3 to 2.6 milliseconds. Kohl's saves money—only paying for the IT resources the company uses and avoiding a big investment in enough hardware to handle maximum spikes in seasonal customer demand. Kohl's can now process twice as many online transactions per second due to a 130-percent increase in overall system capacity and a 160-percent increase in processing speed.

LEGO: Building for a Flexible Future

AN IBM COMPETITOR'S SOLUTION IS TOPPLED BY THE INNOVATION OF IBM ON DEMAND BUSINESS SOLUTION DELIVERABLES

Toy fashions and fads quickly come and go. Creating toys that stand the test of time and gain the approbation of generation after generation is an art. But in a marketplace that is literally brimming with new products and ideas, you also have to capture the imaginations of youngsters and parents with novelty and innovation. As one of the leading players in the toy market, **LEGO** appreciates that it must balance the continued success of the world's most popular construction toy, while responding to marketplace changes with new products.

Why become an on demand business? Just as LEGO bricks can be reassembled into an unlimited array of structures and creations, LEGO wanted to achieve the utmost adaptability for its business. It needed greater flexibility and integration company-wide and saw that it could achieve this by streamlining and simplifying processes and systems, and by introducing variable cost structures.

How and where did LEGO start? LEGO's existing technology infrastructure did not support the flexible and responsive business model it envisaged. Based on 61 HP Alpha servers and 175 HP Compaq servers, it was also expensive to manage. But by working with IBM and business partner TSS to implement a totally new platform based on IBM server and storage technology, LEGO realized it could achieve a far more flexible infrastructure and reap significant cost savings.

What benefits did LEGO achieve? The IBM solution provides LEGO with increased levels of business flexibility, permitting the dynamic addition and reallocation of server and storage capacity across different workloads according to business needs and priorities. LEGO estimates that it can reduce the number of servers by a factor of 10, with all the associated savings in systems management that the reduction will enable. IBM's on demand business solution includes significant additional capacity that can be made available in an instant to handle peak loads. Payment is based on metered usage, permitting LEGO to pay as its business grows.

Metro Group: Supplying the Future of Retail On Demand

PROVIDING RETAIL CUSTOMERS THEIR PRODUCT CHOICE AT THE RIGHT TIME AND PLACE, IN AN EVER-INCREASING ON DEMAND BUSINESS WORLD

Metro Group, a major retailer in partnership with IBM, has opened a futuristic supermarket in Germany to test new retail technologies with customers. These technologies bring the promise of increased efficiency and customer responsiveness, key differentiators in the tough retail market.

Why become an on demand business? A major problem for retailers is keeping track of their stock and being able to provide their customers the right product, in the right color or size, at the right time in the right place. "Shrinkage"—the disappearance of articles by theft, damage, or loss—costs retailers billions every year. Metro Group chose IBM for its radio frequency identification (RFID) technology, which means products can be tracked at any point in the supply chain, eliminating significant inefficiency and shrinkage while better satisfying customer demand.

How and where did Metro Group start? IBM's business consultants began by redesigning the supermarket's supply chain process, integrating a system that now includes "smart shelves" that have the ability to communicate—warning the stock rooms when the razor rack is nearly empty, for example, or triggering promotional advertising on a nearby screen as a customer picks up a shampoo bottle.

The RFID gives a complete picture of where each item is in the store, unlike bar codes, which only track by groups of identical items. RFID tags in shopping carts tell store management how many carts are in the store. If there's an increase, additional checkouts are opened to accommodate the extra volume, bringing an end to delays at the checkout.

What benefits did Metro Group achieve?

1. Reduction in stock-carrying costs through better inventory management
2. Fewer sales losses caused by empty shelves
3. Reduced theft because tags raise alarms at exits unless products are scanned at a checkout
4. More effective staff because PDAs provide real-time information on stock
5. Increased sales through targeted in-store promotion
6. Access to information on customer buying habits, data that can be used to predict demand and automatically be fed back through the supply chain

Mikasa: No More Leaving Business on the Table with On Demand Business

SYSTEMS INTEGRATION ENHANCES CUSTOMER'S ON DEMAND VISIBILITY OF AVAILABLE PRODUCT INVENTORIES

Mikasa, a fine china and dinnerware manufacturer, needed to build communications across its enterprise as part of a drive to provide a more convenient Web-based shopping experience for customers. Mikasa chose IBM to integrate its online store with its backend systems to make search and purchase functions more efficient and reliable. The quick addition of updated Web site features, such as a gift registry, was an added benefit.

Why become an on demand business? Mikasa's lack of integration between its online store and backend systems, such as order processing, made for a clunky online store and potentially poor customer experience—a recipe for disaster. By integrating these systems, customers are able to see what inventory is available at any given time and place an order, which is processed automatically, rather than being re-keyed by Mikasa staff.

How and where did Mikasa start? Citing stability, quick response time, and reliability, Mikasa chose IBM's WebSphere software, including WebSphere Commerce, Commerce Studio, Applications Server, Studio Application Developer, and DB2 to build a new Web site. IBM's e-server iSeries and IBM Intel processor-based servers were selected to host it. IBM business partner Pulver Technologies was tapped to build a fully integrated solution, which was delivered in only 12 weeks.

What benefits did Mikasa achieve? Mikasa's new inventory tracking system has let the company expand the number of products available to customers online, increasing the number of stock keeping units (SKUs) from just 400 to over 20,000. Furthermore, orders are now processed immediately rather than being re-keyed by back-office personnel, improving order accuracy and cutting order processing costs. The new Web site also features a gift registry linked to both the online store and point-of-sale terminals in Mikasa's 167 retail outlets, giving customers added flexibility to shop wherever they want.

National Digital Mammography Archive (NDMA): Empowering Hospitals with X-Ray Access On Demand Business

A POWERFUL AND UNPRECEDENTED ON DEMAND BUSINESS GRID COMPUTING SOLUTION RESULTS IN IMPROVED MEDICAL DETECTION AND DIAGNOSIS

National Digital Mammography Archive (NDMA) emerged in response to the need for a more responsive process that would permit radiologists to access records in real time, enhancing opportunities for improved diagnosis, while saving hospitals millions in X-ray film costs.

Why become an on demand business? Hospital radiologists needed a faster, more reliable way to share medical data to speed up and improve detection and treatment. The challenge was to develop a patient-centric medical record system that could capture—from any location—a full range of healthcare files, including high-fidelity patient medical images, records, and clinical history.

How and where did NDMA start? With IBM's help, by 1999, the National Scalable Cluster Project (NSCP) had successfully created a Grid Computing unit among three universities. The NSCP then decided to direct its efforts toward healthcare and developed the scalable NDMA. They knew the NDMA would have to be a massive distributed computer, delivering computing resources as a utility-like service over a secure Internet connection so all hospitals within the network could access data.

What benefits did NDMA achieve? This complete IBM Grid Computing solution saves hospitals millions of dollars annually in X-ray film costs because of the use of digital imaging. It also permits real-time access to patient data, compared to weeklong waiting under the previous process, and further enhances research capabilities between universities and healthcare facilities. In addition, hospitals can leverage a shared technology infrastructure, avoiding up-front infrastructure costs and ongoing maintenance expenses. Image data is shared with any hospital on the grid because of its open infrastructure.

New York City: Building Inspections On Demand

IBM and LinksPoint Devise an On Demand Business Solution for Wireless Speed Researching and Reporting Related to Building Inspections

The **New York City Department of Buildings** (NYCDOB) oversees about 900,000 structures in the city of New York. That's a huge inventory of physical real estate represented by a very large, centralized set of Lotus Domino databases securely maintained on IBM eServer xSeries computers.

Why become an on demand business? NYCDOB inspectors needed to quickly transfer their observations, opinions, and violation determinations to the city's chief inspector so that they could efficiently make decisions on building closings, repairs, and re-openings. In conducting their inspections, NYCDOB inspectors also needed to rapidly access regulatory information in the databases from the field.

How and where did NYCDOB start? IBM, teaming with its partner LinksPoint, developed a prototype of a mobile application to assist in the preparation and filing of building inspection reports. LinksPoint built the application to run on HP iPAQ handheld PC devices, and developed the synchronization capability for updating NYCDOB's Lotus Domino databases using information from the field. IBM Global Services managed the entire process, provided business process analysis, and designed the Domino databases that serve as the central repository for building inspection information.

What benefits did NYCDOB achieve? The new mobile application saved the NYCDOB about 40 minutes per inspection, increasing efficiency 25–40 percent. Inspectors can efficiently update the enterprise database from the field for rapid information delivery. The wireless infrastructure created to enable this application also supports access in the field to electronic versions of the city's building code and the NYCDOB's *Construction Inspection Manual*, among others. By delivering such information through HP iPAQ handheld PC devices, inspectors are spared carrying around thick binders, or worse, going without the information.

Olex Cables: Answering the Call for Response On Demand

Australian Manufacturer Wanted to Become More Responsive; Now, It Is an On Demand Business Provider with Immediate Responses

Olex Cables, Australia's largest cable manufacturer, could not respond fast enough to customers who were calling in record numbers. Operators needed quicker and easier access to the company's inventory, pricing, and order information. IBM's Web Services and IBM Business Partner Synergy Plus transformed Olex's call center into an on demand business by creating an open standards-based extranet that integrates Olex and customer systems. The staff now can handle more inquiries per hour, and customers can look up product and marketing information and order directly.

Why become an on demand business? Faced with a 50 percent increase in customer calls and a smaller staff because of consolidation efforts, Olex Cables needed help in responding quicker to client requests.

How and where did Olex Cables start? IBM and business partner Synergy Plus, an independent Australian software vendor, built a three-tier Java-based application with WebSphere Application Server that enabled authorized customers to access price checks, inventory levels, and order status from a secured area of the Olex Web site. After considering a proprietary solution, the medium-sized business made a strategic decision to use Java technology to build an open standards-based, flexible on demand Operating Environment infrastructure to ease future integration.

What benefits did Olex Cables achieve? Olex is saving more than $300,000 (USD) annually in labor costs and expects to realize full payback within three years. Operators are serving customers better by handling hundreds of inquiries daily, and customer response time has been reduced to seconds. Olex is also saving money by using the solution to integrate employee, business partner, customer, and inventory systems into a coordinated process.

A Financial Services Company: Streamlining Operations and Expanding Services with On Demand Business

INTEGRATION OF SYSTEMS INTO AN ON DEMAND WORLD ENHANCES CUSTOMER RESPONSIVENESS AND REDUCES OVERALL COSTS

A large financial services company that serves individual and institutional customers worldwide wanted to reduce costs in the administrative areas of its business, while at the same time expand its IT environment by adding applications and systems. This financial services company chose to implement an IBM IT solution to automate the administration of user access rights for more than 65,000 users, manage security of access privileges across hundreds of systems, ensure that security policies were enforced enterprise-wide in an auditable fashion for regulation compliance, and ensure that all private data was adequately protected.

Why become an on demand business? Financial services firms around the world have discovered that it's possible to establish a competitive advantage by offering a broad portfolio of services to customers. As this New Jersey-based financial services company discovered, that advantage can be lost if customers do not have a simple, convenient way to access accounts and policies—on demand. The company understood increasing profits meant becoming more responsive to customer needs, for access to updated information in real time and across accounts while reducing costs.

How and where did the financial services company start? The financial services company decided to improve responsiveness and cut costs by providing customers with the ability to access and manage their accounts instantly, and determined the best solution was a Web portal that would integrate information from an array of sources. Working with IBM Software Services, the financial services company developed a portal based on Java technology that integrated ten disparate systems from nine lines of business in just five months. The solution is built on IBM WebSphere Portal, Application Server, MQ Series and Studio, IBM DB2, IBM Lotus Domino and IBM pSeries and IBM UNIX-based processors.

What benefits did the financial services company achieve? The financial services company now enjoys lower operating costs and improved service levels. This on demand business transformation now allows the company to deliver new products and tools to customers more quickly and securely than before. IBM's solution provided several primary benefits to the financial services company. For example, the portal currently supports 300,000 registered users making 8,000 unique visits. Fifteen percent of exchanges and five percent of redemptions are processed through the site. There are also four very key benefits that are realized by this on demand business transformation:

1. Nearly 155,000 mutual fund statements have been downloaded since inception
2. Self-service capabilities have reduced call center volumes by 23 percent, lowering costs and freeing staff for more value-added tasks
3. The integration of content through open standards, such as Java, XML, and XSL (eXtensible Stylesheet Language) allows for greater flexibility and development efficiency
4. The resilience of pSeries and IBM UNIX-based processors provides business continuance and disaster recovery

Quimica Suiza: Proving that Wireless Is the Right Prescription

A PHARMACEUTICAL DISTRIBUTOR NEEDED REAL-TIME, ON DEMAND ACCESS TO A PLURALITY OF SALES FORCE INFORMATION

Quimica Suiza, a leader in the wholesale distribution of pharmaceutical products in Peru, was losing sales to faster, more responsive competitors. To remain competitive, Quimica Suiza's sales force had to collect payments as well as sell, but its previous sales force automation solution provided little support for either task. IBM, in conjunction with its business partner Synopsis, delivered a wireless solution offering real-time access to the desired information as needed.

Why become an on demand business? Outdated, manual processes meant the company's sales force could not get up-to-date inventory and pricing information. Another challenge was that 70 percent of all orders were handwritten and required a full business day just to go from the customer's office to the Quimica Suiza warehouse. The company realized it needed to transform, automate, and integrate its processes and systems. Quimica Suiza realized that aspects of on demand business operations were in order, to be more responsive and more successful in a very competitive market.

How and where did Quimica Suiza start? Quimica Suiza elected to develop a wireless solution that offered its sales force real-time access to its existing sales and inventory database and order processing transactions. IBM business partner Synopsis delivered the solution in just eight months using IBM WebSphere Commerce, IBM WebSphere Application Server, IBM DB2, and IBM Mobile Connect running on IBM eServer zSeries and xSeries.

What benefits did Quimica Suiza achieve? The solution, called QSMOVIL, permits the sales force to check inventory or account details, answer customer queries, and place orders in real time. As a result, order fulfillment time has improved by 38 percent. This more responsive process has helped to drive double-digit increases in both revenue and profits. QSMOVIL also has reduced order entry errors, has led to a 15 percent increase in sales force productivity, and has cut the costs of sales for Quimica Suiza.

Saks Fifth Avenue: Knitting Business Operations Together for On Demand Business

SAKS FIFTH AVENUE SHOPS AROUND FOR RESPONSIVE IT SYSTEMS AND SELECTS IBM-ARIBA FOR ITS ON DEMAND BUSINESS SOLUTION

After a series of key acquisitions, **Saks Fifth Avenue**, a nationally known retail brand, sought to rein in its procurement expenses to efficiently target suppliers and divest itself of non-core operations. Saks worked with IBM and its strategic alliance partner, Ariba, to gain a single view of all requisition-related activities.

Why become an on demand business? Saks wanted to give employees a corporate information system to track and manage all merchandise on demand. Saks also sought to divest itself of operations not directly related to these core tasks to control spending, reduce the cost of operations, and improve financial performance.

How and where did Saks start? IBM and strategic alliance partner Ariba helped Saks deploy an e-procurement solution that aggregates spend data, electronically reports travel-related expenditures, manages the competitive bidding process, finds suppliers, and negotiates agreements.

What benefits did Saks achieve? The IBM-Ariba solution has given Saks a single view of all requisition activities. The company has realized 12–14 percent in procurement savings, flexible and efficient cost structures, and centrally managed spending operations. It is on track to achieve anticipated ROI.

Telstra: Finally, Obtaining Data That is Accessible, Relevant, and Timely

A WORLD-CLASS TELCO TRANSFORMS PROCESSES TO RESPOND TO CUSTOMER DEMANDS AND ADAPT TO AN EVER-CHANGING MARKETPLACE FOR ON DEMAND BUSINESS SERVICES

Telstra continually strives to be a world-class, full-service, integrated Telco, helping Australian and Asia-Pacific customers and communities prosper through their access to innovative communications services and multimedia products. Following deregulation and partial privatization, Telstra faced serious threats: competition and viewing customer service as a key differentiator.

Why become an on demand business? Telstra was facing a changing marketplace. Deregulation had increased competition and made customer service a key differentiator. This posed a significant challenge for the company: Inefficient processes meant its service technicians spent more time retrieving data than they did solving problems. To be more responsive to customer needs, the company had to update its processes and systems for more efficient service.

How and where did Telstra start? Telstra created a new electronic communications system that gives field agents real-time access to accurate, up-to-date business information. The solution's pervasive infrastructure, based on open standards, supports a wide variety of wireless devices. IBM Global Services Australia designed and developed the integrated, end-to-end solution, which is based on IBM WebSphere Everyplace Server, IBM Host Publisher, and IBM HTTP Server accessing applications on IBM eServer zSeries.

What benefits did Telstra achieve? IBM's solution has streamlined nearly all facets of the dispatch and reporting process, allowing technicians to complete more work in less time and improving customer satisfaction. In addition to improving the efficiency of information access, the solution has improved the quality of Telstra's information. By enabling a broad restructuring of business processes, the wireless solution also has enabled significant cost savings, particularly in the area of support costs. Because it is based on open standards, the solution gives Telstra the flexibility to incorporate new devices or add new applications to respond to changing customer needs.

Twinlab: Proving Vitamin Supplements Are No Longer Kids' Play

UNIFYING DISPARATE BUSINESS PROCESSES TO IMPROVE FOCUS AND FLEXIBILITY AROUND MARKET OPPORTUNITIES

Twinlab is a leading manufacturer of high-quality, science-based nutritional supplements. What started as a "mom and pop" vitamin shop about 35 years ago on Long Island has since expanded into a multinational, $250,000,000 (USD) health and wellness company, with over 500 employees. From vitamins, herbs, and minerals to sports and diet supplements, Twinlab's complete line of products can now be found in health food and nutrition stores throughout the U.S. and overseas.

Why become an on demand business? As a result of Twinlab's enormous growth from a small business vitamin shop to a multi-million dollar international health company, problems surfaced with inventory management, poor communications across five divisions, and the inability to scale to meet customer demand. Complicating matters, the midsize manufacturer, lacking strong technical support, wasted resources on IT administration. To overcome lost business and profits, Twinlab needed to unify its disparate business processes with a fully supported on demand business platform, making it more flexible to take advantage of market opportunities and more focused on core competencies.

How and where did Twinlab start? Twinlab and IBM Global Services integrated all five divisions, implementing a comprehensive inventory management system, consolidating databases, and streamlining operations and communications. This end-to-end enterprise resource planning (ERP) solution utilizes the IBM eServer xSeries, running SAP, to manage Twinlabs' distributor warehouses and an iSeries with DB2 to handle the operating system, database management, and networking. For technical and functional support, Twinlab employed AMS, the IBM Help Desk, and received IBM services, including technical resources and training.

What benefits did Twinlab achieve? This comprehensive solution enabled Twinlab to respond to customer demand with greater speed and more quickly implement aggressive growth strategies to drive revenues. In addition, Twinlab saw improved asset efficiency and inventory management by achieving a real-time look across five divisions in different locations. These improved enterprise-wide communications were made possible by the reliable xSeries, and virtually eliminated system downtime while lowering TCO. Finally, Twinlab can concentrate on its core competencies, rather than IT administration; due to its tight integration with IBM help desk support and on demand business services. IBM's Application Management Services offering provides the safety net for ongoing support and the flexibility to utilize resources in response to fluctuating demands.

Viewpointe: Checking Out Archive On Demand Business Services

TRANSFORMING CORE BANKING PROCESSES AND DELIVERING A NEW IMAGE WITH A STRONG PRESENCE IN THE ON DEMAND BUSINESS WORLD

Breaking new ground in an area that has remained virtually unchanged for years is part of the stellar mission of **Viewpointe**. Viewpointe Archive Services was conceived with the goal of providing leading financial institutions with an innovative, online archiving repository for storing, accessing, retrieving, and sharing check and document images. Viewpointe's service is based on a pay-as-you-go (utility) model. This means banks only pay for what they use, enabling them to improve operational efficiencies, save money by reducing costs, and provide better services for their own customers, including enhanced fraud protection. As such, Viewpointe is driving a fundamental transformation of core banking processes and giving this service level a whole new on demand business image.

Why become an on demand business? Even though online bill-paying services have increased significantly, bank customers have remained loyal to personalized checks, writing approximately 50 billion of them each year. So, finding new ways of reducing check processing costs and reducing the sheer manual effort and time involved have become an industry "Holy Grail." Viewpointe wanted to shift its business to a pay-as-you-go model and sought a technology partner to underpin its business, which was capable of delivering a reliable and secure check imaging service.

How and where did Viewpointe start? IBM is providing Viewpointe with a storage solution designed to support exponential growth, projected at 12–15 petabytes of data (a petabyte equals a thousand terabytes or a million gigabytes!). Viewpointe's flexible infrastructure from IBM enables it to use only the storage resources it needs. But as its requirements grow, it can quickly and easily expand archive capacity, removing the large up-front costs that could have prevented its business from taking off in the first place and giving it precise control over the pace of growth. **What benefits did Viewpointe achieve?** Checks are now processed once and the image is shared by both the receiving and issuing banks, reducing associated operating costs by one-third. The solution lets Viewpointe reduce risk and deliver high performance and reliability to customers, respond faster to bank needs, and reduce the costs and errors associated with manual processing.

Wimbledon: Continuing to Volley in IBM's Court of On Demand Business "High-Volume" Information Processing

A WORLD-CLASS EUROPEAN TENNIS CLUB LEVERAGES IBM ON DEMAND BUSINESS DELIVERABLES FOR FLEXIBILITY AND RESPONSIVENESS TO SCALE RAPIDLY DURING THE WIMBLEDON EVENT: AN EXTREMELY HIGH-VOLUME END-USER EVENTS SCENARIO

Doubles partners at the **Wimbledon** championships since 1990 have relied on IBM's innovative technology strokes to enhance the viewing experience of nearly 1 billion tennis fans each year. This solution suite manages extremely high volumes of networking traffic. As part of this total on demand business solution, IBM is helping our partners scale to 250 times normal IT capacity during the two-week tennis championship, and only paying for the increased demand as and when needed: Utility Computing at its best.
Why become an on demand business? The rapidity of change in how major sporting events are covered, and the ever-increasing expectations of present and virtual spectators, meant the All England Lawn Tennis and Croquet Club could not remain comfortable and maintain its title as the premier tennis championship in the world. To maintain this coveted title, Wimbledon had to find a way to be responsive in fulfilling a myriad of global spectators, players, and media needs—with resilience enough to manage any and all last-minute changes.
How and where did Wimbledon start? IBM turned Wimbledon into an on demand business through a robust infrastructure capable of scaling its normal capacity to 250 times for the two-week event. The total IBM solution integrates 7 operating systems and 24 applications that are brought together by a team of 180 experts from around the world. The solution includes year-round support for the club's back-office functions.
What benefits did Wimbledon achieve? By partnering with IBM, the All England Lawn Tennis and Croquet Club is able to enhance the Wimbledon experience for tennis fans worldwide, utilizing the latest innovative technology of on demand business. This way, Wimbledon can also remain small (in IT infrastructure) while delivering world-class services to a large audience, varied resources, and complex IT infrastructure—only during the two-week Wimbledon championships. By partnering with IBM, Wimbledon can access massive IT support, become an on demand business, without owning, managing, or maintaining the IT infrastructure.

A World-Class Internet Auction Company: Finding On Demand Business Services and Deliverables

A GLOBAL ONLINE AUCTION COMPANY ADOPTS ON DEMAND BUSINESS TO DELIVER RESPONSIVE SERVICES TO 62 MILLION WORLDWIDE BUYERS AND SELLERS

Servicing more than 62 million registered users, this online auction company is the leading online marketplace for the sale of goods and services in more than 18,000 categories. This online auction company needed a world-class on demand business solution to deliver a robust, responsive, resilient, new platform to support its growing customer base. This ultimately ended up allowing the company to focus on the core business issues, and become more responsive to buyers and sellers for their auction needs.

Why become an on demand business? The nature of this online auction company's business dictates that its network services be available 24x7. The company was concerned about whether its IT infrastructure could support its enhanced real-time services, while offering localized Internet auction portals in new regions such as Germany, France, Korea, the UK, and Australia.

How and where did the online auction company start? This online auction company decided to redesign its online auction business by expanding worldwide with country-level auction portals, producing additional revenue. To provide the supporting architecture, this online auction company selected the IBM Corporation because of its business transformation leadership and philosophy of in-depth customer support, allowing this online auction company to focus on functions designed to drive revenue in its core business.

IBM Software Services for WebSphere assisted this online auction company in transforming the IT infrastructure, which is now "powered by" open standards-based IBM WebSphere Application Server, Tivoli software, and resilient IBM eServer xSeries systems. Furthermore, IBM Learning Services also provided a full year of customized education and training, complete with hands-on labs and one-to-one mentoring.

What benefits did the online auction company achieve? IBM's solution provided four primary benefits to this online auction company:

- Maximized growth through the development of new revenue-producing auction portals to further support countries in Europe and Asia.
- More than 99.9 percent online auction availability helps guarantee that customer requests can proceed on plan, 24x7, increasing revenue and boosting user satisfaction.
- The close working relationship with and around-the-clock support from IBM allows this online auction company to focus on developing new real-time auction services for users, strengthening its overall competitive position.
- An investment in open standards-based, resilient systems and software helps manage complexity and resist external threats. This online auction company can quickly respond to new opportunities and add auction features with ease and flexibility.

A Major Healthcare Provider: Keeping a Health Insurer's Business Healthy

A MAJOR INSURANCE PROVIDER NEEDS CLAIMS INFORMATION CHEAPER, FASTER, AND IN AN ON DEMAND BUSINESS, RELIABLE FASHION

A major healthcare provider required new technology and new approaches to handling claims from its more than four million members and 15,000 employer accounts. IBM's Healthcare On Demand for Payer Operations solution was just what the doctor ordered. It not only reduced the customer's IT costs, matching expenses to revenues, but it also provided more real-time information on volume and cost of claims, allowing the customer to better manage its business now and develop new products for customers of the future.

Why become an on demand business? This major healthcare provider wanted to modernize its IT infrastructure and deploy a next-generation claims entry/processing system. The insurer needed to not only reduce its claims processing expense, but also manage its customer information in such a way that it could provide more individualized, customer-focused product offerings.

How and where did the major healthcare provider start? Working with the client and an IBM solutions partner, IBM delivered its Healthcare On Demand for Payer Operations solution. This robust solution incorporates business process transformation, global sourcing, systems integration, IT outsourcing, Utility Computing, and financing.

Now in partnership with this healthcare provider, IBM markets the utility-based solution on the shared infrastructure to other companies in the industry.

What benefits did the major healthcare provider achieve? Rather than a fixed price, the healthcare provider now enjoys a pay-as-you-go model that varies with the number of its members. This solution eliminates up-front technology investment and matches the company's IT expenses with its revenues. A gain-sharing arrangement provides a means of reinvesting the cost savings and upgrading the company's IT infrastructure.

The healthcare provider's claims processing and data warehouse were integrated by design to provide easily accessible and current data on actual volume and cost of claims, at a level of detail formerly unavailable to the company. This more accessible information should provide insights into medical management not before possible as well as allow the customer to generate new product offerings faster.

Savings to the client will be between $286,000,000 (USD) and $346,000,000 (USD) over 10 years.

A Global Air Carrier: Making Self-Service Customer Solutions No Longer "Pie in the Sky"

SELF-SERVICE KIOSKS RESPOND TO AIRLINE CUSTOMER DEMANDS AND IMPROVE THE PASSENGER EXPERIENCE FOR THIS GLOBAL AIRLINE

This global air carrier is widely recognized by major travel organizations for its outstanding customer service. While the carrier remains pleased with its growing passenger base, the airline was concerned that its legacy customer service systems might not be able to accommodate expanding growth rates.

Why become an on demand business? When a major customer survey revealed a need to improve the overall passenger experience, this air carrier looked toward innovative technology to revamp the airline's check-in processes. This air carrier wanted to improve its customer service system by creating self-service kiosks that allowed customers to make dynamic changes during the check-in process. The solution needed to match or exceed the existing level of agent-provided service.

How and where did the air carrier start? This air carrier pioneered self-service kiosks at airports where customers could buy tickets, check in, select seats, and look up arrivals and departures in real time. The kiosks connect with this air carrier's existing electronic ticketing and check-in system, based on IBM Transaction Processing Facility (TPF). This air carrier chose IBM as its technology supplier because of its broad range of solutions, especially in self-service kiosks, business transformation for on demand business services, and experience with the TPF infrastructure. IBM Global Services designed the solution, which includes IBM WebSphere MQ, IBM Kiosk Manager, IBM Consumer Device Services, and IBM Tivoli Distributed Monitoring, running on IBM eServer xSeries and pSeries systems.

What benefits did the air carrier achieve? The self-service kiosks have reduced passenger check-in time by 80 percent, when compared to traditional check-in ticket counters, thereby increasing workforce productivity and customer satisfaction. In fact, 93 percent of passengers surveyed said the kiosks improved their flight experience. This carrier is a role model and pioneer in this method of customer check-in.

Furthermore, the kiosks let staff focus their efforts in other areas, making this air carrier more profitable and responsive to customer needs. For example, this autonomous system monitors each kiosk in real time and can automatically page a technician if one malfunctions, reducing downtime and maintenance logistics, and improving this air carrier's responsiveness to the customer. The solution integrates smoothly with this air carrier's existing infrastructure, making the airline more cost-effective. In addition, the open standards platform provides a faster, more flexible development environment for future on demand business features that can further strengthen this air carrier's competitiveness.

European Catalog Retailer: Saving in Style with On Demand Business Operations

INVENTORY ON DEMAND DELIVERS STRONG SAVINGS, AND AN OVERALL ENHANCED CUSTOMER SHOPPING EXPERIENCE

Staying ahead of fashion trends can be perilous even for the most savvy. This European catalog retailer of clothing and textiles was regularly finding itself either overstocked with too much merchandise, or turning to more expensive suppliers at the last minute due to unforeseen demand. The company asked IBM to re-evaluate its supply chain management processes and find a solution that would reduce its on-hand stock and increase order fulfillment while keeping its costs fashionably low.

Why become an on demand business? With no automated system for anticipating or reacting to customer demand for products, this European catalog retailer frequently found itself facing inventory management difficulties. This was due mainly to the seasonal nature of its business. The company often had to spend more money than necessary reacting to customers' desires, rather than keeping up with which colors, styles, and fabrics were in vogue.

How and where did the catalog retailer start? IBM Business Consulting Services evaluated the company's supply chain management business and proposed key process transformations of its manufacturing resource planning (MRP) system. After a successful pilot of a new MRP system, IBM implemented it across the entire company, accessorizing the solution with troubleshooting and training for more than 180 employees.

What benefits did the catalog retailer achieve? It only stands to reason that improving inventory control also can boost the number of orders filled—you have more of what the customer wants ready in stock. Not only did this European catalog retailer realize savings of 7 million Euros in the first 12 months after the new MRP system was installed, the company also increased its order fulfillment from 84 to 88 percent. In addition, on-hand stock has been trimmed 35 percent. Now, with IBM providing ongoing performance monitoring, this European catalog retailer is targeting 92 percent fulfillment and another 20 percent reduction in stock.

A Large Global Telecom Provider: Keeping Suppliers Handy and Helpful with On Demand Business Services

Global Telecom Provider Seeks to Integrate Supplier Communications and Relationships

A large global telecom provider recently had to adjust its vision, a vision that has worked for the last 126 years of operations. Today, this telecom provider is an industry leader in mobile infrastructure communications, with a keen focus on four major areas: systems, services, licensing, and pervasive handheld devices. IBM offered the advanced technology this telecom provider was looking for—in semiconductors, ASICs[1] (application-specific integrated circuit), and microprocessors. This, in turn, helped it to successfully become the first to market with its third-generation telecom products.

Why become an on demand business? Increasing competition and a weakening telecom market meant this telecom provider needed to reduce procurement and administration costs while improving supplier relations, a task complicated by the dozens of disparate supply chain applications, each with separate access and log-in functions for suppliers.

1. An ASIC (application-specific integrated circuit) is a microchip designed for a special application, such as a particular kind of transmission protocol or a hand-held computer. One might contrast it with general integrated circuits, such as the microprocessor and the random access memory chips in a PC. ASICs are used in a wide-range of applications, including auto emission control, environmental monitoring, and personal digital assistants (PDAs).

How and where did the telecom provider start? This telecom provider chose IBM to implement an integrated on demand business approach to its supplier operations with the development of a supplier Web portal, providing single point-of-entry access to the telecom provider's supply applications. Corporate sourcing suppliers now utilize the portal to get real-time information, and to collaborate with this telecom provider on nine key areas of supply and procurement. Suppliers can become involved at all stages of the production process, including product development.

What benefits did the telecom provider achieve? The supplier Web portal allows this telecom provider to bring key sourcing partners into the value chain earlier, making the company more responsive to the supplier's combined needs, ultimately reducing production costs and creating higher quality products that better meet market demands. The portal also improves this telecom provider's supplier collaboration and supplier relations through:

- Integrated access and SSO to all applications
- Provision of workroom capabilities for supplier collaboration

An Asian-Based Automotive Manufacturer: Strengthening Customer Loyalty with Innovative On Demand Business Solutions

A LEADING AUTOMOBILE MANUFACTURER CREATES PERSONALIZED SERVICES FOR DRIVERS USING INNOVATIONS IN VEHICLES

This automotive manufacturer selected IBM to provide the technology for an IT solution that would allow it to improve customer loyalty through more personalized services while simultaneously lowering the cost of services it offers to customers. This automotive manufacturer chose IBM for its expertise in Telematics technology, as well as its knowledge of the automotive industry.

Why become an on demand business? This automotive manufacturer wanted to increase responsiveness to customers through personalized information and services. In addition, the company needed to improve warranty service, scheduling, and delivery. The solution had to be flexible enough to respond to market changes and integrate with key business processes.

How and where did the automotive manufacturer start? IBM Business Consulting Services designed an end-to-end customer Telematics solution to support and manage delivery of services for a variety of consumer interests, device types, and vehicles. This automotive manufacturer uses Telematics to understand the customer experience over the lifetime of vehicle ownership, provides access to relevant real-time vehicle information, and delivers personalized services to owners of its vehicles.

What benefits did the automotive manufacturer achieve? With this IBM solution, this automotive manufacturer has increased responsiveness to its customers. The solution provides access to real-time vehicle information, which helps this automotive manufacturer make more insight-driven decisions related to customer needs, service details, and vehicle maintenance data. It also allows the company to deliver more personalized services to customers—when they want it and how they want it.

The use of an open standards-based solution with advanced cellular communications enables an effective, lower cost infrastructure. It provides this automotive manufacturer better cost controls through flexible, scalable technology.

A Large Energy Company: Deploying On Demand Business Solutions Brings a New Sense of Responsiveness to Customers

INTEGRATION OF DATA AND FINANCIAL SYSTEMS IMPROVES BUSINESS OPERATIONS ON A GLOBAL SCALE FOR A GLOBAL ENERGY CORPORATION

An integrated energy business spanning multiple continents found its legacy systems were not integrated, costing the company not only time and money, but hampering its international growth goals and damaging its competitive position. This integrated energy business looked to IBM to bring its systems together to optimize its business processes and provide the company with a corporate view of its financial data.

Why become an on demand business? Operating in more than a dozen countries, this integrated energy business had disconnected data and financial systems, preventing it from responding to rapid global changes. Integrating its systems would not only provide improved operational flexibility, but also significantly reduce costs.

How and where did the energy company start? Under a five-year strategic outsourcing contract, IBM hosts this integrated energy business' chosen ERP system to provide the company with the integrated data it was seeking. IBM Business Consulting Services worked with the company to transform its financial and procurement processes, and develop management scorecards to improve the company-wide view and understanding of financial performance.

What benefits did the energy company achieve? Sounding like a page out of IBM's on demand business vision, this company's objectives target the following: interoperability, integration, collaboration, learning, and continuous improvement. This integrated energy business improved business processes and obtained a clear view into its financial systems, translating into quantifiable improvement in its cost of operations. This integrated energy business anticipates 70 percent of the benefits to come from improved decision-making from information access and 30 percent from process transformation improvements. The company's achievements to date include:

- A 1 percent improvement in return on invested capital
- A 9 percent reduction in well maintenance costs, with 15 percent in compression costs and 10 percent in drilling evaluation costs
- A 2–5 percent improvement in procurement
- A 1 percent increase in project rate of return

A Large Japanese Automobile Manufacturer: Improving Dealer Communications in the Automotive Industry

STREAMLINING AND AUTOMATING COMMUNICATIONS PROVIDE MAXIMUM CUSTOMER FLEXIBILITY AND RESPONSIVENESS IN THE JAPANESE AUTOMOTIVE INDUSTRY AS IT DEVELOPS A NEW PRESENCE IN EUROPE

One of Japan's top five largest automobile producers recently achieved a major milestone with the first automobile engineered by its company in Europe. The company plans to produce 40,000 units of this new model annually, to be sold throughout the European markets. To meet this aggressive goal, the European operation had to first overcome an inefficient distribution process that was negatively impacting dealer relationships.

Why become an on demand business? In Europe, this Japanese auto manufacturer had a complicated manual process for dealing with orders placed by European car dealerships. Invoices were centrally printed in batch, and then sent out via express courier to all European distribution centers/dealers. This inefficient approach was damaging dealer relationships and threatening the company's bottom line. This Japanese auto manufacturer's European division needed to further automate its order processing system to streamline business operations and become responsive in real time to its dealers' demands.

How and where did the Japanese auto manufacturer start? This auto manufacturer's European division implemented a content management solution, allowing it to capture, manage, and track crucial purchase orders and invoices. The new solution automatically registers new orders and updates accounts to reflect order changes, guaranteeing that the European arm of this Japanese auto manufacturer always has up-to-date information regarding vehicle order status. The order tracking system is integrated with the production environment, ensuring only requested parts are shipped.

What benefits did the Japanese auto manufacturer achieve? This Japanese auto manufacturer can immediately produce and supply the invoices that its European dealers request. More accurate order processing improves dealer satisfaction, strengthening business relationships between the Japanese business' European car dealerships. With the content management solution, the auto manufacturer's European business division can now easily manage sudden increases in dealer demands. The automated system eliminates the request for reprinting in batch, which was very time-consuming, thus reducing overall total costs.

A Large Travel Services Company: Heading in the Right Direction with On Demand Business

PROVIDING CUSTOMERS WITH REAL-TIME TRAVELER SUPPORT INFORMATION, AROUND-THE-CLOCK, IS A "WALK IN THE PARK" FOR ON DEMAND BUSINESS SOLUTIONS

A travel services company launched an online support network for high-end leisure travelers. A companion company provides real-time traveler support, including customized travel planning, an around-the-clock call center to provide directions, travel reservations, and 24x7 travel planning and emergency assistance. What the startup required was a solution to provide scalability and flexibility to grow as required, but to only pay for this capability as and when needed.

Why become an on demand business? Due to the travel service's strong brand recognition, the new service must guarantee high availability to members, which involves managing daily and seasonal usage fluctuations with severe spikes expected during the summer and major holidays. Initially, the travel services company looked at traditional outsourcing solutions: purchasing hardware and support up-front, and investing in additional technology over time. Then IBM presented another approach: the on demand Operating Environment.

How and where did the travel services company start? Instead of using standalone Web, application, and database servers, the travel services company will run its operations entirely on Linux-based virtual servers, hosted and managed by IBM. As part of this on demand business solution, IBM is providing a scalable, standards-based utility service to provide processing power on an as-needed basis. As a result, the travel services company only pays for the processing, storage, and networking capacity it requires, and can scale its virtual infrastructure up to meet demand spikes. This flexibility frees funds to be invested in improvements to the travel services company's products and marketing, ultimately improving the success of the companion company and other future services.

What benefits did the travel services company achieve? As a result of this on demand business solution, the travel services company expects to realize multiple benefits and achieve the following competitive advantages:

- Minimized up-front capital expense due to flexible, utility-based pricing
- Scalability to match any level of future business growth
- Fast deployment of new virtual servers and applications, resulting in reduced time-to-market
- Simplified capacity planning from additional available server, storage, and network capacity
- Ability to handle workload surges, such as during daily or seasonal Web site usage spikes

Online Movie Rentals: Giving a New Online Business a Jump-Start with On Demand Multimedia Deliveries

A Responsive, Highly Scalable Solution Is Required for an Unpredictable Number of Users Seeking Advanced Forms of Multimedia On Demand

A new online movie rental corporation, recently launched as a joint venture by five major Hollywood studios, wanted to offer a more convenient alternative to traditional video rentals by downloading first-run movies over broadband Internet connections.

Why become an on demand business? The online movie company's main challenge was how to provide a flawless movie-going experience to an unpredictably large number of users without overextending its budget. As a startup, the online movie company also knew it could not manage the technology itself if it was going to be highly responsive to its customers and studio partners.

How and where did the online movie company start? The online movie company sought to provide a highly available, reliable, and secure Web site as a platform for movie downloads. The company searched for a reputable hosting partner that could handle the technical side of delivering large video files, letting the online movie company focus on its core business operations. IBM evaluated and helped deploy the online movie company's Internet-based solution, enabling the startup to handle increasing Web site traffic and deliver movies on demand. IBM's hosting services provide the online movie company's on demand business solutions at two sites. This allows a plurality of services, including Internet connectivity, video file staging, traffic routing, security (including comprehensive intrusion detection and virus protection), backup, archiving, storage management, disaster recovery, load balancing, network monitoring, and much more.

What benefits did the online movie company achieve? Very high hosting availability promotes fast, accurate video downloads for satisfying customer experiences, accommodating approximately 1 million unique visitors per month. The online movie company has the capacity to handle variable transaction levels, including unpredictable surges. The flexible pricing structure lets the online movie company pay for only what it uses of some services, reducing its operating costs. The company is moving toward adopting this pricing model on a broader scale.

Large Petroleum Company: Discovering the Benefits of On Demand Grid Computing

A SUPERCOMPUTING GRID COMPUTING ON DEMAND SOLUTION FOR OIL EXPLORATION THAT DOESN'T COST THE EARTH

A large petroleum company utilizes seismic surveying, while producing three-dimensional images utilized for locating oil and gas deposits. This requires a massive amount of computing power. IBM's supercomputing on demand offering was the answer, enabling the petroleum company's Global Data Processing division to buy supercomputing service units—equal to a specified amount of processing power per week—according to its needs.

Why become an on demand business? The petroleum company's data processing organizations wanted to explore new and more cost-effective ways of providing computing power for seismic surveying to its gas industry clients. Simply by the nature of its business, the company wanted a more flexible way of accessing additional high-performance computing power for short periods of time. The petroleum company needed the responsiveness to ramp up capacity fast in real time—according to fluctuating needs—without spending excessive amounts of money.

How and where did the petroleum company start? IBM's on demand business solution is enabling this company to take advantage of the latest Linux cluster technologies. This solution can serve up the massive amounts of computing power required for seismic surveying using a cost-effective Intel/Linux platform. IBM is providing this petroleum company with 384 x335s running Linux, as well as some x345 systems for a three-month project in the Gulf of Mexico.

What benefits did the petroleum company achieve? This petroleum company will have the flexibility to scale up infrastructure fast in response to peaks in demand, being ultra-responsive to business opportunities while reaping significant savings on systems management and maintenance. By opting for high-performance computing power on demand, this petroleum company predicted it could potentially save more than $1.5 million (USD) per year, simply by utilizing this innovative Grid Computing on demand business solution.

Global Logistics Company: Getting It There by Sea, Land, or Air

A GLOBAL LOGISTICS BUSINESS NEEDS TO SEE SHIPPING MOVEMENTS AS AN ON DEMAND BUSINESS, A SIMPLIFIED VIEW

A German integrated logistics business spans the globe—its complexity and size are both a virtue and a challenge. IBM is helping this company meet its customers' demands for speedy quotes and reliable service.

Why become an on demand business? As a leader in integrated logistics services, this company prides itself on the ability to provide "one-stop-shopping." But in an increasingly complex global environment, the company was finding it increasingly difficult to provide quick answers to customer requests for information, price quotes, and delivery schedules. The company anticipated that it could establish a competitive edge by integrating processes and systems to offer a more responsive, single-face-to-the-customer approach.

How and where did the logistics business start? IBM Global Services specialists became part of the integrated logistics business team by defining a sophisticated e-fulfillment system supporting a self-service Web portal. The integrated logistics business team elected to develop a platform that would allow communications among customers, transportation partners, and employees through a single interface.

What benefits did the logistics business achieve? The IBM solution helps to enable integration with customer supply chains as well as provide a "single face" to the customer. This integrated logistics company believes that these are decisive factors in maintaining competitiveness. The solution also provides this integrated logistics business with a more consistent view of its business processes, with the latest accurate information always available. By making accurate data available in an easy-to-access fashion, the solution can help recognize potential transportation problems at an early stage, and enable alternative approaches to be developed in cooperation with customers. The modular design enabled by WebSphere supports the development of new services and the integration of new partners and customers.

Hospital IT Company: Winning and Realizing a Small-to-Medium Business (SMB) Success

A HOSPITAL IT COMPANY PROVES ONE IS NEVER TOO SMALL TO NEED ON DEMAND BUSINESS CAPABILITIES

A hospital provides radiology services to medical institutions. Traditionally, radiology administration requires significant investment in medical equipment, medical process redesign, hardware, software, and services, which can total up to $2,000,000 (USD), even for smaller institutions. A hospital wanted to offer a new set of flexible services that would help its customers—local hospitals with limited budgets—maximize patient care while streamlining operations. The hospital selected IBM for its Utility Computing services model.

Why become an on demand business? This hospital sought to expand its marketplace by providing customers with access to radiology services on a pay-per-use basis. This approach would position this hospital as being responsive to the tight budget constraints of its client base and allow it to increase its customer base beyond in-house capabilities.

How and where did the hospital IT business start? IBM is performing fully managed Web hosting services to provide a reliable, powerful infrastructure in support of this hospital's Web-based applications, which schedule appointments, manage medical reports, and store digital images for the hospital's radiology department.

What benefits did the hospital IT business achieve? IBM provided this company with a Utility Computing services solution and model that allows hospitals to tap into radiology resources on a "cost per use" basis. This hospital customer reduces costs by paying for only what it uses and avoiding up-front hardware costs, improves efficiency by allowing 24-hour Web-based access to all aspects of operations, and eliminates the need to replace legacy systems by using standards-based technology easily integrated with existing medical software and hardware.

A Large Utility Company: Finding a Mobile Service Scheduling Solution That Provides a "Jumbo" Advantage

COMPANY BECOMES MORE RESPONSIVE TO FAST-CHANGING SERVICE NEEDS

A large utility company provides metering and billing services for over 2.5 million gas, electric, and water consumers in northern England. Facing stiff competition after industry deregulation, the utility company looked to employ new technology to solve the old problem of not having real-time information in the hands of those who needed it most.

Why become an on demand business? Recent deregulation of utilities in the UK created strong competitive pressure to reduce costs and provide better customer service. By replacing its inefficient paper- and phone-based dispatching system, the company found a way to be more responsive to customer demands for fast service and maximize its productivity, thus boosting profitability.

How and where did the utility company start? This utility company deployed a mobile service scheduling solution that responds in real time to changing needs for meter inspection, repairs, and provisioning of new customers, dynamically rearranging job lists according to customer availability. The solution, JUMBO, for Joint Utility Meter Business Operator, developed by an IBM business partner, lets service technicians log on at any time and download work orders in their geographic area, or upload repair reports and new orders. The JUMBO solution is based on IBM WebSphere MQ, connecting to the central database server at headquarters, and handheld devices using IBM WebSphere MQ Everyplace.

What benefits did they achieve? The wireless JUMBO solution makes technicians more responsive to fast-changing service needs because work orders are updated in real time. It also provides greater mobility and flexibility, increasing the number of jobs utility workers can handle in one day and making them more productive. This utility company now spends less time and money on dispatching, allowing it to focus more on its metering and billing services. Heightened responsiveness to customer needs has boosted its satisfaction, growing the customer base from 2.5 million to 10 million accounts.

Summary

In this chapter, we explored the industry factors driving new on demand business transformations across many vertical industries. We profiled customer stories that describe several successful transformations that serve as examples of what can be accomplished today, and as an inspiration for what can be accomplished in the future. *Can you see it?*

Conclusions 9

In this book, we have explored several strategy and technology perspectives of on demand business. We have also discussed a wide variety of topics related to on demand business: the on demand Operating Environment, Autonomic Computing, and Grid Computing. We explored the fact that the on demand business involves business transformation. It is an evolution, not a revolution or a temporary state of operations: a state of operations that you should not simply sit back and watch other businesses achieve. We have explored key aspects of the service providers delivering these types of advanced services for on demand business operations.

The more in-depth portions of this book unveiled the Autonomic Computing Blueprint, Autonomic Computing and Grid Computing strategies, and the future of Grid Computing. We explored key points with concise, hard-hitting explanations of how these technologies intersect with on demand business. We explored the fact that on demand business includes Autonomic Computing, and sometimes, these solutions will incorporate Grid Computing into more advanced problem-solving solutions. We discussed the On Demand "Journey." This Journey is not unique to any industry, rather, a common mission that all industries must consider, if they have not already.

We explored many key examples of customers amid the transformation of becoming an on demand business enterprise. We intersected with some industrial notions of what is involved in this transformation.

As we conclude this book, let's shift our focus onto the industries and contrast everything we have discussed with exactly what is key to consider from within certain industries, as it relates to on demand business transformations. We will not explore all global industries as we conclude, just a few key industries that have recognized the need and taken significant steps in approaching or executing their own Journey.

By now you might be asking, how is the industry in which my business is a part of affected by all of this? Or, is it? Are there other industries that are more affected than the one I am a part of, and if so, which ones are they? When will I know it is time to begin the Journey for my business?

When your company becomes an on demand business you will absolutely realize the change. This on demand business state of operation quickly demonstrates many new capabilities to respond, unlike never before; you will recall that we discussed in previous chapters that it will allow for you to:

- React to many marketplace dynamics in a positive and effective manner
- Focus on the core business strategies, and outsource the rest to valued partners
- Develop and leverage flexible, variable cost structures
- Increase operational resiliency and safeguard critical information

We discussed that the advantages of achieving an on demand business state of operations and efficiencies are extremely beneficial, but it may not be obvious as to how these new benefits play out across the certain industries. IBM's teams of on demand business global experts are working with many companies to deliver actionable and measurable improvements towards establishing and executing their own On Demand Journey.

Let us take a conclusive look at how on demand business is addressing trends and issues in a few key industries. These markets are:

- Finance
- Banking
- Media & Entertainment
- Electronics
- Automotive

- Insurance
- Retail

Market Perspectives

The following discussions address the global industries, and perspectives that describe some interesting dynamics, from a practitioner's point-of-view.

Financial Markets

Due to the declining economic situation over the last three years, and the relatively stagnant financial markets across the board, revenue streams have yielded less than landmark profits for many financial services firms. Mergers and IPOs are down dramatically in the stock markets; broker-dealer commissions are just now beginning to stabilize; and asset management firms, once rich with a constant flow of assets in the 1990s, have actually seen assets under management shrink.

Financial Services Firms Transform Themselves to On Demand Business

After enduring three years of stagnant growth, watching fee-based businesses decline and reliable revenue sources such as commissions and clearing become commodities, investment banks and asset management companies seem to be at the mercy of a stressed market. Although the industry has been conditioned to expect a certain amount of volatility, the past few years have challenged securities firms more unexpectedly than anyone can ever remember.

In an environment where the only constant is uncertainty, financial market institutions will gain more control over their own destinies through a variety of on demand business environments.

For the most part, securities firms have responded to this market downturn as they always have, which is by cutting costs wherever possible. They have reduced budgets, consolidated infrastructures, and reduced the workforce by (in some cases) thousands of employees. This movement has seemingly been to little advantage, since even after such drastic measures these firms have still been very challenged to hold their margins.

It is obvious that these previous types of business practices are no longer yielding world-class economic results. Financial services firms need to continue seeking a new on demand business vision, discard the less-than-satisfactory approach of cost reductions, and focus on creating new value propositions. Financial firms can do this by embracing an entirely new business transformation model, integrating on demand operating models, unlocking new levels of productivity, thus causing a discovery of new value positions across the markets.

Banking: As a global industry, banking is transitioning from a set of independent, vertically integrated financial services institutions to a network of affiliated financial firms. Banks are starting to specialize in consumer services, by embracing a specific industry role of product development, distribution, and transaction processing. These decisions are made according to what best matches their strengths, then turning to other companies or other units in their own companies to execute the rest of the transition.

ON DEMAND BUSINESS IS CHANGING BOTH BANKS AND THE BANKING INDUSTRY

With recent value creation strategies somewhat stalled, banks are undertaking fundamental changes to their business structures and consumer value propositions; meanwhile, the industry itself is taking on a new shape. As banking institutions reconstruct and the industry itself deconstructs, these two paths will very naturally converge on the road to full service on demand business banking environments.

In an environment that is touching the pulse of almost every global consumer, expect the continued introductions of innovative on demand business environments. This will very soon be the case across the board, as banks become more flexible while rendering a full suite of on demand business consumer services.

Meanwhile, individual enterprises are reconstructing their own business models. They are decomposing product towers into smaller, independent components that can be shared across the enterprise. These enterprises are also beginning to craft an integrated view of their customers. However, these financial channels remain for the most part product-centric and managed by business units. This movement is creating unnecessary complexities and rigidity of this arrangement, and often overwhelms any long-term benefit that could be derived from these shared processes.

These developments have, therefore, introduced innovative on demand business environments where banks can become flexible enough to respond rapidly to many customer demands, market opportunities, or external threats.

Media & Entertainment

It is no surprise that the media and entertainment (M&E) industry features so many doctors, lawyers, and therapists in movies and on television sitcoms. The industry has, in fact, always been a manic mix of high-risk investments with unpredictable rewards. And let us not ignore the unpredictable audiences, of which we are all a part, where everything is always riding on the next big "hit." But now many viewers will tell you that it is more disorderly and seemingly uncontrolled than ever.

ON DEMAND IS A BIG SUCCESS WITH MEDIA AND ENTERTAINMENT FIRMS

A recent study by MTV revealed that their typical television viewer lives a 30-hour day. That's right, a 30-hour day. No, this does not mean that America's youths have given up sleep. It means that they are living in an on demand business world: They surf the Internet, view DVDs, play MP3s, send instant messages, download movies, and sometimes even watch a little TV. Our youths are doing enough of this multi-tasking work, simultaneously to add up to 30 hours of daily, à la carte media consumption. And these multitasking teenagers are not alone: Today audiences of all kinds are revolutionizing how they access media and where and when they consume it.

In an environment that enters the lives of virtually all individuals, entertainment is expected to continue with brilliant introductions of on demand business types of media events. Entertainment corporations are now gaining more control over their own destinies, through a variety of on demand business operational environments

Many new and innovative digital technologies have introduced a wide variety of media devices, delivery formats, and methods. Digital networks enable customers to create customized products and services, rendering what they want to see or do, exactly when, where, and how they want it. As consumers of M&E focus their attention across these many options, "hits" are becoming increasingly more difficult to predict and revenue from these offerings continue to decline.

However, there is a promising future for M&E on the horizon. Embracing on demand business services and methods, many M&E firms are quickly gaining the ability to sense and respond to market change and unexpected consumer situations, while staying focused on providing high-quality M&E, which is what they know how to do best. The M&E industry is proving itself to be able to handle threats with confidence, while remaining flexible in how it organizes itself and structure costs in a new virtual world.

Electronics

The electronics industry is constantly striving to succeed by managing innovation, while the industry's products are increasingly becoming commodities. In an attempt to increase revenues and reduce costs, firms are more and more reaching towards cutting costs in research and development. How do you resolve that paradox?

ON DEMAND BUSINESS PRACTICES SOLVE STHE PARADOXES FACED BY ELECTRONICS FIRMS

Electronics companies are creating new business models and network services that can flex with the unknown and the unpredictable, resulting in an ability to manage earnings volatility in a quarter. This means they must continue building more resilient and Autonomic Computing capabilities for globally dispersed operations.

As we have described in this book, we call this on demand business, and consumers are seeing strong interests in this across the electronics industry, as they continue to manage innovation.

An ever presence of readjusting the planning cycles is a constant challenge. Sustaining communications and business controls across a wide geographical network of suppliers in North America, Europe and Asia is an ongoing endeavor. Being able to immediately cut costs when revenues take a sudden plunge is another difficult challenge, and a seemingly never-ending task. Constantly faced with chaos and forms of uncertainty, electronics firms spend entirely too much time reacting and not enough time managing innovations, at least as not as much as they would like.

Some of the global electronic firms' thought leaders believed that mergers and acquisitions would perhaps solve these challenges. However, in most cases, this world of mergers and acquisitions became a bit fractal throughout the recent years. One of the causal impacts of these mergers and acquisitions is suggesting that the firms are acquiring more complex versions of the problems they were trying to resolve, prior to the merger. Costs remain excessive; innovation in some cases seems to exceed demands; consumers are less than happy with the buying process, and all too often, product values. Enterprise operations are still locked in vertical towers; and the global supply chain is still lacking coordination.

Automotive

Despite a number of mergers, business alliances, and consolidations that were supposed to help auto companies expand and compete, it is not apparent that all is well across the overall industry. For example, the time-to-market for a new car model now averages three years. The industry has a large number of plants. Some have said that supply not exceeds demand. Per-vehicle warranty costs now average $700 per vehicle. Complex arrays of government standards and regulations have increased the complexity and cost of doing business.

ON DEMAND AUTOMOBILE COMPANIES ARE ON THE ROAD TO SUCCESS

New imperatives are driving specialization, changing the automotive blueprint from the traditional silo view to a dynamic, on demand business networked ecosystem. In this new ecosystem, virtual value nets lead to new partnerships that will be formed quickly in response to market needs. Competencies and players in other industries will become integrated as parts of the automotive ecosystem, and local and regional ecosystems will begin to transform into a single global system. In this environment, information can flow in real-time throughout the value net.

In an environment that holds much promise, consumers can expect to realize a variety of changes as the industry and its products transform to on demand business environments.

Aside from the fact that some of the auto companies have made great strides in the past few years, the strongest players in the industry will need to become even more dynamic, more collaborative in how they deliver products and introduce new innovative value creation. Automotive companies will have to become much more responsive to the industry threats, more flexible and adaptable, and reach out to partners while resisting the urge to produce everything in-house.

The automotive industry, however, is now beginning to catch on and is pursuing more specialized roles based on business needs, regional rules, crossover products, and extremely dynamic market opportunities. This is a welcome change to both manufacturers and consumers.

Insurance

The insurance industry is facing a serious disadvantage. Transformation and enhancements of critical business processes are in need of taking place during one of the most severe IT budget crunches in years. The traditional approach of purchasing IT solutions in order to resolve a particular business problem no longer works because it does not address the pressing need for wider integration of new and legacy systems in the industry. Conversely, the current environment is being questioned as to whether or not it is sustainable or affordable over the long term.

Insurance Firms Are Placing a High Premium on Becoming an On Demand Business

On demand business is a new way for insurers to operate their businesses, and better manage the costs of large investments. On demand business enhances execution of their current strategies and expands their strategic options. This is accomplished by providing a broader range of business and operational strategies, which are economically, strategically, and technologically possible.

By creating an on demand Operating Environment, designed with open standards for ease of integration, insurers can implement new capabilities, simpler consumer processes, and accomplish this all with significantly less effort and cost than that of the past.

Insurers are realizing that on demand business offers an affordable way to create increased shareholder values, while modernizing and integrating almost every business process insurers utilize. This is occurring both internally and externally of the insurers' organization. That is because on demand business centers around value creation at the business process level, as opposed to focusing on the selection of a particular technology.

Once an On Demand Readiness Assessment demonstrates that the enterprise meets the expectations of the consumers, insurers can be much more responsive, developing innovative revenue-generating services, new product offerings and simpler policies for consumers. The insurers' core systems can now scale up (or down) based upon customer demand and actual on demand utilization. Ultimately, the insurer can focus on transforming differentiating market capabilities, underwriting, claims adjudication, policy administration, just to name a few of the benefits. The insurance enterprise can now respond rapidly to change, reconfiguring and continuing to transform itself as required.

Retail

The retailers of today face some very steep challenges, for example, empowered customers, decreasing workforce productivity, highly competitive and saturated markets, and sweeping technological change. Add in economic and political instability, and retailers are being forced as never before to concentrate on three strategic priorities: transforming the customer experience, making employees more capable, and cutting costs.

RETAILERS AND CONSUMERS ARE REALIZING THE BENEFITS OF ON DEMAND

With increasing competitive pressures more demanding on retailers then ever before, and with consumers' buying trends shrinking profit margins, retailers are tiring and some are even retiring from the chase. Knowing that the future promises an even more frenzied pace, it is time for retailers to consider trading in common practices, for on demand business processes that provide the speed and flexibility.

Retailers must now stand out in the crowded retail marketplace, offering consumer unique value propositions and buying opportunities.

Yes, the e-commerce boom of recent past caused impacts across several markets, but this did not stop the Internet from growing and providing more and more online retail outlets. Today, the Internet is pervasive in our homes, our cities, and our daily lives. Even the brick-and-mortar retail chains have been forced to join the Internet e-commerce solution space. Financial services companies (e.g., PayPal) have made it very simple for consumers to pay for goods and services over the Internet. Shopping can be conducted at the click of a mouse, while in the comfort and privacy of the consumer's home, or preferred public access point.

Savvy retailers are reconsidering their business retail models, developing a new "sense and respond" approach to retailing. That is, instead of buying goods, placing them up for sale, and then promoting and discounting them until they finally disappear, the on demand business retailers put the emphasis on the consumer. They do this by determining the consumer wants and needs, and this is then captured and internalized throughout the retail enterprise. This, in turn, seems to respond to that need of the consumer in a very quick and efficient manner.

By simply integrating business processes and using the best technologies as we have discussed in this book, on demand business retailers have proven that they can transform the way they did business, achieving new optimum customer service and cost-effectiveness goals. This is a welcome change to the consumers, and is continuing to improve.

Closing Thoughts

In closing, we have discussed (across several dimensions) that on demand business is a complex state of operational efficiencies, with concepts, methods, and techniques: It is process-centric, enabled by technology.

Performing as an on demand business is something that almost all major businesses, worldwide, will have to confront sooner or later. As a business leader, do you feel or have you observed the strong industry pressures to transform? Do you feel the economic factors related to considering on demand business solutions, and the need for transformational activities across your enterprise?

Throughout this book, we have explored technical and strategic concepts, which in some cases go well beyond what those skilled in the traditional art of business operations have yet considered or practiced. This new operational state of on demand business indeed requires carefully planned transformation activities, and in most cases, across many dimensions. This is inviting us all to continuously strive to achieve and maintain a new competitive edge. This includes tighter business partnerships intersecting with important autonomic operations, plus new innovative ways of thinking about how to manage the business. The on demand business strategy has indeed set precedence for a new plateau of conducting business operations.

Consumers and service providers, alike, are rapidly growing to expect more and more advanced services, and they want them to be available—24x7—365 days a year. As consumers, we also have a strong desire for more pervasive forms of conducting business and living our daily lives; and, we want this while utilizing a plurality of devices that are available in a multitude of form factors. We seem to desire to be continuously networked to innovative, advanced Web services; as an example, we are now beginning to connect the day-to-day appliances in our homes.

The global (and domestic) telecommunications firms delivering these on demand business services through "pipes" of a sort are working very diligently to bring the best and most advanced services to us. Likewise, many business enterprises are now able to serve in a role as service providers (i.e., Web content service providers) while competing to deliver the "killer applications" that we will all want to utilize in our everyday lives. These content providers are also partnering with other service providers to deliver the best possible forms of services to the general public. Networking service providers are leveraging best of breed networking capabilities in new ways to ensure that the on demand Operating environments are the richest, most responsive and reliable environments possible.

Government institutions of the world, biomedical explorations in medicine, science, research, and academics are all trying to determine how to best leverage these powerful on demand business services. As we have explored throughout the discussions in this book, some of the world's most powerful computing environments have been applied to the world's most difficult problems in Grid Computing solutions.

As with any evolution, it is simply time itself that is required for us to see a whole new world of transformed business operations and cultural practices. Some of us are just seeing it now, and some of us have already achieved and experienced it; however, this is not by any means a signal for us to become complacent.

The Internet continues to advance daily in the capabilities it provides our global societies, which in turn are rendering a very powerful international on demand business landscape. Business leaders are discovering new, innovative ways to enhance their strategic positioning across their global and localized markets. Industries strive to increase time-to-market, cut costs of both operational expenditures and capital expenditures, while at the same time, deliver enhanced forms of advanced Web services.

We are discovering new ways in which to simplify our daily lives, thanks to some wonderful technological innovations. We are also developing new ways to enrich our learning efficiencies in schools as students and educators. Oftentimes this is approached by integrating many forms of pervasive devices into daily routines that yield many forms of on demand information and problem-solving methods. Information is available around every corner, every channel, and every Web page we visit. We are able to obtain this instantaneously, thanks to some outstanding on demand business leadership and precision execution of delivery strategies.

In fact, we now expect that all products and services deliver richer and more robust features than ever imagined. Why? The answer is straightforward: We expect this, now, simply because it is possible to deliver vast amounts of information on demand. Our global cultures and daily routines are transforming, as are our societies, in ways that not even 10 years ago we had ever imagined. Let us not forget that in 1995, it was hardly even possible to conduct a "secured" transaction on the Internet. Today, this medium, enabling many of the on demand business capabilities, seems to be something that some now just accept as a common way of operating.

On demand business is indeed an operational state of achievement, one that continues to refine itself each and every day: It is a transformation. These perspectives on business transformation, strategy, and technology integration, implemented across any enterprise, will enable this state of on demand business operations. It is one that any business can ascertain once the enterprise understands the points of entry, their tactical plan for execution and delivery, and the incredible values yielded from such a transformation to both the consumers and the enterprise.

Can you see it?

Appendix A
IBM On Demand Developers Conference

This Appendix section describes several of the developmental activities related to many IBM on demand business global initiatives. These activities were presented at a recent IBM On Demand Developers Conference, and are highlighted in this Appendix.

The Author would like to extend a special thank you to the contributing editor of the information in this Appendix, David Singer (IBM Distinguished Engineer, Almaden Services Research), who co-chaired this unique On Demand Conference event for IBM. Additionally, the author would like to thank Dave Ehnebuske (President, IBM Academy of Technology, and Senior Vice President and Group Executive) for his support and sponsorship of this event, and encouragement to publish the event information as described in this Appendix.

The following information identifies many (but not all) of the respective contributors to the subject event, and gives brief descriptions of their accomplishments.

Alvarez, Guillermo A.; Chambliss, David D.; Pandey, Prashant; Jadav, Divyesh; Xu, Jian; Menon, Ram

SLEDS: Autonomic Performance Virtualization

Previously isolated pools of data, typically stored in a myriad of direct-attached storage units, are currently being consolidated into large data centers for total aggregate capacities of hundreds of terabytes to petabytes. That trend bodes well for on demand business computing, as consolidated centers serve the simultaneous needs of multiple clients, and can reap the benefits of sharing high-end storage devices and networks. The needs of on demand business are easier to satisfy by shifting collocated resources from one client to another in a nimble manner, while keeping hardware and management costs low.

However, previously unrelated clients now interfere with one another, by competing for resources such as disk drive actuators, SAN links and end-points, switch back planes, controller processors, data caches, system buses, and SCSI interconnects. The resulting coupling weakens the assurance of acceptable performance to any given application, because of competition with others outside of its control. The Quality of Service (QoS) experienced by a client may degrade to an arbitrary degree because of increased demands placed on the storage system by other clients outside of its control.

Today's virtualization technologies have been widely successful at aggregating capacity from multiple storage devices. But sufficient capacity doth not a true on demand business system make: Applications need not only performance isolation, but also guaranteed minimum levels of performance lest costly consequences ensue. That may stem from application-specific constraints (e.g., response times for online transaction processing) or from general administrative considerations (e.g., a backup having to complete within a time window).

We believe that platforms for on demand business should offer access to virtual data containers with guaranteed capacity AND performance, thus isolating end clients and system administrators from the concerns of interference and inadequate resource provisioning. Only truly autonomic solutions can satisfy those requirements in a cost-effective way—manual action taken by system administrators is not fast or accurate enough to respond to failures and fluctuations in the workload.

In this paper we present SLEDS, a production-capable outboard storage controller that manages client workloads to meet QoS goals. SLEDS provide virtual slices of system performance, as governed by per-client, demand-dependent, statistical QoS guarantees. By relying on periodic workload monitoring and introspection, SLEDS introduces the autonomic capability of self-optimization to the storage system, not only "within the box" but also with a view to the best achievement of system performance objectives.

SLEDS controls the load imposed by each client on the system, and throttles it to a reduced throughput whenever necessary to ensure that all guarantees are met. Throttling does not normally reject I/O operations, but delays their substantive processing. It is also transparent to client hosts and back-end devices (e.g., RAID storage servers, disk enclosure controllers, or even individual hard disks) as the performance control is achieved by inserting a gateway device. The system is rapidly responsive to changes in workload and system conditions. A leaky-bucket throttle discipline maintains accurate control of resource use even when I/O submission rates are highly variable. This contrasts with a strategy of performance management by data placement only, which may require hours or days to perform data migrations in response to overloads. Because SLEDS makes autonomic resource allocation decisions based on global knowledge of the system's state, it can detect and unclog any bottleneck in the system, be it collocated or not with the QoS enforcement point.

We present the architecture of the system and discuss our design decisions. Our current prototype of SLEDS consists of approximately 48,000 executable lines of C and Java code running in user mode on the Linux operating system. We evaluated the prototype's performance on a mid-range storage system with heterogeneous hosts, operating systems, and devices with very encouraging results. Despite the fact that SLEDS does some processing for every single I/O in the system, we found that it adds less than 1.3 microseconds to service times, with a penalty of 2.5% or less in the form of decreased maximum throughput. In addition to not significantly decreasing performance, the prototype provides very good performance isolation and responsiveness to changes in the workload. In particular, the dynamic throttling decisions made by SLEDS without human intervention turned out to result in almost no QoS violations, unlike static limits like a system administrator might apply by hand.

Bellwood, Tom

Web Services Failure Recovery— A Framework and Demo in the ETTK

This session will discuss follow-on development activity on the Active e-Biz project, which was sponsored by the SWG Emerging Technology group last year. Active e-Biz combines the Grid, Web services, and Autonomic computing technologies to provide a service provisioning Hosting Service which supports self-healing of the services it manages. It employs a Distributed Network Agreement (aka "DNA") mechanism, which is essentially a map of how to reconstruct and provision each service, managed by the Hosting Service should one or more fail. It also allows for collection of state information applicable to the service to assist with re-provisioning. The Hosting Service monitors the health of the services it hosts and automatically heals, initially only via re-provisioning them after failures. Hosting Services are Grid services, which share the DNA information for each service hosted by all other Hosting Services in the network, across machines such that re-provisioning could occur in any connected environment in the network.

Among other things, Active e-Biz also includes capabilities such as defining of affinity groups, which essentially identify a group of Hosting Services between which DNA information (fragments) for hosted services are shared. This creates a convenient boundary within which healing actions for services managed by each of the Hosting Services in the group can be performed. Any Hosting Service within the group can heal the services of an individual Hosting Service.

Since the initial project, the strategy we've adopted is to extract the unique and useful features from the Active e-Biz work for development and packaging such that they can be used to augment existing provisioning products and technologies. The follow-on work is known as On Demand Failure Recovery (ODFR). The ODFR Framework is a generalized solution, which is intended to allow other provisioning technologies and products to acquire the self-healing behavior for themselves as well as the services they manage, with minimal changes.

During 1H03, we delivered the initial version of the ODFR Framework available with connectors for the first client we are focusing on—Service Domain. Service Domain is essentially a Web service aggregation technology, which is currently offered via the Web Services Toolkit (WSTK) on developerWorks. We plan to offer interface implementations for Service Domain, allowing it to be managed by our ODFR Framework.

The ODFR Framework will also support an Administration Console. It will display the grid of all related ODFR Frameworks with information on the services each supports. This will eventually behave as something of a healing environment explorer, displaying relevant information on healing and re-provisioning events, highlighting problem areas based on recurrent healing events, etc.

Clearly, one could also use this self-healing behavior to perform On Demand provisioning based upon policy and SLA criteria. In the future, we intend to extend the work in this direction, taking advantage of other efforts in this space where possible. We also are looking at optimization capabilities for managing self-healed environments. Follow-on work may include integration with other technologies and products as well.

Biddle, Edd; Clark, Duncan G.

OGSA Portals

The session described work being undertaken within Hursley Services & Technology organization to examine and demonstrate how a portal might operate in a on demand business environment.

The investigation concentrated on how the emerging standards associated with grid and portals can be effectively combined to meet typical customer portal needs. Standards include WSRP/WSIA (Web services for remote portlets), OGSA (Open Grid Services Architecture) and the portlet API.

Issues to be investigated include the granularity of portlet services in a grid environment, the nature of the WSRP container/portal, approaches to virtual portals and alternative means of aggregating content on the client. These clients would include native Java and Flash rendering as well as conventional browsers.

Interoperability and the ability to work with heterogeneous technology are central to the work although the emphasis is to ensure compatibility and easy integration with Websphere Portal Server.

A presentation will provide a narrative of the work undertaken to date together with issues and findings while a demonstration will show progress and demonstrate how grid based portals would actually work in an on demand business environment.

Binding, Carl

Web Services for Mobile Devices

We present IBM's current efforts in extending the reach of Web Services technology to the pervasive and mobile device space. After a brief review on Web Services, we present our implementation of the JSR172 J2ME Web Services. Potential extensions of this work to asynchronous transport protocols and programming models are presented.

Boss, Greg

PC Grids—Potential Value and Cultural Issues

The value of a consumer/employee PC Grid can be measured in many ways. One study shows PCs account for 70% of the average company's computing resources. What priorities should be placed on creating an IBM PC Grid? What types of applications could run on such a grid? In what ways could such a PC Grid be used? We discuss these questions and current plans for extending the IBM intraGrid to include vast resources of PCs.

Consumer/employee PC Grids will produce cultural issues that must be addressed. Users will initially have concerns about performance degradation, hardware, and battery life. What is the best way to educate users on these issues and encourage participation? Ideas like using the IBM intranet portlet to show real-time contribution levels on individual and team basis are also discussed. Lessons learned from the IBM Smallpox Research Grid project that began in 1Q03 with 5000^{+} IBM participants are also discussed.

Boutboul, Irwin; Zhou, Nianjun (Joe); Meliksetian, Dikran S.

A Download Service over IBM intraGrid

We present a download service (named downloadGrid). We undertook this project with two main objectives in mind. First, we wanted to deploy a scalable and reliable download service over the IBM intraGrid. Second, we wanted to derive and develop a method for other grid service-based applications.

The downloadGrid system consists of three components:

1. A management center, which provides a download service, a distribution service, a monitoring service and an administration service
2. FTP servers, which provide the download service
3. Clients that download files using an advanced download algorithm

A typical usage flow of the process is as follows:

1. An administrator uploads a file through the administration service, with a specific policy. The policy contains the priority of the file, the mode of deployment (public or private), and geographical location. For private files, the file is first locally encrypted before being uploaded, along with the encryption key.
2. The distribution service takes care of distributing the file on a set of FTP servers in order to satisfy the above policy. The file is also distributed to

more or less FTP servers based on its priority. The distribution is enhanced in two ways:

- The distribution is done step by step. The file is first distributed to a set of servers. Then others servers obtain the file from this first set, and so on. We have a logarithmic distribution. FTP servers use the download service to get the file.
- The algorithm avoids network bottlenecks. The selection of the servers in each step is performed in such an order as to span networks.

3. A client sends a download request to the download service specifying a file with QoS requirements. Currently, the only QoS requirement considered is the download rate. The management service creates an optimized plan based on the location of the client and the current status and usage information of the FTP servers. An optimized plan contains the size of the requested file and a list of FTP servers with associated download rates.

The client starts the download from the first FTP server. It then adds step-by-steps additional sessions with other FTP servers, as long as it detects an increase of the download rate after adding a new session. Each FTP server is allocated a chunk of data depending on its speed, and will abide the to the download rate specified in the optimized plan.

The whole download process is an adaptive parallel FTP process from multiple servers. If a specific server is too slow to provide the download rate as planned, the client will assign to other FTP servers, that have completed their assigned chunks, additional chunks from the original chunk assigned to the nonconforming server. We have implemented our own FTP server to be able to control the download rate of each FTP session on the server side.

During the download, the client sends feedback to the management center about FTP servers, which are not accessible. At the end of the download, the client sends feedback about the complete download process (list of chunks/ FTP servers).

Dynamic and autonomic content delivery network At any time, the management center is aware of the current load and performance of the FTP servers, thanks to the feedback depicted above. The distribution service is able to replicate the file as needed (depending on current demand of the file, availability of FTP servers etc.). The monitoring service continuously monitors feedback and checks/restarts/shuts down FTP servers accordingly. The distribution of the file is continuously automatically adjusted to take into account the current status of the grid to provide a reliable download service.

The downloadGrid is the first Grid application deployed over the intraGrid. The methodology developed for this application can naturally extend to other applications. We envision that any practical grid application would have service providers, consumers, and brokers. The broker, implemented as the download service for downloadGrid, takes the responsibility of providing the roadmap of the service, securing the application, and maintaining audit records.

Brody, Andy

An Introduction to Topic Maps

Topic Maps are now one of the hottest topics within the Semantic Web and Knowledge Management research areas. They enable meaningful representation, navigation, and structuring of unstructured information (without altering it), and allow the deployment of information sets in different environments with different requirements.

We explain what a Topic Map is and how real-world concepts can be represented within the Map to provide a knowledge model that can span both digital and physical business assets.

Performing useful knowledge modeling with digital business assets is a difficult challenge. Digital assets can be held on anything from a DB/2 instance running on a mainframe to an Apple iPod. Physical assets, such as paper documents or people, are even harder to model. Businesses can make better decisions if they can construct semantic links, rather than just views, between their heterogeneous business assets. Topic Maps provide semantically rich navigation paths to access all types of business assets as a considerable enhancement to the usual simplistic linking of data.

Topic Maps assist with achieving the requirements of an on demand business. Consider the definition of an on demand business: "An enterprise whose business processes, integrated end-to-end across the company and with key partners, suppliers, and customers can respond with flexibility and speed to any customer demand, market opportunity, or external threat." The ability of Topic Maps to create rich semantic relationships between heterogeneous assets allows the business to identify and react to the changing environment, by letting the business understand what they have and the relationships between them.

Topic Maps have grown out of the ISO SC34 working group, which is responsible for producing standard architectures for information management and interchange based on SGML (ISO 8879).

The current Topic Map standard is set by the XML Topic Map specification v1.0 (XTM 1.0), an evolution of the ISO 13250 standard that defines both the model and the XML serialization format of Topic Maps. These are now being integrated back in to ISO process via The Standard Application Model for Topic Maps, Topic Map Constraint Language (TMCL) and Topic Map Query Language (TMQL) drafts.

The Standard Application Model defines a conceptual and formal data model for Topic Map applications. This will eventually play a critical role in the evolution of Topic Map technology as it forms the basis of both TMCL and TMQL.

Carlson, Robert (Bob)

WebFountain Technology Drives Business Transformation

As Internet data grows, so does the knowledge gap between what companies know and what they do not know. This translates directly into lost opportunities and missed revenue. This knowledge gap is driving a trend to transform passive, reactive processes and organizations into companies that can become proactive and agile. These business transformations enable companies to first sense and then quickly respond to real-time internal and external changes.

To implement such processes, enterprises need automated tools to help them access comprehensive, relevant, and timely information. Currently, the vast majority of all information/data cannot meaningfully be used without substantial human intervention. This has created a demand for tools that help enterprises obtain relevant, actionable information that can create value for the enterprise by helping to improve decisions and reduce time needed to discover, assimilate, analyze, and synthesize information.

WebFountain (WF) is an on demand business Innovation Services solution that collects, stores, and analyzes large amounts of unstructured and semi-structured text. It is built on an open, extensible platform that enables the discovery of trends, patterns, and relationships from the data. WF applications identify patterns, trends, and relationships in text. This information gives executives and business leaders insight into the core opportunities and challenges that are affecting their business so they can proactively and confidently develop competitive and deliberate strategies to successfully manage their enterprises.

Carpenter, Brian

Why On Demand Business Needs IPv6

On demand business requires:

- Unlimited scaling of the network.
- The ability for any service requester to connect securely to any service.
- Ease of configuration as resources are added, moved, or removed throughout the network.

These are exactly the areas where IPv6 has strategic advantages over IPv4:

- The essentially unlimited address space sets no practical scaling limit, avoids ambiguous addressing, and allows universal connectivity.
- Universal connectivity allows a simple approach to end-to-end network security, if required, without preventing conventional firewall, proxy, and intrusion detection techniques. This simplifies applications-level security such as Web Services Security.
- Both stateless neighbor discovery and DHCP-style automatic configuration are available for IPv6, as well as Mobile IPv6 for roaming devices.

Thus, we see IPv6 as a major and significant enabling technology for the large-scale deployment of on demand business.

Coyler, Adrian

Aspect Oriented Software Development

Aspect Oriented Software Development (AOSD) is an exciting new direction in software engineering. Going beyond OO (Object Oriented), AOSD can capture and modularize crosscutting concerns in a system: Easy examples are things such as tracing, logging, and first-failure data capture (i.e., error and exception handling policies); however, it can also do so much more than this. A feature of aspect-oriented techniques is support for "unanticipated evolution"—the ability to non-invasively modify existing behavior, or to extend the feature set, of an existing application. When this ability is coupled with load-time (or even later) support for the application of aspects, one has the perfect environment for cleanly encapsulating behavior and dynamically applying it into an execution environment. Thus, we can adapt applications "on-demand" and over time in a modular fashion.

Chun, Jen-Yao; Jeng, Jun-Jang (JJ)

On Demand Business Process Monitoring and Management

To transform a business into an adaptive enterprise demands effective monitoring and management of its business processes. A successful solution of monitoring and managing business processes requires the following characteristics:

1. Dynamic instrumentation
2. Anomaly detection
3. Adaptive analytics
4. Intelligent control

We have created an adaptive platform, coined as BPMM (Business Process Monitoring and Management), for managing business process centric solutions. A business process centric solution affects many enterprise entities: multiple business processes, different organizations, various execution platforms, multiple trading partners, and a constantly changing business environment. It is an uneasy task to create such solutions, especially in a context with increasing complexity in both business and IT worlds.

The BPMM platform is the response to this challenge by creating an adaptive platform so that developers can leverage it to build management applications. In particular, BPMM supports on demand business monitoring and management via both infrastructure support and policy framework. BPMM has in fact been applied to various domains such as insurance, supply chain, logistics, and banking; and has proven to be an effective platform to achieve on demand business monitoring and management. The BPMM infrastructure presented in this talk covers the following subjects: BPMM Foundation, Policy Framework, BPMM Architecture and Implementation, and BPMM Use Case Scenarios.

The BPMM infrastructure can also be treated as an autonomic business activity management platform supporting a complete functionality to sense, interpret, predict, automate, and respond to business activities. This infrastructure is aimed to decrease the time it takes to make the business decisions. Actually, there should be almost zero-latency between the cause and effect of a business decision. The BPMM architecture enables:

1. Analysis across corporate business processes
2. Notification of the business of actionable recommendations
3. Automatically triggering of business operations effectively

Therefore, BPMM can literally close the gap between business intelligence systems and business processes. The presentation will focus on demonstrating how on demand business monitoring and management can be realized via the BPMM infrastructure in an autonomic fashion.

Curtis, Michael

Applying Grid and Autonomic Capabilities to Benefit Government Agency Missions

This presentation is a discussion of opportunities for enhancing the value of information systems for large-scale and highly diverse government organizations. This discussion will focus on large agencies' efforts to integrate data and processes across technically, politically, jurisdictionally, and geographically dispersed virtual entities. Several examples of initiatives and programs in the USA federal arena are also mentioned to illustrate the environment and needs, to stimulate discussion of IBM's ability and approaches to enhance those agency's mission capabilities by applying grid and autonomic solutions.

Deen, Glenn; Kauman, James; Lehman, Toby

OptimalGrid and Building Massive Online Games

OptimalGrid is a research project from Almaden Research, which makes creating grid applications easy by hiding the complexity of the underlying grid. An ExtremeBlue project during the summer of 2003 applied OptimalGrid to the task of extending the Quake II server using grid designs. The result was a game server infrastructure, which grew from a single server many-player game, to a massive online world using dozens of servers and supporting hundreds of players.

Dias, Daniel M.

Cayuga On Demand Manager

The IBM High-Volume Web Site (HVWS) team and IBM Research are developing a technology, code-named Cayuga, on IBM WebSphere that provides dynamic resource allocation and workload balancing for multiple mixed workloads in a grid infrastructure. The solution supports multiple transactional and parallel numeric applications, each of which has a service level agreement (SLA) defined; the SLAs specify the service objectives, such as the transaction rate and response time objectives, for each application. Internally, each application is assigned a priority level, with a minimum and maximum number of processors that are allocated to each application. As traffic of a higher priority application increases to such a level that the SLA objective cannot be met, the system either drains the lower priority applications to other servers in a local or remote grid, reduces the number of servers allocated to lower priority applications up to a minimum number, or configures additional servers that are then allocated to the applications. As the traffic of higher priority applications decreases, the system re-assigns additional servers to the lower priority applications. This is accomplished in a reactive mode, as the demand occurs, or in a predictive mode by forecasting demand and reconfiguring the system in advance of violating SLAs.

The entire system is controlled by an autonomic MAPE (Monitor, Analyze, Plan, Execute) controller. Based on SLA definitions for each application, an SLA monitor determines when SLAs are violated or predicted to soon be violated by a forecaster. The analysis component then estimates the number of servers needed by each application to meet the SLAs. A planner then determines the re-allocation of servers to each of the applications, based on the SLAs, priorities, and the maximum and minimum servers able to be allocated to each workload. The execution component then reconfigures the load balancers and servers appropriately. Prototypes for interacting MAPE controllers at geographically distributed sites, have also been built, where remote resources are negotiated and configured when local resources have been exhausted.

Everhart, Craig

Global Name Space and Advanced File System Function

The integrated Data Grid can benefit from a convergence of technologies including NFS version 4 and CIFS, SAN.FS, and the span of volume level copy services provided by the sweep of block devices, brokered by OGSA. Benefits start with a uniform file name space for enterprise-wide resource location and grow to exploit copy services to provide enterprise-class availability and reliability. Cooperative enterprises, such as universities, have a natural way to join their name spaces and contribute to a global service.

Ford-Hutchinson, Paul V.

What's the Difference Between a Software Product and a Utility Service?

There is much interest in the provision of "e-utility" services; however, our experience demonstrates that this is a deal more complex than buying (or making) a bit of software and putting it on the Internet. Customers come to utilities for reliability, security, availability, support, and a host of other non-functional reasons.

After over 20 years of delivering EDI services (arguably the first e-utility) IBM, currently in the form of IGS e-business Hosting Services, has a wealth of experience in this field.

This session will look at what qualities, both technical and organizational, are required, to deliver services and not just host software.

Giangarra, Paul

Uses for Grid Computing and On Demand Business in Public Sector

This presentation discusses requirements and potential uses and/or needs for on demand business and Grid Computing Technologies that we have encountered while working with public sector (especially Federal Government, Military, Intelligence, and Defense) customers.

This discussion addresses the potential applicability of Grid Computing in the areas of distributed provisioning and resilience. A set of these requirements is also presented and discussed.

This discussion addresses specific requirements in on demand business computing that are less often encountered in private sector, discusses how they specifically apply to public sector, and shows how they can result in significant future business opportunity.

Gibson, Christopher

Applying Reputation Systems to On Demand Businesses

The ability to respond flexibly, efficiently, and rapidly to a changing business environment characterizes an on demand business. Take, for example, a supply chain. Maximizing the choice of service providers in a supply chain allows a business to select those who best fit its immediate business needs. These needs can and do change rapidly; in the morning low cost might be important, in the afternoon it could be lead-time.

There are an increasing number of online service providers that can be leveraged in support of this model. However, real flexibility can only be achieved when a business has access to all available service providers and not just the inevitably small set with whom contractual agreements are in place. Anxiety about the reliability and security of unpremeditated business engagements—a reflection of the real risks involved—inhibits both choice and the ability to react.

In this context, reputation is the empirical measure of a service provider's likelihood of delivering a service to an agreed standard. Likewise, assurances are promises or guarantees intended to inspire consumer confidence. Taken together they help a business to form an opinion of a service provider's (un)reliability. The ARIES (Assured Reputation Information Exchanged Securely) framework provides online access to such information, enabling business to select new service providers quickly and with confidence.

ARIES services can be incorporated into existing service acquisition models including automated contract negotiation. One of the principle advantages of ARIES is that it can be used both manually (for example, a Web browser plug-in) and automatically (for example, an online procurement system) allowing the fully automated, agent-based use of assured online services.

ARIES does not provide reputation and assurance information itself. Rather, it is a method for querying information providers, and securely incorporating the results into a transaction flow. The distributed components of ARIES choreograph an exchange between the service requester, the service provider, and one or more reputation authorities (trusted organizations to which the service requester delegates the acquisition of service provider reputation and assurance information).

The application of reputation in online business communities is a nascent field of study, but a field that is beginning to bear fruit. ARIES is characterized by its ability to support the practical application of reputation information in the decision-making process. In summary, its features include:

- The automatic acquisition of reputation and assurance information for any online offering, artifact, or entity that could attract a rating.
- The use of reputation and assurance information to maximize service provider choice (i.e., its use in service provider look-up and selection).
- Secure delivery of the information, specifically the confirmation of its origin and rendering it tamper-proof.
- The use of reputation and assurance information to automate service acquisition.

Bi-directional queries, both the service requester and the service provider seeking reputation information and assurances, are addressed.

Goering, Ron

eServer Platform Provisioning and ThinkControl Integration

The ability to quickly create, allocate, provision, re-provision, configure, and install server and OS images (including servers, blades, LPARs, and VMs) is an important element of many on demand business solutions. The eServer

Platform Provisioning work (ePP) will provide standard Web service interfaces to the eServer series provisioning capabilities.

The ThinkControl technology that is now part of the IBM portfolio as a result of the Think Dynamics acquisition provides provisioning orchestration capabilities for many logical and physical data center resources.

This discussion addresses ePP interfaces, capabilities, and the work we have been doing to integrate ePP into ThinkControl. This integration provides the ability to provision IBM eServers and coordinate these activities with provisioning of other logical and physical resources to enable dynamic addition and removal of capacity to a business application in response to changing application needs.

The first delivery of ePP/Think integration was delivered in 2003, as a fully supported product.

Goldszmidt, German

Oneida

Oneida is one of the incubator projects of the On Demand Application Environment, with focus on business process integration. It demonstrates the applicability of IBM middleware and tools to enable firms to interoperate and share in collaborative business processes.

Oneida will support interactions between hosted and in-house Web services, and demonstrate the ability for businesses to rapidly deploy new business processes by integrating existing services, across a virtual value chain.

In a generic scenario, several enterprises decide to collaborate on a business venture partnership, which may include participation from other firms. The partners define a set of business process templates, business objects, and applications.

The BPEL (Business Process Execution Language) for Web services is also used to define the business processes. Business rules that are germane to the offering may be described in Business Rule Beans (BRBeans).

The business objects are manifested as variables and containers. The applications are also rendered as Web services, to provide access to the offering. The customers will work with the partners' service teams to customize the process templates, data structures, and message formats to meet their specific business policies.

This discussion addresses multiple *use case scenarios*, and the design of some of the specific solutions for a *Proof of Concept* prototype.

Heisig, Steve

Using Machine Learning for Resource Consumption Characterization

Workload Characterization can provide a model, which humans as well as autonomic functions can query to compare normal application behavior with current or abnormal states. This will allow abnormal behavior such as program loops, stalls, and memory leaks to be detected programmatically. It can also enable generalizations to be made about common factors in performance problems.

This discussion presents a model based on clustering of application performance data, which allows humans to explore the data, as well as proposing a program API that would allow autonomic functions to take actions based on comparisons between observed values and model predictions. This approach is currently being tested in two dissimilar domains; one is a large bank running legacy work on IBM z/Series software and the other is a GRID-based prototype of a proposed national mammography database.

Jania, Frank

IBM Community Tools Providing Collaboration On Demand

IBM Community Tools is a suite of applications for interacting with various communities through instant and broadcast messaging. The power of client-side-filtered broadcast messaging is exploited to allow users to locate answers, start impromptu discussions, and to alert and survey large groups of people instantly, in real time. With 23,000 average monthly users, IBM Community Tools is a success story written with a number of complementary technologies. Web services, instant messaging, broadcast messaging, and Java are all leveraged to create a set of services and applications, which allow people to collaborate on demand. At the heart of the system is the broadcast Web service. This service allows authenticated broadcast to all the members of an entire community at once. The IBM Community Tools client provides a GUI front end to these services, and a rich interface to the receipt settings and notification of broadcasts.

The client consists of a number of plug-in applications that provide on demand collaboration functionality. The *w3alert* allows the user to broadcast messages to one's community, along with an optional URL for additional information. *TeamRing* is a tool that allows users in a community to view the pages of a Web-based presentation in concert, without the overhead of application or screen-sharing programs. *PollCast* gives the user the capability to send out a survey to one's community and get the results back, displayed in graphical form, in real time. *FreeJam* allows the user to broadcast a topic for discussion out to the community and start a just-in-time chat room for that topic. *SkillTap*, the most widely used application, allows the user to broadcast out a request for information or help out to the entire community and start a one-one instant messaging session with another user who can provide the requested information. *SkillTap* also offers a persistence mechanism to store, at the discretion of the responder, the responses to frequent queries in a searchable database.

There are two releases of IBM Community Tools currently in use. The IBM internal release has had an average monthly user base of 23,000 users. Since its inception the *SkillTap* application has been the enabler for 20,000 answered queries, all of which were answered by people throughout IBM with an average response time of about one minute. This on demand business solution represents an approximate savings of $2,000,000 USD worth of help desk calls.

The tremendous on-going success of the internal implementation of IBM Community Tools sparked the creation of an external implementation focused at the iSeries iNation community. iSeries customers and professionals from around the world use this implementation to collaborate in real time. This external implementation has been extended to the large alphaWorks community, which will allow members of both the alphaWorks community and the iSeries nation to collaborate with their own communities, or across communities.

Kataoka, Katsuhisa; Kako, Naoko

Rich Client for Enterprise Business Systems

In recent years, HTML clients have been widely used in Web environments regardless of the system types or the target users, but they have not completely satisfied all industrial customers' requirements. One of the common problems is the efficiency in inputting data. This discussion addressing the customer pains for the client of the enterprise business systems, shows our approach to relieve them, discusses what a good client is, and proposes a method to develop it.

Enterprise systems were running on centralized main frames, then shifted to the client-server systems, and finally have changed to Web systems. The client for Web is browser. Web browsers are made for browsing documents, not for inputting data. End-users, who have been familiar with rich user interface of client-server systems, do not satisfy.

We have developed several enterprise business systems for Japanese customers, such as an order entry system for an industrial manufacturer, a financial assessment system for a banking customer, an electronic form application system for a governmental organization, etc. Our experiences tell us that from the end-users' viewpoints, efficiency is the most important characteristic for enterprise business systems. Enterprise business systems are developed for specific business operations. Their efficiency is linearly related to their business performance. On the other hand, from the developers' viewpoints, ease of development and maintenance are the most important characteristics, because the development budget and the schedule are always limited. We need a client system with efficient input as well as easy development and maintenance. Otherwise, we cannot develop responsive and useful enterprise business systems.

Our solution is a Java-based client system called BAP (Business Adaptive Panel). We (1) made a common foundation to assemble the customizable and reusable components for business forms, and then (2) arranged the components that can be used immediately. This allows application developers just to assemble components without writing new Java codes. We also defined an XML vocabulary to define screen images and made Java executables to interpret XML into Java component instances. The screen definition XML is distributed using HTTP like HTML, and the data exchange is done separately. BAP runtime can be used as applet and is easily maintained and distributed to the clients from the central site. The data fields are kept on the client and can be transmitted to the server when required; the server need not retain session information and this reduces the server workload.

BAP enables people who aren't familiar with Java to develop client systems. BAP also supports to make prototypes easily. Almost all customers require screen images from the early phase and discuss their business processes through them. From users' view, client user interface seems to drive their business processes. BAP affords this design method and common understanding for both users and developers. Additionally, by defining client data model first, look and feel of the BAP client can deal with change requests from users flexibly. We believe BAP and client user interface driven design will greatly improve user experience of the enterprise business system.

Kephart, Jeff

Unity: A Prototype Autonomic Computing System

IBM's autonomic computing initiative is aimed at solving a fundamental problem: the complexity of integrating and maintaining large-scale systems, which oftentimes threatens to overwhelm the capabilities of humans. IBM is, indeed, building systems that are self-configuring, self-healing, self-optimizing, and self-protecting.

Recognizing the need to demonstrate these types of systems that exhibit the desired "self-" properties, and to show that the autonomic computing architecture is absolutely on the right track, IBM Research established a milestone by developing and demonstrating a robust, responsive, holistic autonomic computing system in 2003. We describe *Unity*, a medium-scale autonomic computing system prototype that has been delivered from autonomic elements in accordance with the IBM autonomic computing architecture and is based on one of the on demand business scenarios. We discuss how, by integrating IBM technologies that are appropriately imbedded within the architecture, one can achieve self-configuration, self-optimization, and self-healing, while at the same time, learn some important lessons about the autonomic computing architecture.

Knox, Alan

eDiaMoND—A Grid for Screening and Diagnosis of Breast Cancer

Oxford University has joined with IBM, Mirada Solutions, and a group of clinical partners to build a Grid that will support diagnosis of breast cancer and provide medical professionals with more information to help treat the disease. The project, which represents a joint investment of approximately USD $6 million by IBM and the UK Government e-Science Program, has been named "eDiaMoND" by Oxford researchers. eDiaMoND will be one of the first Grids built entirely with commercially available technology, including state-of-the art software developed by Mirada Solutions to standardize new and existing mammogram x-ray images.

Krantz, Steve

Infrastructure On Demand

IBM corporate strategy is to optimize operations to dynamically respond to the needs of customers, employees, partners, and suppliers; that is, to sustain ourselves as an on demand business enterprise. The resultant infrastructure has become integrated, open, virtualized, and autonomic. The strategy also considers user satisfaction and productivity, cost optimization, the establishment of a showcase(s) for IBM and partner products, and support for our business processes.

IBM offers a broad set of products and services, many of which provide features of an on demand Operating Environment. Plans are in place to continuously improve and enhance these products/services, as well as to develop new ones in support of new on demand business features. Many of these products and services fit into the evolving IGS on demand business Universal Management Infrastructure (UMI) framework.

A forward-looking strategy is required that encompasses succeeding years to address technology, internal processes, organization, and culture. The technology strategy is to evolve a rules-based, virtualized infrastructure utilizing both private and public utility assets. With the private utility, infrastructure is dedicated to IBM internal requirements, with on demand business acquisition of additional resources, when needed from the public utility that is shared with other IGS customers. Internal processes improve to include deployment automation based on encoded rules (e.g., standards, policies, and service-level agreements). Metered billing also provides process improvements. In terms of continual transformation of our organization(s) and culture, risk tolerance analysis continues to be performed on end-user groups, and the results used to foster the rate and pace of change and communications as new technologies are introduced.

As we continually transform the IBM IT infrastructure, we measure our progress. New high-level, end-to-end metrics have been developed to address this.

Looking at the business value of the prospective investments and their accompanying change, significant cost reductions and cost avoidance should occur throughout several tactical periods of time.

The transformation of the infrastructure (as literally defined by the term "transformation") is an evolutionary process. The CIO's office continually addresses several key steps to assist in business transformation. For example, this office has defined new metrics and supports the IBM Global Services (IGS) On Demand Services (ODS) computing platforms with several key applications. The CIO's office has planned a number of new infrastructure investments for the future, to help realize and strengthen the strategy, such as accelerated production use of the IGS UMI and ODS environment(s). Investments, such as use of a public utility and automated application provisioning, are included.

This discussion pays full treatment to these very important aspects in any on demand business transformation.

Kuchhal, Manu

Pervasive P2P Grid

An end-to-end uniform services-oriented paradigm on all classes of devices, ranging from servers, PDAs, network devices (e.g., routers, switches, etc.), to smart devices will enable a Common Resource Management model. This model is critical to ensure smooth realization of on demand business. In such an environment, all resources are made available as utilities, available on demand, and bound by QoS (Quality of Service) and SLAs (Service Level Agreements). It can also serve as a basic infrastructure to implement the architectural blueprint for Autonomic Systems, as suggested by IBM.

Motivation: Pervasive Computing is about "a billion of devices interconnected and interoperable with each other; exchanging information anytime, anywhere, anyplace." P2P talks about sharing of resources between multiple parties, at one-to-one levels, without any centralized controls. A Grid standard, like OGSA (Open Grid Services Architecture), provides a mechanism for uniform abstraction, aggregation, federation, and integration of resources, for example, like compute cycles, data storage, and managed networks as services.

Smart devices (are connected to networks and respond to requests for specialized services over networks) can be made accessible, manageable, and controlled over a common bus of OGSA compliant protocols, ensuring compatibility with all vendors and clients. Examples include lab-instruments, accelerometers, biosensors, telescopes, wind tunnels, security appliances, entertainment devices, office, and health care equipment. If one were to consider Grid as a very large meta-computer, smart devices are its I/O.

Various applications such as automated software provisioning and management, file-sharing, and CPU cycle scavengers like @SETI could run as specialized OGSA services, thereby reducing code size and complexity (by unifying common functionalities among them).

Most of the features of P2P (Peer-to-Peer) frameworks like resource abstraction as services, virtual peer groups, directory services, discovery and lookup mechanisms, extensible service description, and location transparency, already exist in OGSA, thereby making it an excellent choice for P2P computing disciplines.

Solution: OGSA running on all classes of devices is desirable in order to enable access, usage, and monitoring of resources anytime, anywhere, anyplace; and allow interoperability and interconnectivity between these devices with the ability to engage in P2P interactions.

A lightweight OGSA node is suggested that could run on PCs, as well as on small form-factor devices, such as: PDAs, network devices, smart devices, and Game Consoles. When used in conjunction with the other features of Extension Services for WebSphere® Everyplace™ (ESWE) platform, this enables an incrementally installable, very manageable GRID node utilizing very small amounts of resources.

The ESWE platform provides a component-based Java™ runtime environment that implements the OSGi (Open Services Gateway Initiative), an open-standards platform supporting the development, deployment, and management of services to pervasive devices. OGSi already provides some of the requirements of OGSA, including Web Services; they just need to be wrapped into a solution complying with OGSA specifications.

Legg, Steve

Storage Virtualization—An On Demand Business Enabling Technology

Storage Area Network (SAN) technology was introduced around five years ago. One of the achievements that SAN has accomplished is to bring to the open systems arena some of the storage management advantages that have been familiar to mainframe system administrators for several years. Customers are demanding more, and Storage Virtualization enables the next big step in disk storage management.

This discussion positions Storage Virtualization as an enabling technology. The discussion will go on to describe IBM's product announcement: the IBM 2145 SAN Volume Controller. This product, developed in Hursley and Almaden IBM Research Centers, is IBM's strategic platform for Storage Virtualization. This discussion illustrates many of the design decisions considered in order to bring this product to market. One is able to explore comparisons with competitive offerings from other vendors, as well as to speculate on future directions for this key technology area.

Levine, David W.

A Toolkit for Autonomic Computing

IBM is delivering a toolkit for building autonomic managers. The toolkit focuses on providing the "instruments" required for building the required parts of an autonomic element. This includes a set of tools and technologies, which aid in the development of the specialized portions of autonomic managers.

This discussion focuses on *how* this toolkit is structured, including *how* it aids in the rapid deployment of the grid portions of the autonomic element. This discussion defines the various sources of technology in the toolkit, including the Globus Grid toolkit and the Agent Building and Learning Environment (ABLE), as well as other related technologies. One will explore some very key strategies for building autonomic elements, including policy-based computing. Finally presented are future directions for the autonomic manager toolkit.

Loader, Becca

Extreme Blue

The Extreme Blue™ program is IBM's incubator for talent, technology, and business innovation. The Extreme Blue program challenges project teams of technical and MBA interns (along with their technical and business mentors) to start something BIG by developing new high-growth businesses starting with a promising idea, market, or technology.

In just twelve weeks, Extreme Blue project team members are expected to develop both technology and the business plan for an emerging business opportunity in areas as diverse as on demand business, Grid computing, pervasive computing, and many others. This dedicated small team approach is unique combining high-potential business and technical interns with early career IBM professionals, as well as IBM's most distinguished technical and business leaders, scientists and visionaries. Extreme Blue interns face the challenge of making a real impact in the industry by leveraging IBM's unmatched resources (people, technology portfolio, and infrastructure) in a leading-edge innovation work environment worldwide.

The Extreme Blue program is committed to the professional career development of its interns and early career professionals. Professional staff is on hand every day for mentoring and training in technical and business disciplines. Team members network both formally and informally with all levels of IBM leadership across business and technology both for their projects and for their own career development.

For more information about Extreme Blue, visit, ibm.com/extremeblue.

Luniewski, Allen

DB2 Information Integrator: Bringing On Demand Data to the Grid

The Grid/On demand environment is a highly distributed, heterogeneous collection of sources of data. Applications need to access and integrate data across this collection of data sources. Some of this access is anticipated and static in nature. Other access patterns are also unplanned. And access patterns will change over time. The DB2 Information Integrator product (DB2/II) provides the facilities to address this need through its replication, caching, and federation capabilities. The Grid/On demand environment presents challenges beyond those addressed by the current DB2/II product. We will discuss these problems and how DB2/II will need to evolve to address them.

Magowan, James

OGSA Data Access and Integration

The Open Grid Services Architecture—Data Access and Integration (OGSA-DAI) project is building an efficient Grid-enabled middleware reference implementation of the components required to access and control data sources and resources in an OGSA environment. As well as being a technical challenge in its own right, this project is closely linked to the standards processes of the Global Grid Forum (GGF) and to the Globus team providing an implementation of the Open Grid Services Infrastructure (OGSI). The project involves team members from IBM, the National e-science centre (NeSC), Edinburgh Parallel Computing Center (EPCC), University of Manchester, University of Newcastle, Oracle, and University of Southampton. See: http://www.ogsadai.org.uk for more information.

Nadalin, Anthony

Security Challenges in an On Demand Environment

Securing an On Demand environment brings its own challenges. It brings up the need to provide a comprehensive security infrastructure—architecture and roadmap. This is in order to support, integrate, and unify popular security models, mechanisms, protocols, platforms, and technologies in a way that enables a variety of systems to interoperate securely and dynamically. It also suggests the need to integrate the model between the operating environment and the application environment to provide a manageable security environment. This discussion focuses on some of these challenges and possible approaches to address those based on existing and emerging security technologies, standards, and products.

Nash, Simon

WebSphere SDK for Web Services

IBM WebSphere SDK for Web Services (WSDK) is an integrated kit for creating, discovering, invoking, and testing Web services. WSDK is designed to address the needs of experienced Java programmers who want to quickly learn how Web services can be created using existing Java components and achieve seamless integration with disparate systems.

WSDK combines the industry-leading expertise of IBM in Web services with the productivity of the Eclipse IDE framework and the power and functionality of the market-leading IBM WebSphere Application Server to offer a low-risk, affordable entry to Web services. WSDK enables developers to create and test Web services that conform to the Web Services Interoperability Organization (WS-I) Basic Profile 1.0.

Supporting the latest specifications for Web services including WS-Security, SOAP, WSDL, and UDDI, WSDK supports the following key capabilities and benefits:

1. Provides everything necessary to create and test Web services in a tightly integrated package including Eclipse IDE plug-ins for Web services development, a simplified Web application server, a private UDDI registry, tools, documentation, and sample applications.
2. Interoperates with other vendors' implementations—specific samples are provided for Microsoft .NET clients.
3. Delivers an out-of-the-box solution with quick installation for building and testing Web services.
4. Simple integrated tools speed the creation and testing of Web services, including hot deployment to the application server.
5. Provides comprehensive documentation, samples, tutorials, and an online newsgroup.
6. Migration is available to enterprise-scale deployment with IBM WebSphere products.

Novaes, Marcos

TAGSS Grid Scheduler

The Topology Aware Grid Services Scheduler (TAGSS) is a new technology capable of dynamic scheduling requests to stateful Grid services. TAGSS also introduces a new programming model for Grid computing, called the Parallel Array Language. The presentation includes an overview of the TAGSS architecture, the PAL programming model, and a demonstration of an application developed using TAGSS and PAL. The talk will also include customer case studies and a review of recent activities related to TAGSS, including its relationship to other scheduling strategies within IBM.

Perry, Riggs

Disconnected Application Architecture

The migration from Notes-based technology to Web-based has removed a standardized technique for providing disconnected functionality for mobile users. The result of this change is that application owners/developers are engineering their own custom local data stores and synchronization mechanisms. This produces interoperability issues, duplicate development efforts, etc. The larger issue in this area is the growing need to allow IBM employees to access their data and applications On Demand wherever and whenever their job requires it.

This presentation will discuss the W3 Proactive Portal project, which provides off-line access to selected content on IBM's intranet. The goal of this work is to create a framework of characteristics for Web-based applications and a "playbook" of in profile, standardized technologies that can be used to provide off-line capabilities. Using this playbook, application owners will spend less time creating unique solutions and have a better idea of the cost and performance associated with providing disconnected functionality.

IBM will benefit by having a standard set of solutions that can be implemented and managed in the infrastructure instead of ad hoc implementations created by individual application owners.

Pollitt, Daniel

Agile Software Development Within IBM

With the "On Demand" era, the speed that IBM needs to react to changing requirements in order to deliver business value to its customers is increasing.

Agile Software Development methods such as eXtreme Programming can work within IBM to meet these challenges, following the simple value statements of focus and preference presented by the Agile Alliance (http://www.agilealliance.org) in the Agile manifesto (http://www.agilemanifesto.org).

A project identified as "agile" and developed using Test-Driven Development (TDD) implies a great deal about its quality and ability to "embrace change." For more information, please see http://www.extremeprogramming.org, http://www.xpdeveloper.org.

Raghavachari, Mukund

A Programming Model for the On Demand Era

The acceptance of on demand business computing depends in no small part on a programming model that simplifies the development of robust programs for this flexible and dynamic environment. Current programming models fall short as a result of two characteristics of the on demand business era. First, On Demand computing blurs the distinction between clients, servers, and databases (the classical 3-tier architecture). The trend of blurred boundaries has, in some sense, been visible in the application-database interface as the programming model has evolved from batch-oriented processing to transaction processing monitors to application servers. With On Demand computing, the distinction between applications and databases vanishes as the database becomes yet another service used by the application. Our contention is that in this era, concepts such as transaction support and data integrity constraints that have long found use in database systems must be introduced into the programming model to support the development of robust, maintainable applications.

The second trend of the on demand business computing era is the pervasive use of XML as a data exchange format. Current programming models have little support for XML both in terms of integrating XML data into an application as well as efficient algorithms for processing XML. To access XML data from an application, a programmer might write code that utilizes DOM or SAX to operate on XML data. Not only is the code complex and error-prone, but also the programmer gets little assistance from the programming environment (such as static type-checking) to ensure that the developed programs are correct. Efforts such as WDO (WebSphere Data Objects) alleviate the problem by providing an easier model for accessing XML data. Our thesis, however, is that as XML becomes ubiquitous, it is important to support XML as a first-class data type in the programming model, that is, programmers may declare variables of XML type and get the guarantees, such as static type checking and optimizations, that have long been expected from a programming language. Furthermore, the programming language should support standard XML mechanisms such as XPath for the navigation of memory-resident XML data.

The DOMO project at the Thomas J. Watson Research Center is focused on developing a programming model for the On Demand era. We are developing abstractions that introduce transactions, data integrity constraints, XML Schema types, XPath navigation, etc., as first-class programming constructs. In this presentation, we shall address two specific issues. The first is XJ, an extension to Java that introduces XML Schema types as Java types. Programmers may "import" XML Schemas and work with the types defined within schemas in an intuitive manner. The XJ compiler takes an XJ program and generates a Java program that uses DOM (or WDO) to access the XML data. XJ shields a programmer from the idiosyncrasies of these APIs allowing him or her to program in a manner natural for XML. We shall also describe algorithms that the compiler may use to check schema compliance, and to efficiently implement XPath accesses in a streaming fashion.

The second topic of this presentation is the introduction of data integrity constraints as a first-class construct in a programming language. The programmer may specify constraints (such as, an employee's salary must be less than his or her manager's) on objects or on database schema objects. The programmer may also specify when constraints must be enforced and actions that should be taken to repair inconsistencies. The system takes these declarative specifications and analyzes application code to ensure that data integrity constraints are satisfied. The system may generate code automatically to verify constraints that cannot be verified statically.

Read, Paul

Smart Use of DB2

This presentation shows the new autonomic and SMART features in DB2 UDB V8.1.

We will look at the technology for reducing human intervention/cost in the operation of a DBMS that includes one or more of the qualities of automation, decision-making, or expert advice.

We will explain:

- What we mean by SMART/Autonomic
- What new functions and tools are available
- How to use the new functions and tools
- What will be the focus in the future releases

Sheftic, Rick; Vijayan, Geetha

Autonomic Computing—Key Enabler of the IBM Global Account On Demand Transformation

IBM has made the strategic decision to lead the on demand business era. Autonomic Computing has been identified as one of the four characteristics and critical enablers of the on demand Operating Environment with the promise of significant long-term business value. Therefore, IBM is also an early adopter of autonomic computing technology in its on demand Operating Environment. The Gartner Group predicts it will take eight to ten years to realize the full potential of autonomic computing and that the maturity process are evolutionary across multiple dimensions of technical capability and business value. Clearly, IBM is at the very early stages of a complex evolutionary journey.

To be an effective early Autonomic Computing technology adopter, it is important to develop an Autonomic Computing capabilities roadmap. Creation of this roadmap requires a repeatable method that will enable business units and IT architects to assess the potential benefit of Autonomic Computing technologies applied to specific environments. This pre-deployment assessment identifies the cost/benefit tradeoffs required to optimize the business value realized from autonomic computing deployments.

During the first half of 2003, the IBM Business Transformation/CIO Enterprise Systems Management team conducted a preliminary assessment of the IBM Global Account Autonomic Computing capabilities and developed a set of high-level recommendations and a tactical roadmap proposal for autonomic computing adoption in the IGA over the next 18 months. The assessment scope included the Tivoli, Websphere, DB2, Lotus Notes, and Storage IBM products. Key results from this preliminary assessment are:

- Value is derived from Autonomic Computing in three dimensions: Return on Investment (Total Cost of Ownership), Quality of Service and Resiliency, and Time to Value.
- The potential incremental value an organization may realize from the adoption of Autonomic Computing technology is a function of the current autonomic maturity within that organization. To that end, IBM's Integrated Technology Services (ITS) has developed an Autonomic Computing Adoption Model to help organizations understand their current system management issues and future Autonomic Computing requirements.
- {Other assessment items have not been listed, for IBM internal reasons}

This paper addresses key Autonomic Computing adoption questions in the context of the IBM Global Account, and will discuss the preliminary assessment recommendations and tactical IGA Autonomic Computing adoption successes and roadmap.

Smith, Andy J.

Executive Dashboard

The Executive Dashboard is an intranet Web application that brings together customer satisfaction and quality data from disparate sources and presents it in a unified way. The data sources are managed by different parts of the corporation, and vary considerably in structure, organization, and concept, and each has its own mechanisms for administration, access, and analysis. The Executive Dashboard integrates these data sources to enable comparisons and analyses that were not previously feasible.

This application enables executives and product managers to leverage the wealth of information which IBM collects for its products in order to quickly focus on exceptional situations and make informed decisions. Trend views quickly highlight the areas of concern, and drill-down allows them to be examined in more detail. Users are able to track performance against targets for groups of products using predefined groups or groups that they create for themselves.

The Executive Dashboard application incorporates a data analysis engine, which delivers data via a Web service to a Web application. JSP and servlet technologies are used to provide the user interface via a Web browser. A DB2 database accessed via JDBC enables the Web application to support personalization and administration. We will explain how the use of J2EE technologies has helped us to partition and simplify the complex task of presenting heterogeneous distributed data in a consistent and dynamic way to produce the required user experience.

Tan, Yih-Shin

On Demand Application Environment Scenario—Service Domains

Service Domain is a way of aggregating and sharing of multiple Web services of various origins by presenting a logical view of single services. At the type level, a service domain SD is a set of port types, i.e., SD = {pT1, ... ,pTn}. At the instance level sd, a service domain is a collection of ports, i.e., sd = {p11, ... ,p1k(1), ...,pn1, ... pnk(n)}, Here, each port pji sd is of port type pTj SD. With this technology a service broker can be established easily that manages a heterogeneous group of service suppliers to provide services to multiple groups of consumers. The China Software Development Lab is exploiting this technology to engage customers in China by offering an On Demand Service Grid solution for e-science and e-service applications, as an On Demand customer pilot project that is later announced as the China Education and Research Grid. This technology has the main role of broker and marketplace for services in the on demand Operating Environment scenarios. The talk will present the concept and a demo for the virtualization of an On Demand Service Grid.

The China Education and Research Grid enables universities across the country to collaborate on research, scientific, and education projects. This is one of the world's largest implementations of grid computing—which takes untapped application service, data, and computing resources from different computing systems and makes them available where and when they're needed, resulting in a single, virtual system. The Grid was launched in October of 2003 initially with six universities, and will link more than 200,000 students and faculty members at nearly a hundred universities across China when the project is completed. When phase one of the project is completed in 2005, the grid will perform more than six teraflops, or trillions of calculations per second, and eventually will be capable of more than 15 trillion calculations per second.

Thompson, Lance

Verification On Demand Implementation Strategy

Engineering and Technology Services (E&TS) is one of IBM's newest endeavors to create a services organization that applies the best technology and the best engineering talent to bring lasting value to our customers. IBM's verification technology has long been an advantage that IBM chip and system designers have enjoyed. In addition, IBM verification engineers have a significant history of solving complex verification problems. This paper describes the E&TS team's thoughts on making IBM's verification technology and engineering skills available on demand with the concept of Verification On Demand.

The paper begins with a brief introduction to the tools and methods that make up IBM's verification advantage. The discussion includes the topics of the engineering design automation grid, cycle simulation tools currently in use, and the importance of applying IBM Verification Engineers to provide high-value service offerings.

The focus of this paper is the technologies that E&TS applied to form the offering. These technologies range from the queue manager in the electronic design automation grid to the graphical user interface presented to the customer. A highlight is the high degree of reuse achieved by E&TS by utilizing an existing internal design automation infrastructure. Some of the elements used to create the Verification On Demand offering are:

- Verification Client, which interfaces with the user to collect tool data, parameters, and commands to send through the DropBox API to the Grid.
- DropBox acts as our secure communications middleware facility. DropBox has an application program interface (API) that facilitates a client/server-programming environment in which the client is customizable to appropriately control data flow through the DropBox to the server (Grid).
- Job Submission Daemon runs on the Grid and retrieves data, parameters, and commands via the DropBox API and launches appropriate pre-submit wrapper.
- Pre-submit and Runtime Wrappers prepare the data and environment and runs the tools.
- LoadLeveler/Scubed performs queue management and automated job submission. Scubed also includes some self-healing technology to the mix.

While still a work in progress, we are confident we have laid the foundation for a unique on demand business offering that enables IBM to provide high-value Engineering Design Automation service offerings.

Van Heuklon, Jeff

Use of ABLE for On Demand Computing

ABLE (Agent Building and Learning Environment) is an Autonomic Computing core technology. This cross-platform framework can be used as a foundation for On Demand computing on IBM eServers. To truly follow the on demand business vision, eServers must be able to self-configure, self-heal, self-optimize, and self-protect (self-CHOP). Intelligent agents built on the ABLE framework have the ability to do all of these. This presentation will consist of two main topics. First, there is a high-level overview of the ABLE technologies, showing the structure of the framework and its capabilities. Secondly, ABLE is already being investigated by multiple development organizations across IBM. Descriptions of some of these possible projects, as well as ideas for future projects, are also covered.

ABLE Technologies The ABLE framework, component library, and distributed agent platform (collectively called the ABLE Runtime) are all written in Java. Individual agents that do work are written as Java beans. This makes the technology portable enough to run on any server platform and operating system. The ABLE library includes AbleBeans for reading and writing text and database data, for data transformation and scaling, for rule-based inference capability using Boolean and fuzzy logic, and for machine learning techniques such as neural networks and decision trees. This presentation will show how developers can extend the provided AbleBeans or implement their own custom algorithms. An agent editor is provided that developers can plug these new or updated ABLEBeans into. From there, properties can be changed, bean input can be displayed, and output can be viewed as machine learning progresses. There will also be a discussion on how rule sets created using the ABLE Rule Language can be used by a number of different inference capability engines. Some of these engines are included in the ABLE component library, and they range from simple if-then scripting to heavyweight Artificial Intelligence algorithms.

These ABLE technologies are available to be downloaded from IBM's alphaWorks Web site (http://www.alphaworks.ibm.com/tech/able). In addition, ABLE is shipped with OS/400 on the iSeries server. As ABLE becomes more pervasive, additional methods of distributing this technology will likely be made available.

Applications For the initial releases, focus is being put on agents that help customers with planning and analysis. Ideas for the initial ABLE agents include:

1. Setting up the right kind of software trace to troubleshoot a problem
2. Monitoring the health of a server, and when there are warning signs, automatically make the required changes to improve the health.
3. Monitor utilization of system resources and use prediction capabilities to purchase additional resources on demand to satisfy SLAs.
4. Analyze TCP activity and take protective actions when intrusions are detected.

Vashaw, Bart

WebSphere Scalability On Demand

On demand business is delivered in the WebSphere Software Platform. IBM's new set of on demand business offerings, such as IBM Server Allocation for WebSphere Application Server and IBM Web Server Provisioning

Offering are designed to help customers get started on their journey to become a on demand business, driving down costs, increasing organizational productivity, and enabling more flexible business operations.

In this work, several new on demand business initiatives are outlined and demonstrated. The particular initiatives discussed at this are focused on techniques to strengthening one's environment utilizing WebSphere application scalability and resiliency. Creative techniques for one's consideration in the intelligent workload management, and an exploration of the high-performance Web services delivered are at the core of these initiatives.

For example, the "on demand business partnership" between IBM and Akamai is spotlighted as an instance of the integration of the IBM WebSphere application platform with the Akamai distributed computing platform, paving the way to truly enable on demand business. Akamai EdgeComputing *Powered by WebSphere* products from IBM provides companies with a way to tap into virtual, Internet-based resources whenever they require them. The solution permits companies to extend applications built on WebSphere to Akamai's distributed network of more than 15,000 servers. Instead of paying for spare capacity that will remain under-utilized (much of the time), companies can now meet their demands on a real-time basis; this is a "*pay-as-you-go*" type of utilization environment.

Walker, Lance

Enterprise Component Business Architecture (ECBA)

The Enterprise Component Business Architecture (ECBA) is a new business architecture for IBM's on demand business enterprise. Its mission includes the development of a more fully integrated relationship among the designs of processes, applications, and data. To promote this integration, the modularization of business processes and IT functions is a strategic element that is also one of the fundamental topics discussed during this session.

A formalized business process framework is a key ECBA ingredient that prescribes a modular process construct for the rapid design of solutions. Its advantages are discussed to explain how it promotes the development of a solution through the assembly of smaller process elements, called Enterprise Process Modules, which incorporate modular IT constructs. Modular IT components are not really a new idea, and for the purposes of this architecture, components are defined as software elements combining business function and data that can be independently developed and plugged in when needed. Specific Enterprise Process Modules and Enterprise Components are being defined to meet IBM's business interaction and transaction needs -- both internal and outwardly facing. Examples of these elements are Order, Customer, and Contract.

With the above constructs understood, business analysts can assemble process models by incorporating predefined Enterprise Process Modules that are associated to predefined Enterprise Components. In addition to Enterprise components, workflow and services are other IT elements needed to complete the ECBA picture. Process models can be operationally instantiated as *workflows* to replace large applications, and then be designed to "thread" through components. Components are exposed as multiple services that can be choreographed into workflows as needed, to rapidly provide assembled solutions for new requirements that may result from sudden on demand business changes in tactical directions. Tools and products to create and use this framework were also described.

In summary, the ECBA architecture will result in smaller modules of more manageable components (Enterprise Process Modules and Enterprise Components). This can then be independently and easily generated, and updated. These component-based functions that directly align with specific IBM business processes, respond more quickly to evolving customer demands. Smaller and more granular functions (Enterprise Component services) are associated to the Enterprise Process Modules, and used in various combinations to produce variable cost solutions, as opposed to the creation of large monolithic and fixed applications that are often more narrowly defined in scope, and oftentimes difficult to change. Web Services, as a standards-based interface, is designated as a primary method to access the Enterprise Components, which promotes a Service Oriented Architecture (SOA) to allow for a more open interaction between business functions.

White, Steve; Whalley, Ian

Self-Management of the Grid

Much of the work in defining the "Grid," to date, has focused on functional characteristics, how to provide services of various kinds, and how these services can be bound together to create *virtual organizations*. As the Grid matures, ever-larger collections of services are bound together to deliver more complex functions, and these functions will then be dynamically configured and reconfigured.

In this work, we suggest that the resulting systems can become the largest, most complex, most dynamic systems ever built, and that traditional systems management may not adhere well to the challenges that they will encounter. We propose a unified approach for *self-management* of Grids, in which each Grid service is responsible for internal self-management. Establishing and enforcing negotiated relationships with other Grid services accomplish this. Policy-based control is used to raise the semantic level of the management interface and to simplify management. This design pattern is repeated at successively higher levels of abstraction in the system, resulting in self-managing domains, self-managing virtual organizations, etc. We review lessons learned in prototyping systems that use this approach.

Wissenbach, Jens

Automation Tool

Automation Tool is a configuration, reporting, and scripting tool that allows a customer to handle a number of Enterprise Storage Servers (ESS) from one desktop. It uploads configuration, monitoring, and performance data of all attached ESSs in dBs with a JDBC interface. It sends automatic reports in ASCII format via email. It has a command line interface via a server client SSL interface.

Wright, Roger P.

Designing "On Demand" Business Applications

This abstract summarizes an approach to building applications that can change "On Demand." Much has been made of the On Demand Operating Environment.

In the opinion of the author, "applications," as described below, have not been sufficiently considered in the discussion. Many of the characteristics of the "On Demand Business" were defined in 1999 in Stephan H. Haeckle's book, "Adaptive Enterprise" (Harvard Business School Press, 1999), which he called a "sense-and-respond organization" at that time. For example: "A sense-and-respond organization does not attempt to predict future demand for its offerings. Instead, it identifies changing customer needs and new business challenges as they happen, responding to them quickly and appropriately, before these new opportunities disappear or metamorphose into something else" (page 3).

Haeckle's book did not attempt to show how IT systems might be constructed to meet these requirements—it was about Business Organization, Management Strategy and new approaches to running a business. It did recognize the need for IT to support the "sense-and-respond" organization and several of the case studies featured custom-built applications that supported the specific needs of those organizations. However, we can see that with the recent emphasis of *On Demand* approaches to business, it is necessary to determine how IT systems may be constructed to meet the premise.

There has been a considerable amount of work lately looking at *On Demand* IT infrastructure, typically referred to as the "on demand Operating Environment." This asks the question "what are we operating?" and leads to the title of this work. In essence, what sort of applications do we need to support an On Demand Business? Here, "applications" are defined as those parts of the IT infrastructure, which actually embody one's organization. They are NOT the Web Browser, the Application Server, the Message Transformation hub, the Workflow Manager, the Content Manager, etc. They ARE the Order Management system, the Warehouse system, the Logistics system, the Billing system, the Seat Reservation system, the General Ledger system, etc. These applications used to be custom-built and kept many of our colleagues gainfully employed for many years.

Today, package vendors like SAP and Siebel (and so) often provide them. Generally speaking, IBM no longer develops these kinds of applications and has recently divested itself of such of these applications as it had developed in the past. These observations are based on this author's {*Wright, R.*} approaches used in deliverables of his current engagement as Chief Architect on a major design, build, and run contract in the Communications Sector.

Wynter, Laura; Liu, Zhen

Managing On Demand Infrastructure Through Yield Management—Framework and Architecture

Currently, resource management systems for on demand computing centers aim to satisfy constraints associated with the computing needs of the current clients of the system.

However, the management of on demand computing centers must take into account the varying requirements of different customers, arriving in real time, and the different levels of service that are promised to each.

The real-time nature of on demand requests, as well as the underlying assumption that not all on demand requests are equally valuable (to the client or to the provider), adds new and significantly more complex requirements and revenue-gaining possibilities to the currently used resource management systems.

All components of a resource management system should be revisited in the context of on demand. These components include the subscription and billing services that interact with customers, the resource allocation and scheduling components that link customer requests to the computing center hardware, software, and network infrastructure, and the monitoring tools that evaluate the throughput and response times of each job, and in more recent systems, keep track of those parameters and possibly others with respect to service-level promises.

In particular, what is missing from this framework and what can serve to tie the existing system with the on demand context, is a yield management component, and in particular, a yield management reservation system.

The yield management components are based on a fine segmentation of customer demand using data on price/service-level elasticities that relate levels of demand to price and service levels offered to the clients and link these demand-side characteristics with the supply-side through a yield management scheduler and arbitrator and the yield management reservation system. In other words, a yield management component links sophisticated models of "what customers want and are willing to pay" with what the on demand center can offer, and does so at each instant in time, as the state of the on demand center changes. One may use the analogy of airline reservation systems to understand the nature of an on demand yield management-based system.

This presentation presents the ideas, methodology, architecture, and examples of the yield management framework for on demand computing centers.

Yamamoto, Naoya

Applying ARIMA Model to Transaction Data and Forecasting Server Capacities

This paper will suggest a methodology of estimating system resource capacities by forecasting and analyzing actual data monitored in the system. Before system reinforcement, a customer must decide the date and the size of it. A reinforcing needs a lot of time for preparing, purchasing, testing, etc. So earlier and more exact forecasts of the date when the system reaches the limit are required.

Today in most cases the forecast of server resource capacities is accomplished by hand, or no forecast is made. This causes delayed actions or a wayward plan with poor accuracy. This costs a lot of human resources and money. Current capacity planning methods usually need some access scenarios, which most users overlook. And they need CPU usages for each scenario in key machines. Then, by inputting the ratio of these scenarios and the estimated transactions (or some special objectives such as the number of concurrent users), they specify the machines required.

The point here is these data are usually prepared at the constructing time, and include many suppositions. These scenarios and the respective ratios may be differing from the suppositions, and will oftentimes change as time progresses.

The method this work describes utilizes actual data from the online production system. This method aims to automatically generate these data. And then, to forecast the result, the data is fitted to the ARIMA model with the "Box Jenkins" approach. These analyses are accomplished utilizing long-term actual data.

This method provides long-term and more exact forecasting capabilities, utilizing this detailed data. This is especially useful for Web application servers, and capacity planning programs. This method also allows customers to make better investment planning assumptions. Furthermore, in the short-term, this method is useful for an autonomic computing engine that actualizes a self-configuring system for Autonomic Computing and on demand Operating Environment.

Yih, Jih-Shyr

IBM Value Chains Store Construction and Information Integration

IBM's own sales and distribution on demand business infrastructure is a key proving ground for IBM's on demand business services and software offerings. IBM Research was given the opportunity to prototype the Common Commerce Engine for store consolidations, and was then invited to help the end-to-end business process transformation, down to tasks in physical server architecture for deployment.

This work covers experiences and lessons learned in the value chain constructions, with respect to tool/application integrations and various data feed integrations. We also found that behind every great store, there is a fulfillment, which supports from quote to cash. In this work, the authors will explore information integration methodology utilized in the current value chains; and problem opportunities in areas such as data quality, duplication, propagation delay, transformation, the handling of large documents, and so on. The work presents solutions using a Websphere Business Integrator (WBI) clustering technique, and enhancements in WBI adapters for active persistent streaming data operations. A cut-through method using a "double data schema annotation" technique is to be discussed on how data is handled and manipulated amongst application specific access methods, a built-in persistent operational data store, and collaboration brokers.

Yusuf, Larry

Who Are You Today?—Through the Customer's Eyes, a Unique Approach to Solution Test

You have been told to build a portal for your customers to your business applications. It must provide access to dozens of major business applications and related data spread throughout your business units. And also you have been asked to add your business partners' applications to the portal. Integration also will mean developing processes, since there is some logic to the sequence of these assembled applications.

You have twenty (20) business units and a dozen business partners. Your portal must be available on the Web 24 hours a day, seven days a week. You have been given a staff of six (6) developers, including yourself, of course, and you have four (4) months to deliver it into production.

Is it as easy as it sounds? Using the breadth of our tools, the various integration components on the number of available platforms, can this be as easy as it sounds for the developers? What experiences will these developers have? How effective are products when put together in addressing such challenges?

Who are you today? ... is a presentation that:

1. Introduces the Solution Test approach to persona-based testing.
2. Provides a description of the roles adopted by the team and its impact on the test approach, content, and execution.
3. Through a case study, examines application development, deployment, and infrastructure considerations when testing customer-like solutions.
4. Reflects on the effectiveness of this approach in unleashing integration and consistency issues.
5. Discusses the effectiveness of persona-based testing in the on demand era.

Glossary

Apache AXIS An implementation of the Simple Object Access Protocol (SOAP) specification. It defines a core technology framework to build SOAP message enveloping and message transport.

Autonomic Computing A computing environment with the ability to manage itself and dynamically adapt to change in accordance with business policies and objectives. Self-managing environments can perform such activities based on situations they observe or sense in the information technology (IT) environment rather than requiring IT professionals to initiate the task. These environments are self-configuring, self-healing, self-optimizing, and self-protecting.

Business Web A collection of businesses that dynamically comes together on the Internet in a unique way to sell new products and services, build new channels, and create new business models.

Common Information Model (CIM) A model that describes management information and offers a framework for managing system elements across distributed systems.

Common Management Model (CMM) A set of common management functions that applies to the resources in a grid. CMM is an abstract representation of real IT resources such as disks, filesystems, operating systems, network ports, and Internet Protocol (IP) addresses. CMM defines a set of common management interfaces by which these manageable resources are exposed to external management applications for the purpose of managing these resources.

Computational Grid An infrastructure framework that provides dependable, consistent, pervasive, and inexpensive access to high-end computational capabilities, including computing power, hardware, and software.

Data Grid Deals with all aspects of Grid Computing data, including data location, transfer, access, management, and security.

Distributed Computing A method of computing in which very large problems are divided into smaller tasks that are distributed across a computer network for simultaneous processing. Individual results are then brought together to form the total solution.

Generic Security Service Application Program Interface (GSS API) The GSS API is defined in the Internet Engineering Task Force's (IETF's) request for comment (RFC) 1508. The GSS-API provides a way for applications to protect data that is sent to peer applications. This data might be from a client on one machine to a server on another. The current Globus Toolkit uses the GSS API for security programming.

Global Extensible Markup Language (XML) Architecture Builds on the current XML Web services baseline specifications of SOAP, Web Services Description Language (WSDL), and Universal Description, Discovery, and Integration of Web Services Universal Description, Discovery and Integration (UDDI)[1]. This architecture is a design-based principle of creating modular, integration-capable, general-purpose, and open standard architecture solutions for Web services.

Grid Computing A form of distributed computing over a network (public or private) using open standards (e.g., Open Grid Services Architecture (OGSA)). Includes the aggregation and harnessing of un-utilized computational and data resources across heterogeneous, geographically dispersed environments. Grid Computing enables the virtualization of distributed computing and data resources such as processing, network bandwidth, and storage capacity to create a single system image, granting users and applications

1. UDDI specifications form the necessary technical foundation for publication and discovery of Web services implementations both within and between enterprises.

seamless access to vast IT capabilities. Just as an Internet user views a unified instance of content via the Web, a Grid Computing user essentially sees a single, large, virtual computer.

Grid Job Scheduler Schedulers are responsible for the management of jobs such as allocating resources needed for a specific job, partitioning jobs to schedule parallel execution, data management, and service level management.

Grid Resource Allocation Manager (GRAM) A Globus-provided resource manager, GRAM provides a set of standard interfaces and components to collectively manage a job task and provide resource information, including job status and resource configuration. This information can be used for resource allocation for a specific job.

Grid Resource Broker Provides matchmaking services between service requester and service provider. This matchmaking enables the selection of the best available resource from the service provider for the execution of a specific task.

Grid Security Infrastructure (GSI) GSI is based on public key encryption, X.509 certificates, and the Secure Sockets Layer (SSL) communication protocol. Extensions to these standards have been added for single sign-on (SSO) and delegation. The Globus Toolkit uses the GSI to enable secure authentication and communication for participants in a grid.

GWSDL The Grid Web Services Description Language is a description language for Grid Computing services. This language is modeled after WSDL; however, it tries to overcome some of WSDL 1.1's drawbacks on inheritance and open content extensibilities. Related to defining a *service provider interface*, it should strongly consider WSDL, since that is the language used in the Web Services architecture to describe service interfaces. However, it is not strictly necessary to directly write the WSDL code. For example, one could start out with a Java interface, and generate the WSDL code from the Java code.

IBM Software Strategy for on demand business Designed to meet the needs of users in heterogeneous development environments, the Application Framework for on demand business is an architecture that enables continued on demand business transformation through the use of open standards-based on demand business platforms, the patterns of on demand business, and a comprehensive product set. It is based on proven, open, multi-platform, and multi-vendor industry standards like hypertext markup language (HTML), Transmission Control Protocol/Internet Protocol (TCP/IP), Java, and XML. The applications offered provide the availability, reliability, security, and performance necessary for on demand business settings. It also guarantees compatibility, interoperability, and connectivity with the latest scalable operating system architecture.

Index Service A Globus-provided high-level service responsible for managing static and dynamic state data for grids.

Model-Driven Architecture (MDA) A platform-independent modeling architecture for building interoperable solutions on standards such as the universal modeling language (UML) and meta-object facility (MOF). This is the core architecture for future computing and the Open Management Group drives it.

.NET Framework A Microsoft Corporation framework (and toolset) to develop Common Language Runtime (CLR) solutions and XML Web services.

On Demand Business An enterprise whose business processes—integrated end-to-end across the company and with key partners, suppliers, and customers—can respond with agility and speed to any customer demand, market opportunity, or external threat.

On Demand Operating Environment The new computing architecture initiative from IBM designed to help companies realize the benefits of on demand business. The on demand Operating Environment exhibits four essential characteristics: It is integrated, open, virtualized, and autonomic.

On Demand Operating Infrastructure Defines a set of core components and solutions to meet on demand Operating Environment characteristics.

Ontology Used to describe collections of information like concepts and relationships that can exist between resources and objects. This taxonomy defines classes of objects and the relations among them.

Open Grid Service Architecture (OGSA) The open standards architecture for next-generation Grid Computing services to enable the creation, maintenance, and integration of Grid Computing services maintained by virtual organizations (VOs).

Open Grid Service Infrastructure (OGSI) A core component of the OGSA, which provides a uniform way to describe Grid Computing services and define a common pattern of behavior for all Grid Computing services. In short, this architecture defines Grid Computing service behaviors, service description mechanisms, and protocol binding information by using Web services as the technology enabler.

Patterns for on demand business A repository of successfully implemented on demand business designs and a source of information for architects and developers. It is the distillation of the collected wisdom of IBM, partners, and customer IT architects, containing code samples and representative implementations.

Peer-to-Peer (P2P) Computing The sharing of computer resources and services by a direct exchange between systems. These resources and services include the exchange of information, processing cycles, cache storage, and disk storage for files. P2P computing takes advantage of existing desktop computing power and networking connectivity, allowing economical clients to leverage their collective power to benefit the entire enterprise.

Resource Description Language (RDL) A standard for the notation of structured information, recommended by the World Wide Web Consortium (W3C) for meta-data interoperability across different resource description communities.

Resource Specification Language (RSL) A Globus Grid Computing Toolkit-defined XML standard for describing a job.

Semantic Web The next-generation Web, in which information is given well-defined meaning and computers and people are enabled to work in cooperation. This enables the data in the Web to be defined and linked together by computing agents for automatic decision-making.

Service Level Agreement (SLA) An agreement between a service provider and service requester to provide certain quality-of-service (QoS) guarantees.

Service-Oriented Architecture (SOA) The architecture to define loosely coupled and interoperable services/applications, and to define a process for integrating these interoperable components. In an SOA, the software system is decomposed into a collection of network-connected components/services, and applications are composed dynamically from the deployed and available components in the network.

Standard Generalized Markup Language (SGML) (1) An International Organization for Standards (ISO) language for defining a computer formatting task in a text document. Widely used in the publishing industry, an SGML document uses a separate Document Type Definition (DTD) file that defines the format codes, or tags, embedded within it. Since SGML describes its own formatting, it is known as a "meta-language." SGML is a very comprehensive language that also includes hypertext links. HTML is an SGML document that uses a fixed set of tags, while XML is a simplified version of SGML. (2) A standardized, generalized markup language for describing the logical structure of a computer document.

Simple Object Access Protocol (SOAP) The definition of the XML-based information that can be used for exchanging structured and typed information between peers in a decentralized, distributed environment. This includes message packaging and message exchange scenarios.

Software Architecture An abstraction of the runtime elements of a software system during some phase of its operation. A system may be composed of many levels of abstraction and many phases of operation, each with its own software architecture [Fielding].

Universal Description, Discovery, and Integration of Web Services (UDDI) The building block that will enable a business to quickly, easily, and dynamically find businesses and transact business with each other. This is based on open standard UDDI specifications that provide seamless interoperability and integration capabilities. There are many directory services, both Internet-/intranet-related, based on UDDI, which can be treated as global registries for Web services.

Universal Modeling Language (UML) Modeling is the process of designing software applications. UML provides a standard and industry-driven methodology for software modeling. This modeling language is defined by the Open Management Group and is not limited to software. It can form the basis for designing non-software systems, too.

Utility Computing The network delivery of IT and business process services by Utility Computing providers.

Virtual Organization (VO) Coordinated resource sharing and problem-solving in a dynamic, multi-institutional organization. Sharing is with direct access to computers, software, data, and other resources by a range of collaborative problem-solving and resource brokering strategies. Sharing is controlled with resource providers and consumers defining what is shared, who is allowed to share, and the conditions under which sharing occurs. The set of individuals and/or institutions defined by the sharing rules forms the VO.

Web Service Description Language (WSDL) An XML language for describing Web services. This language provides a model and XML format. WSDL provides facilities for service developers to separate abstract definition of the service (interface, message, and schema definitions) from concrete definition (service, port, and binding definitions).

Web Services (1) A Web service is a software system identified by a universal resource identifier (URI) whose public interfaces and bindings are defined and described using XML. Its definition can be discovered by other software systems. These systems may then interact with the Web service in a manner prescribed by its definition using XML-based messages conveyed by IPs. (2) Self-contained, modular applications that can be described, published, located, and invoked over a network (generally the Internet). Web services go beyond software components because they can describe their own functionality and look for and dynamically interact with other Web services.

Web services provide a means for different organizations to connect their applications with one another to conduct dynamic on demand business across a network, no matter what their application, design, or runtime environment.

Web Services Architecture (WSA) Provides a common definition for Web services. This is a standard initiative from the W3C. WSA tries to define and relate the necessary technologies (XML, SOAP, and WSDL) to construct interoperable Web services. It will define how to construct interoperable messages, message exchange patterns, enveloping mechanisms, and a definition mechanism for interoperable services. In addition, the WSA will define the manageability, security, and correlation aspects of Web services.

WebSphere Application Server (WAS) A Java II Enterprise Edition/Java II Server Edition (J2EE/J2SE) built application server framework from IBM. It provides reliable, high-volume transactions for customers. It forms the basis for the IBM on demand Operating Environment framework with built-in support for Web services and OGSA-based Grid Computing services.

eXtensible Markup Language (XML) A meta-language written in SGML that allows one to design a markup language for the easy interchange of documents on the World Wide Web.

Reference Materials

[ACBLUEPRINT] IBM Corporation Software Group. "An Architectural Blueprint for Autonomic Computing," IBM Technical White Paper, April 2003.

[Boisseau 01] Marc Boisseau and Craig Fellenstein. "Services Brokerage and the IBM Services Provider Delivery Environment (SPDE): An IBM Framework for Service Providers," IBM Technical White Paper, June 2002.

[Boisseau 02] Marc Boisseau, Craig Fellenstein, and Noel Luddy. "An Introduction to the IBM Services Provider Delivery Environment (SPDE): An IBM Framework for Service Providers," IBM Technical White Paper, September 2002.

[Fellenstein01] Craig Fellenstein and Ron Wood. "Exploring e-Commerce: Global e-Business and e-Societies," Upper Saddle River, NJ: Prentice Hall, December 2000.

[Ferreira] Luis Ferreira, Bart Jacob, Sean Slevin, Michael Brown, Srikrishnan Sundararajan, Jean Lepesant, and Judi Bank. Portions extracted from the IBM Redbook document, "Globus Toolkit 3 Quick Start," REDP3697, September 2003 (*www.ibm.com/redbooks*).

[Foster01] Ian Foster and Carl Kesselman. Chapter 2, "Computational Grids," *The Grid: Blueprint for a Future Computing Infrastructure*, Morgan Kaufmann, San Francisco, California. 1986.

[Foster02] Ian Foster, Carl Kesselman, and Steven Tuecke. "The Anatomy of the Grid:" (*www.globus.org/research/papers/anatomy.pdf*). 2001.

[Foster04] Ian Foster, Carl Kesselman, Jeffrey M Nick, and Steven Tuecke. "The Physiology of the Grid—An Open Grid Service Architecture for Distributed Systems Integration," (*www.globus.org/research/papers/ogsa.pdf*). 2002.

[Ganek01] Alan Ganek. "Computers must become smarter," *Silicon Valley Business Ink*, April 18, 2003.

[Ganek02] Alan Ganek and Thomas Corbi, "The dawning of the autonomic computing era," *IBM Systems Journal*, Vol. 42, No. 1. 2003.

[IBM01] "Autonomic Computing," (*www-3.ibm.com/autonomic/index.shtml). 2003.*

[IBM02] "On demand business," (*www-3.ibm.com/e-business/index.html). 2003.*

[Joshi01] Joshy Joseph. "A developer's overview of OGSI and OGSI-based GRID computing," 2003.

[Kendall] Samuel C. Kendall, Jim Waldo, Ann Wollrath, and Geoff Wyant. "Note on Distributed Computing," (*research.sun.com/techrep/1994/smli tr-94-29.pdf). 1994.*

[OGSA-WG] OGSA Architecture Working Group, (https://forge.gridforum.org/projects/ogsa-wg). 2003.

[Roure01] David De Roure, Mark A Baker, Nicholas R. Jennings, and Nigel R Shadbolt. "The Evolution of Grid," (*www.semanticgrid.org/documents/evolution/evolution.pdf*). 2003.

[WEB] "Architecture of the World Wide Web," (*www.w3.org/TR/webarch/). 2003.*

[Wladawsky-Berger] Irving Wladawsky-Berger. "Turning points in information technology," *IBM Systems Journal*, Vol. 38, Nos. 2 and 3.

[WSA] Web Services Architecture Working Group. (www.w3.org/2002/ws/arch/). 2003.

Acknowledgements

First and foremost, I would like to thank my family, who are definitely more important to me than on demand business. It was my family who put up with my many late-night hours, allowing for the development and delivery of this book. I would like to also thank my wife, Elizabeth, who spent many long hours editing this book, as the content was being developed, arranged, and rearranged.

I would also like to thank several of the IBM senior executive leaders for their support, professional guidance, and counsel in providing key information on some of the complex topics addressed in this book. Attribution goes to Paul Horn, Alan Ganek, Irving Wladawsky-Berger, and John Gotsch, who all deserve a special thank you for their contributions, support, and encouragement.

I would like to thank Jaclyn Vassallo for her editing support. I would also like to thank Marie Chow and Teressa Jimenez for their outstanding assistance and support in the area of Autonomic Computing; they provided me with important information that has been included in this book.

I would like to especially thank Joshy Joseph, a co-author with me on *Grid Computing* and a colleague of mine at work. I would like to thank Susan Visser, Dave Bartlett, Mark Ernest, Donald Castle, Peter Andrews, Ric Telford, Bart Jacob, David Singer, Richard Murch, Craig Barbakow, Alex Cabanes, and Chris Reech for their weeknight/weekend hours and guidance, bringing insight and expertise to both the Autonomic and Grid Computing topic areas. I would like to thank Kevin Twardus and Gordon Kerr for sharing their telecommunications industry experience and supplying service provider insights.

A very special thank you is due to my Pearson publisher, Jeffrey Pepper, and his support team, including Linda Ramagnano, Kathleen M. Caren, Jim Keogh, and the entire team at Pearson. This team provided excellent support and endless patience in the many activities encountered throughout the entire development and production of this book.

Without all these individuals, this book would not have been produced in its timeframe, depth, breadth, and quality.

Craig Fellenstein

Index

A

N

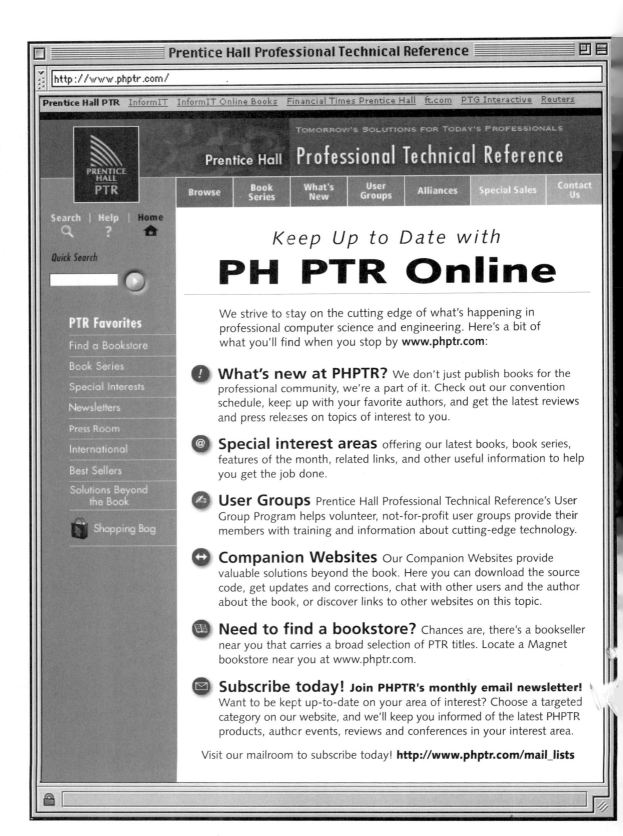